Titles in This Series

Volume

8 **J. Feldman, R. Froese, and L. M. Rosen, Editors**
Mathematical quantum theory II: Schrödinger operators
1995

7 **J. Feldman, R. Froese, and L. M. Rosen, Editors**
Mathematical quantum theory I: Field theory and many-body theory
1994

6 **Guido Mislin, Editor**
The Hilton Symposium 1993
Topics in topology and group theory
1994

5 **D. A. Dawson, Editor**
Measure-valued processes, stochastic partial differential equations, and interacting systems
1994

4 **Hershy Kisilevsky and M. Ram Murty, Editors**
Elliptic curves and related topics
1994

3 **Rémi Vaillancourt and Andrei L. Smirnov, Editors**
Asymptotic methods in mechanics
1993

2 **Philip D. Loewen**
Optimal control via nonsmooth analysis
1993

1 **M. Ram Murty, Editor**
Theta functions
1993

Volume 8

CRM PROCEEDINGS & LECTURE NOTES

Centre de Recherches Mathématiques
Université de Montréal

Mathematical Quantum Theory II: Schrödinger Operators

J. Feldman
R. Froese
L. M. Rosen
Editors

The Centre de Recherches Mathématiques (CRM) of the Université de Montréal was created in 1968 to promote research in pure and applied mathematics and related disciplines. Among its activities are special theme years, summer schools, workshops, postdoctoral programs, and publishing. The CRM is supported by the Université de Montréal, the Province of Québec (FCAR), and the Natural Sciences and Engineering Research Council of Canada. It is affiliated with the Institut des Sciences Mathématiques (ISM) of Montréal, whose constituent members are Concordia University, McGill University, the Université de Montréal, the Université du Québec à Montréal, and the Ecole Polytechnique.

American Mathematical Society
Providence, Rhode Island USA

The production of this volume was supported in part by the Fonds pour la Formation de Chercheurs et l'Aide à la Recherche (Fonds FCAR) and the Natural Sciences and Engineering Research Council of Canada (NSERC).

1991 *Mathematics Subject Classification.* Primary 81T08; Secondary 81V70.

Library of Congress Cataloging-in-Publication Data

Mathematical quantum theory II: Schrödinger operators / J. Feldman, R. Froese, L. Rosen, editors.
p. cm. — (CRM proceedings & lecture notes, ISSN 1065-8580; v. 8)
"Proceedings of the Canadian Mathematical Society Annual Seminar on Mathematical Quantum Theory held in Vancouver, August 4–14, 1993"—Pref.
Includes bibliographical references.
ISBN 0-8218-0366-2
1. Schrödinger operator—Congresses. 2. Quantum theory—Congresses. I. Feldman, Joel S., 1949– . II. Froese, Richard Gerd. III. Rosen, Lon M., 1944– . IV. Canadian Mathematical Society Annual Seminar on Mathematical Quantum Theory (1993 : Vancouver, B.C.) V. Title Mathematical quantum theory 2. VI. Series.
QC174.17.S3M38 1995
530.1′44—dc20 95-6092
CIP

∞ The paper used in this book is acid-free and falls within the guidelines
established to ensure permanence and durability.
♻ Printed on recycled paper.

This volume was typeset using AMS-TeX, and AMS-LaTeX,
the American Mathematical Society's TeX macro systems,
and submitted to the American Mathematical Society in camera-ready
form by the Centre de Recherches Mathématiques.

10 9 8 7 6 5 4 3 2 1   00 99 98 97 96 95

Contents

Preface

These articles constitute the proceedings of the Canadian Mathematical Society Annual Seminar on Mathematical Quantum Theory held in Vancouver August 4–14, 1993. The meeting was run as a research level summer school concentrating on two related areas of contemporary mathematical physics. The subject for the first session was quantum field theory and many-body theory and is covered in Volume 1 of these proceedings. The second session dealt with Schrödinger operators and is covered in Volume 2. Each session featured a series of mini-courses, consisting of approximately four one hour lectures, designed to introduce the student to the state of the art in a particular area. In addition, about thirty speakers gave one hour lectures of an expository nature.

The meeting could not have taken place without the sponsorship of the Natural Sciences and Engineering Research Council (NSERC) and the Centre de Recherches Mathématiques (CRM) of the Université de Montréal. We also wish to acknowledge the assistance of the Canadian Mathematical Society and the Department of Mathematics of the University of British Columbia, in particular Nina Einarsson and Margaret Harney.

Joel Feldman
Richard Froese
Lon Rosen

August 1993

Centre de Recherches Mathématiques
CRM Proceedings and Lecture Notes
Volume 8, 1995

Topics in Spectral Theory of Schrödinger Operators on Non-Compact Riemannian Manifolds with Cusps

Shmuel Agmon

1. Introduction

In this paper we shall discuss some basic topics in the spectral theory of Schrodinger operators on certain non-compact Riemannian manifolds with cusps. The model for such manifolds (of arbitrary dimension) are certain non-compact hyperbolic manifolds of finite volume. The spectral theory in question was first developed for the Laplace-Beltrami operator on non-compact hyperbolic surfaces of finite area. The theory in this case is of a special interest because of the many connections that exist between spectral properties of the Laplacian on hyperbolic surfaces and problems in number theory. By way of introduction we dwell briefly on this important special case.

Recall that hyperbolic surface is a 2-dimensional Riemannian manifold whose Gaussian curvature is -1. Such a manifold is isometric to a Riemann surface $\mathbb{H}^2/\Gamma$ where $\mathbb{H}^2$ denotes the upper half-plane with the hyperbolic metric and where Γ is a torsion free discrete subgroup of $PSL(2,\mathbb{R})$. The subgroup Γ is said to be cofinite if the surface $\mathbb{H}^2/\Gamma$ is not compact but its area is finite. The interest in the spectral theory of the Laplacian on $\mathbb{H}^2/\Gamma$, for Γ which is cofinite, arose in number theory when it was observed by Maass [**8**] that Eisenstein series are generalized eigenfunctions of the Laplacian on $\mathbb{H}^2/\Gamma$. This connection was utilized by Roelcke [**13**] in the case of the modular group to obtain a spectral representation of the automorphic Laplacian. The work of Selberg [**15, 16**] on the trace theorem gave a special impetus to subsequent study of spectral properties of the Laplacian on automorphic forms. The following are some of the many contributions added to this study during the last twenty five years. Roelcke [**14**], Faddeev [**5**], Elstrodt [**4**], Patterson [**10**], Venkov [**18**], Lax and Phillips [**7**], Colin de Verdière [**3**], Hejhal [**6**], Phillips and Sarnak [**11, 12**], Wolpert [**19**] (for additional contributions see [**6**]).

In recent years various mathematicians studied the spectral theory of the Laplacian on certain non-compact hyperbolic manifolds of higher dimension. In particular relevant to our paper is the work of Müller [**9**] on spectral properties of the Laplacian on higher dimensional Riemannian manifolds with cusps. However, most papers in the higher dimensional case consider the spectral theory of the Laplacian on hyperbolic manifolds of *infinite* volume. This interesting (and difficult) case of the theory is beyond the scope of our paper.

1991 *Mathematics Subject Classification.* Primary: 58G25; Secondary: 30F35, 35J10.

Partially sponsored by the Edmund Landau Center for research in Mathematical Analysis, supported by the Minerva Foundation (Germany).

This is the final form of the paper.

This paper is a written version of lectures given at the Mathematical Quantum Theory Conference held at the University of British Columbia, Vancouver, in August 1993. It is intended as an introduction to the spectral theory of Schrödinger operators on non-compact Riemannian manifolds with cusps. A new method based on L^2-weighted estimates is used. We believe that this approach simplifies the presentation of the theory even for the classical case of the Laplacian on non-compact hyperbolic surfaces of finite area.

Because of certain constraints on exposition and publication some topics of the basic theory (such as the scattering operator) were omitted in this paper. We also omitted from the paper the topic: "Instability of embedded eigenvalues", which was discussed in the original lectures. We plan to supplement these omissions in another publication.

2. The Schrödinger Operator

We denote by (M, g) a complete n-dimensional C^∞ Riemannian manifold. We assume that M has the following decomposition:

$$M = M_0 \cup Z_1 \cup \cdots \cup Z_m \tag{2.1}$$

where M_0 is a compact manifold with a smooth boundary and where (Z_j, g), $j = 1, \ldots, m$, is isometric to a Riemannian manifold $(\widetilde{Z}_j, \tilde{g}_j)$ of the form

$$\widetilde{Z}_j = X_j \times [a_j, \infty), \quad j = 1, \ldots, m, \tag{2.2}$$

with $a_j > 0$ and

$$\tilde{g}_j = y^{-2}(h_j + dy^2).$$

Here X_j denotes a compact (boundaryless) $n-1$ dimensional manifold with a metric h_j and y is the coordinate on $[a_j, \infty)$. We refer to Z_j as a cusp (or a cusp neighborhood) and we refer to the pair (x_j, y), where $x_j \in X_j$ and $y \geq a_j$, as the canonical coordinates on Z_j.

The class of manifolds (M, g) defined above was introduced by Müller [**9**]. It contains a large class of hyperbolic manifolds of a finite volume. In particular any non-compact hyperbolic surface with a finite area is in this class. Such a surface has a decomposition (2.1) with $X_j = S^1$ and $\tilde{g}_j = y^{-2}(dx^2 + dy^2)$.

We shall denote by z the generic point on M and by $d\mu$ the measure induced by the metric g on M. $L^2(M)$ denotes the class of L^2 functions on M with respect to $d\mu$ and $\|\cdot\|$ denotes the norm in $L^2(M)$. $L^2_{\text{loc}}(M)$ denotes the class of functions on M which are locally in L^2. $H^k_{\text{loc}}(M)$, $k = 1, 2$, denotes the (local) L^2-Sobolev class of functions of order k on M.

Weighted L^2 spaces will have an important role in our considerations. To define these we choose a weight function $\rho(z) > 0$, $\rho \in C^\infty(M)$, such that in canonical coordinates:

$$\rho(z) = y \text{ in } Z_j \tag{2.3}$$

for $j = 1, \ldots, m$ (ρ will be fixed throughout). For any real α we introduce the class of functions

$$L^{2,\alpha}(M) = \{u \in L^2_{\text{loc}}(M) : \rho^\alpha \mu \in L^2(M)\}. \tag{2.4}$$

For $u \in L^{2,\alpha}(M)$ we set

$$\|u\|_\alpha^2 := \|\rho^\alpha u\|^2 = \int_M \rho(z)^{2\alpha}|u(z)|^2 d\mu. \tag{2.5}$$

We also set

$$L^{2,\infty}(M) = \bigcap_\alpha L^{2,\alpha}(M), \quad L^{2,-\infty}(M) = \bigcup_\alpha L^{2,\alpha}(M). \tag{2.6}$$

We consider $L^{2,\infty}(M)$ and $L^{2,-\infty}(M)$ as linear topological spaces with their natural topologies.

We introduce two notions of rapid decay in a cusp. If u is a function defined in a cusp Z_j, we write

$$u = O(\rho^{-\infty}) \text{ in } Z_j \tag{2.7}$$

if for any $\alpha \geq 0$ there exists a constant c_α such that

$$|u(z)| \leq c_\alpha \rho(z)^{-\alpha} \text{ for } z \in Z_j. \tag{2.7$'$}$$

We write

$$u = O_{L^2}(\rho^{-\infty}) \text{ in } Z_j \tag{2.8}$$

if for any $\alpha \geq 0$ we have

$$\int_{Z_j} |u(x)|^2 \rho(z)^{2\alpha} \, d\mu < \infty. \tag{2.8$'$}$$

We denote by Δ the Laplace-Beltrami operator on M. At the cusp Z_j we have in canonical coordinates

$$\Delta = y^2 \frac{\partial^2}{\partial y^2} + (2-n) y \frac{\partial}{\partial y} + y^2 \Delta_j' \tag{2.9}$$

where Δ_j' denotes the Laplace-Beltrami operator on X_j in the metric h_j. Also,

$$d\mu = y^{-n} \, d\mu_j' \, dy \text{ on } Z_j \tag{2.10}$$

where $d\mu_j'$ denotes the measure on X_j in the metric h_j.

We consider a Schrödinger type differential operator P on M of the form

$$P = -\Delta + V \tag{2.11}$$

where V is a real function which decays rapidly:

$$V(z) = O(\rho^{-\infty}) \tag{2.12}$$

in all cusps. It is our object in this paper to study various spectral properties of P.

We start by observing that since V is bounded it follows by standard results that P has a unique self-adjoint realization in $L^2(M)$. We denote this realization by $\widetilde{P}$. It is clear that $\widetilde{P}$ is bounded from below. The basic results concerning the spectrum of $\widetilde{P}$ are described in the following

THEOREM 2.1.

(i) *The essential spectrum of* $\widetilde{P}$ *is given by*

$$\sigma_{\text{ess}}(\widetilde{P}) = \left[\left(\frac{n-1}{2}\right)^2, \infty\right). \tag{2.13}$$

(ii) *The point spectrum* $\sigma_p(\widetilde{P})$ *consists of an increasing (possibly empty) sequence of eigenvalues of finite multiplicity having no finite limit point. In particular* $\sigma_{\text{dis}}(\widetilde{P})$ *is a finite set.*

We shall prove here only statement (i). (ii) is proved at the end of Section 5. To prove (2.13) we use

LEMMA 2.2. *Suppose that there exists a function* $u \in C^2(M)$ *such that*

$$u > 0 \text{ and } (P - \lambda)u \geq 0 \text{ in } M \setminus K$$

for some $\lambda \in \mathbb{R}$ *and some compact set* K. *Then:*

$$\inf \sigma_{\text{ess}}(\widetilde{P}) \geq \lambda. \tag{2.14}$$

Lemma 2.2 is valid for an operator $P = -\Delta + V$ on any complete Riemannian manifold X where V is a real bounded function. A proof of the lemma for $X = \mathbb{R}^n$ is given in [**2**, Theorem 2.1]. The same proof applies in the general case.

PROOF OF (2.13). A simple computation shows (using (2.9), (2.3) and (2.12)) that for any $\varepsilon > 0$

$$\left(P - \left(\frac{n-1}{2}\right)^2 + \varepsilon\right)\rho^{(n-1)/2} > 0 \tag{2.15}$$

in $M \setminus K_\varepsilon$, where K_ε is some compact. Hence, applying Lemma 2.2, it follows from (2.15) (since ε is arbitrary) that

$$\inf \sigma_{\text{ess}}(\widetilde{P}) \geq \left(\frac{n-1}{2}\right)^2. \tag{2.16}$$

Next let $\lambda \in \left[\left((n-1)/2\right)^2, \infty\right)$ and write $\lambda = \left((n-1)/2\right)^2 + t^2$, $t > 0$. Set $s = (n-1)/2 + it$ and observe that $\lambda = s(n-1-s)$. Define:

$$u_\varepsilon = \sqrt{\varepsilon}\rho^{s-\varepsilon}, \quad \varepsilon > 0.$$

An easy computation shows that the u_ε are functions in $\operatorname{Dom}(\widetilde{P})$ such that

$$\begin{aligned} &\|u_\varepsilon\| \geq c > 0 \text{ for } 0 < \varepsilon \leq 1, \\ &\|(P-\lambda)u_\varepsilon\| = O(\varepsilon) \text{ as } \varepsilon \to 0, \end{aligned} \tag{2.17}$$

where c is some constant. It follows from (2.17) that $\lambda \in \sigma(\widetilde{P})$ and thus we conclude that

$$\sigma_{\text{ess}}(\widetilde{P}) \supset \left[\left(\frac{n-1}{2}\right)^2, \infty\right). \tag{2.18}$$

Combining (2.18) with (2.14) we obtain (2.13), proving (i). □

In the following we shall discuss properties of solutions of the differential equation $Pu - \lambda u = f$ and properties of the resolvent operator $(\widetilde{P} - \lambda)^{-1}$. In these investigations it is convenient to replace the spectral parameter λ by a parameter s related by $\lambda = s(n-1-s)$. Note that the map: $s \to \lambda$ is one to one from the half-plane $\operatorname{Re} s > (n-1)/2$ onto the slit plane $\mathbb{C} \setminus \left[((n-1)/2)^2, \infty\right)$. The map can also be viewed as a one to one conformal map of the s-plane onto the double sheeted Riemann surface consisting of two sheets of the λ-plane joined along the half-line $\left[((n-1)/2)^2, \infty\right)$.

3. Some Asymptotic Results

In this section we digress to consider properties of solutions of the differential equation

$$Pu - s(n-1-s)u = f \text{ on } M \tag{3.1}$$

where s is a complex number. A basic property of solutions of (3.1) is that they possess a simple asymptotic expansion at the cusps if it is known that u increases moderately and f decreases sufficiently rapidly at the cusps. In particular we have

THEOREM 3.1. *Let u be a solution (3.1) with $u \in L^{2,\alpha}(M)$ for some α and $f \in L^{2,\infty}(M)$. The following asymptotic expansion holds at the cusp Z_j, $j = 1, \dots, m$:*

(i) *If $s \neq (n-1)/2$, then*

$$u = c_j \rho^{n-1-s} + b_j \rho^s + O_{L^2}(\rho^{-\infty}). \tag{3.2}$$

(ii) *If $s = (n-1)/2$, then*

$$u = c_j \rho^{(n-1)/2} + b_j \rho^{(n-1)/2} \log \rho + O_{L^2}(\rho^{-\infty}) \tag{3.2$'$}$$

where c_j and b_j are some constants. Moreover the derivative $\partial u/\partial y$ at the cusp Z_j has a similar asymptotic expansion obtained by differentiating both sides of (3.2) (resp. (3.2′)). Finally, if f verifies the pointwise rapid decay condition: $f = O(\rho^{-\infty})$ then the asymptotic expansions (3.2)-(3.2′) (and the corresponding expansions for $\partial u/\partial y$) hold with the L^2 error term $O_{L^2}(\rho^{-\infty})$ replaced by the pointwise error term $O(\rho^{-\infty})$.

Before describing the proof of Theorem 3.1 we introduce the following terminology. Let v be a function defined in the cusp $Z_j = X_j \times [a_j, \infty)$, v belonging to the Sobolev class H^1_{loc}. We say that v satisfies the "zero-mean" condition in the variable x_j, at the cusp Z_j, if

$$\int_{X_j} v(x_j, y)\, d\mu'_j = 0 \tag{3.3}$$

for all $y > a$ for some $a > a_j$. Here $d\mu'_j$ is the measure induced by the Riemannian metric h_j on X_j.

The proof of Theorem 3.1 is based on the following

THEOREM 3.2. *Let v be a function in $L^{2,\alpha}(M) \cap H^2_{\text{loc}}(M)$ for some α, satisfying the differential equation*

$$-\Delta v + qv = f \text{ on} M \tag{3.4}$$

where q is a bounded function on M and $f \in L^{2,\infty}(M)$. Suppose, moreover, that v verifies the zero-mean condition (3.3) in all cusps. Then,

$$(|v| + |\nabla v|) \in L^{2,\infty}(M). \tag{3.5}$$

Here $|\nabla v(z)|$ *denotes the norm of the gradient vector* $\nabla v(z)$ *in the induced metric* g.

The proof of the theorem is given in the Appendix.

PROOF OF THEOREM 3.1. Consider the function u in some fixed cusp $Z_j = X_j \times [a_j, \infty)$. Denote by (x_j, y) the canonical coordinates in Z_j. Averaging $u(x_j, y)$ with respect to x_j in the cusp, we set:

$$\tilde{u}_j(y) = \omega_j^{-1} \int_{X_j} u(x_j, y)\, d\mu_j' \tag{3.6}$$

for $y > a_j$ where $\omega_j = \mathrm{vol}(X_j)$.

From (3.6), (3.1), (2.11) and (2.9) it follows readily that $\tilde{u}_j(y)$ is a solution of the ordinary differential equation

$$y^2 \tilde{u}_j''(y) + (2-n) y \tilde{u}_j'(y) + s(n-1-s)\tilde{u}_j(y) = h_j(y) \tag{3.7}$$

where

$$h_j(y) = \omega_j^{-1} \int_{X_j} (V(x_j, y) u(x_j, y) - f(x_j, y))\, d\mu_j'. \tag{3.8}$$

It follows from our assumptions on u and (3.6) that $\tilde{u}_j \in H^2_{\mathrm{loc}}$ on the half-line $y > a_j$ (and thus $\tilde{u}_j$ is a C^1 function). It also follows that $h_j(y)$ is a function in L^2_{loc} which decays rapidly in the sense that

$$h_j(y) = O_{L^2}(y^{-\infty}) \text{ as } y \to \infty. \tag{3.9}$$

By an elementary representation formula for solutions of the Euler equation (3.7), using the rapid decay condition (3.9), it follows readily that $\tilde{u}_j(y)$ has the following asymptotic expansion as $y \to \infty$, for $s \neq (n-1)/2$:

$$\begin{aligned} \tilde{u}_j(y) &= c_j y^{n-1-s} + b_j y^s + O(y^{-\infty}), \\ \tilde{u}_j'(y) &= (n-1-s) c_j y^{n-2-s} + s b_j y^{s-1} + O(y^{-\infty}), \end{aligned} \tag{3.10}$$

where c_j and b_j are constants (note that the asymptotic expansions (3.10) are pointwise). For $s = (n-1)/2$ we have

$$u_j(y) = y^{(n-1)/2}(c_j + b_j \log y) + O(y^{-\infty}) \tag{3.10$'$}$$

as $y \to \infty$ and a corresponding expansion for u_j' obtained by formally differentiating (3.10$'$).

Next, define in Z_j the function

$$v_j(x_j, y) = u(x_j, y) - \tilde{u}_j(y). \tag{3.11}$$

It follows from (3.11) and (3.6) that v_j verifies the zero-mean condition (3.3) in Z_j. Furthermore, from (3.11), (3.1) and (3.7) it follows that v_j is a solution of the differential equation

$$\Delta v_j + s(n-1-s) v_j = k_j \text{ in } Z_j \tag{3.12}$$

where

$$k_j(x_j, y) = V(x_j, y) u(x_j, y) - f(x_j, y) - h_j(y). \tag{3.12$'$}$$

From (3.12$'$), (3.12), (3.9) and our assumptions on u and f it follows that

$$k_j = O_{L^2}(\rho^{-\infty}) \tag{3.13}$$

as $\rho(z) \to \infty$ in Z_j.

Now introduce a cutoff function $\zeta_j(z) \in C^\infty(M)$ such that $\operatorname{supp}\zeta_j \subset Z_j$ and such that $\zeta(z) = 1$ in Z_j outside some compact set. Define

$$w_j(z) = \begin{cases} \zeta_j(z)v_j(z) & \text{in } Z_j, \\ w_j(z) = 0 & \text{in } M \setminus Z_j. \end{cases} \tag{3.14}$$

It follows from (3.14), (3.12) and (3.13) that w_j is a function on M verifying all the conditions of Theorem 3.2. Hence applying the theorem to w_j, using (3.14), we conclude that

$$|v_j| + |\nabla v_j| = O_{L^2}(\rho^{-\infty}) \tag{3.15}$$

as $\rho(z) \to \infty$ in Z_j. Combining (3.10)-(3.10′), (3.11) and (3.15) it follows that u verifies (3.2)-(3.2′) and that $\partial u/\partial y$ verifies the corresponding differentiated asymptotic formula. This proves the main part of Theorem 3.1.

Finally, we sketch the proof of the last statement of Theorem 3.1. Hence suppose that u satisfies the conditions of Theorem 3.1 where f verifies a *pointwise* rapid decay condition at the cusps:

$$f = O(\rho^{-\infty}). \tag{3.16}$$

Using equation (3.1) and the boundedness of f, recalling that $u \in L^{2,\alpha}(M)$ for some α, it follows with the aid of standard interior elliptic L^p estimates (using Sobolev's embedding theorem and an iteration argument) that u is a continuous function of polynomial growth:

$$u = O(\rho^N) \text{ at the cusps.} \tag{3.17}$$

Hence, it follows from (3.17), (3.16), (3.8) and (2.12) that the function k_j (defined by (3.12′)) satisfies

$$k_j = O(\rho^{-\infty}) \tag{3.18}$$

as $\rho(z) \to \infty$ in Z_j. Applying once more the interior elliptic L^p estimates, this time to the function v_j (solution of (3.12)), it follows from Sobolev's embedding theorem via an iteration argument that

$$|v_j| + |\nabla v_j| = O(\rho^{-\infty}) \tag{3.19}$$

as $\rho(z) \to \infty$ in Z_j. Combining (3.19), (3.11) and (3.10)-(3.10′) we obtain the pointwise asymptotic formulas analogous to (3.2)-(3.2′) for u and the corresponding formulas for $\partial u/\partial y$. □

We conclude this section by defining a class of solutions of equation (3.1) which satisfy certain "boundary conditions" at the cusps. This class (termed s-outgoing solutions) will play an important role in the sequel. In this connection we introduce a first order differential operator D_ρ defined by

$$D_\rho = \rho^{-1}\nabla\rho. \tag{3.20}$$

Note that in the cusp $Z_j : D_\rho = y\partial/\partial y$ for $y > a_j$.

DEFINITION. A solution u of equation (3.1), $u \in H^2_{\text{loc}}(M)$, is said to be "$s$-outgoing" if $f \in L^{2,\beta}(M)$ and

$$\|(D_\rho + s - n + 1)u\|_\beta < \infty \tag{3.21}$$

for some $\beta > (n-1)/2 - \operatorname{Re} s$.

One can replace the "radiation condition" (3.21) by a condition on the asymptotic behavior of u at the cusps. In particular, if f decays rapidly at the cusps, we have

THEOREM 3.3. *Let u be a H^2_{loc} solution of equation* (3.1) *with $f \in L^{2,\infty}(M)$. Then u is s-outgoing iff u has the following asymptotic expansion*

$$u = c_j \rho^{n-1-s} + O_{L^2}(\rho^{-\infty}) \tag{3.22}$$

in Z_j for $j = 1, \dots, m$, c_j a constant. If $f = O(\rho^{-\infty})$ then the conclusion of the theorem holds with the remainder term in (3.22) *replaced by $O(\rho^{-\infty})$.*

PROOF. Suppose that u is a s-outgoing solution of (3.1) with $f \in L^{2,\infty}(M)$. It follows readily from (3.21) that $u \in L^{2,\alpha}(M)$ for some α. Applying Theorem 3.1 it follows that u possesses the asymptotic expansion (3.2) (or (3.2$'$) if $s = (n-1)/2$) and a corresponding expansion for $\partial u/\partial y$. A simple computation shows that condition (3.21) implies that the constant b_j in the expansion (3.2) (or (3.2$'$)) must vanish. This yields the expansion (3.22). That the remainder term $O_{L^2}(\rho^{-\infty})$ in (3.22) can be replaced by $O(\rho^{-\infty})$ follows from Theorem 3.1. That a solution u of (3.1) with expansion of the form (3.22) is s-outgoing is also an easy consequence of Theorem 3.1. □

4. Meromorphic Continuation of the Resolvent

We shall use the notation $B\big(L^{2,\alpha}(M), L^{2,\beta}(M)\big)$ to denote the (topological) space of continuous linear operators from $L^{2,\alpha}(M)$ to $L^{2,\beta}(M)$. Here: $-\infty \le \alpha$, $\beta \le +\infty$.

We shall denote by $G(s)$ the (modified) resolvent of $\widetilde{P}$ defined by

$$G(s) = \big(\widetilde{P} - s(n-1-s)\big)^{-1}. \tag{4.1}$$

$G(s)$ is an operator valued function of s with values in $B(L^2(M), L^2(M))$. More precisely, it is a meromorphic function defined in the half-plane $\operatorname{Re} s > (n-1)/2$ with poles (of the first order) at the sequence of real points $\{s_j\}$ such that $s_j(n-1-s_j) \in \sigma_{\text{dis}}(\widetilde{P})$, $s_j > (n-1)/2$.

It is clear that as a function with values in $B\big(L^2(M), L^2(M)\big)$ the function $G(s)$ cannot be continued analytically beyond the half-plane $\operatorname{Re} s > (n-1)/2$. However, this situation changes if one modifies the definition of $G(s)$ and consider it (in a natural way) as a function with values in $B\big(L^{2,\alpha}(M), L^{2,-\alpha}(M)\big)$ for some $\alpha > 0$. We shall see that under this modification $G(s)$ admits a meromorphic continuation to the half-plane $\operatorname{Re} s > (n-1)/2 - \alpha$. Before discussing this result we make the following:

OBSERVATION. Let $u = G(s)f$ where $f \in L^2(M)$ and $\operatorname{Re} s > (n-1)/2$, s not a pole of the resolvent. Then u is a s-outgoing solution of (3.1).

Indeed, by a well known result any solution u of (3.1) with u and f in $L^2(M)$ satisfies:

$$|\nabla u| \in L^2(M).$$

Thus, in particular, we have:

$$(D_\rho + s - n + 1)u \in L^2(M) \tag{4.2}$$

which implies that u is s-outgoing (condition (3.21) holds with $\beta = 0$).

The following is a basic result.

THEOREM 4.1. *Consider $G(s)$ as an operator valued meromorphic function with values in $B\big(L^{2,\alpha}(M), L^{2,-\alpha}(M)\big)$ for some $\alpha > 0$. Then $G(s)$ admits a meromorphic continuation to the half-plane $\operatorname{Re} s > (n-1)/2-\alpha$. Moreover, the following holds.*

(i) *The coefficients of the principal part in the Laurent expansion of $G(s)$ about any pole in the half-plane $\operatorname{Re} s > (n-1)/2 - \alpha$ are finite rank operators.*

(ii) *Let $f \in L^{2,\alpha}(M)$ and set $u = G(s)f$ ($\operatorname{Re} s > (n-1)/2 - \alpha, s$ not a pole). Then u is the unique s-outgoing solution of the equation*

$$Pu - s(n-1-s)u = f. \tag{4.3}$$

In the following $G_W(s)$ will denote the (modified) resolvent of the operator $-\Delta + W$. Theorem 4.1 is derived from the following theorem.

THEOREM 4.2. *Let α, R be positive numbers. There exists a constant $c_0 = c_0(\alpha, R) > 0$ such that setting $W = ce^{-\rho}$, where c is a constant $\geq c_0$, the following holds:*

(i) *$G_W(s)$ is analytic for $\operatorname{Re} s > (n-1)/2$. Moreover, as a function with values in $B\big(L^{2,\alpha}(M), L^{2,-\alpha}(M)\big)$ $G_W(s)$ admits analytic continuation to the domain:*

$$\mathcal{D}_{\alpha,R} = \left\{ s : \operatorname{Re} s > \frac{1}{2}(n-1) - \alpha, |\operatorname{Im} s| < R \right\}. \tag{4.4}$$

(ii) *Let $f \in L^{2,\alpha}(M)$ and set $u = G_W(s)f$, $s \in \mathcal{D}_{\alpha,R} \cup \{s : \operatorname{Re} s > (n-1)/2\}$. Then u is the unique s-outgoing solution of the equation*

$$-\Delta u + Wu - s(n-1-s)u = f. \tag{$*$}$$

The proof of Theorem 4.2 is technical and quite long. In the Appendix we give some details of the proof. In particular we establish certain *a priori* estimates for s-outgoing solutions of ($*$) (for c large). These estimates form the core of the proof of Theorem 4.2.

PROOF OF THEOREM 4.1. Fix a number $\alpha > 0$ and consider $G(s)$ as a function with values in $B\big(L^{2,\alpha}(M), L^{2,-\alpha}(M)\big)$. We wish to show that $G(s)$ admits a meromorphic continuation from the half-plane $\operatorname{Re} s > (n-1)/2$ to the half-plane $\operatorname{Re} s > (n-1)/2 - \alpha$. To this end pick a number $R > 0$ and then chose a function $W = ce^{-\rho}$, c a large positive constant such that Theorem 4.2 holds for $G_W(s)$ (relative to the constants α, R). Thus in particular the function $G_W(s)$ with values in $B\big(L^{2,\alpha}(M), L^{2,-\alpha}(M)\big)$ is analytic in the domain $\mathcal{D}_{\alpha,R} \cup \{s : \operatorname{Re} s > (n-1)/2\}$. By the resolvent equation we have

$$G(s)\big(I + T(s)\big) = G_W(s) \tag{4.5}$$

where

$$T(s) = (V - W)G_W(s). \tag{4.6}$$

Here $T(s)$ is considered as a function with values in $B(L^{2,\alpha}(M), L^{2,\alpha}(M))$ and I is the identity operator on $L^{2,\alpha}(M)$. Equation (4.5) holds for all s in the half-plane $\operatorname{Re} s > (n-1)/2$, $s(n-1-s) \notin \sigma_{\text{dis}}(\widetilde{P})$. Now, it follows from (4.5) that for such s the operator $I + T(s)$ is one to one on $L^{2,\alpha}(M)$. Observing that $T(s)$ is also a compact operator on $L^{2,\alpha}(M)$ (this is an easy consequence of the rapid decay of $V - W$ at the cusps and interior elliptic regularity) it follows from the Fredholm alternative and (4.5) that $\big(I + T(s)\big)^{-1}$ exists in $L^{2,\alpha}(M)$ and that

$$G(s) = G_W(s)\big(I + T(s)\big)^{-1} \tag{4.7}$$

for all s in the half-plane $\operatorname{Re} s > (n-1)/2$, $s(n-1-s) \notin \sigma_{\text{dis}}(\widetilde{P})$. Now, it follows from (4.6) that $T(s)$ is defined and is analytic function of s, with values in $B(L^{2,\alpha}(M), L^{2,\alpha}(M))$, in the domain $\mathcal{D}_{\alpha,R} \cup \{s : \operatorname{Re} s > (n-1)/2\}$. Hence, by the *analytic* Fredholm theorem [**17**] it follows that $\big(I + T(s)\big)^{-1}$ is a well defined meromorphic function of s in the same domain. It follows from the last remark and (4.7) that the function $G(s)$, with values in $B(L^{2,\alpha}(M), L^{2,-\alpha}(M))$, admits a meromorphic continuation to $\mathcal{D}_{\alpha,R}$ (given by the r.h.s. of (4.7)). Since R is arbitrary it follows, moreover, that $G(s)$ admits a meromorphic continuation to the half-plane $\operatorname{Re} s > (n-1)/2 - \alpha$.

In conclusion we observe that statement (i) of Theorem 4.1 follows from the analytic Fredholm theorem applied to $\big(I+T(s)\big)^{-1}$ and combined with (4.7), while statement (ii) of the theorem follows easily from assertion (ii) of Theorem 4.2 and formula (4.7). □

5. Resonance and Eigenfunctions and Poles of the Resolvent

DEFINITION. We shall use the notation $\Lambda(P)$ to denote the set of poles of the resolvent $G(s)$ of P in the whole complex plane. Here $G(s)$ is considered as an operator valued function with values in $B(L^{2,\infty}(M), L^{2,-\infty}(M))$.

REMARK. In the following when we write $G(s)$ for s varying in the half-plane $\operatorname{Re} s > (n-1)/2 - \alpha$, $\alpha \geq 0$, we tacitly assume that $G(s)$ is a function with values in $B(L^{2,\alpha}(M), L^{2,-\alpha}(M))$ obtained by analytic continuation of the L^2 resolvent of P in the sense of Theorem 4.1.

Observe that for any $s \notin \Lambda(P)$ the operator $G(s)$ is symmetric in the sense that

$$\langle G(s)\varphi, \psi\rangle = \langle \varphi, G(s)\psi\rangle \tag{5.1}$$

for any φ, $\psi \in L^{2,\infty}(M)$. This follows by analytic continuation on s since (5.1) clearly holds for large positive s (in view of the self-adjointness and reality of $G(s)$). More precisely, in view of Theorem 4.1 the same analytic continuation argument shows that (5.1) holds for $\varphi, \psi \in L^{2,\alpha}(M)$ for any non-negative α such that $\alpha > (n-1)/2 - \operatorname{Re} s$. Here and in the following we set

$$\langle f, g\rangle = \int_M f g \, d\mu \tag{5.2}$$

for any two functions f, g on M for which the r.h.s. of (5.2) is a well defined finite integral.

We shall need the following lemma.

LEMMA 5.1. *Let u_i, $i = 1, 2$, be a s_0-outgoing solution of the (respective) differential equation*

$$Pu_i - s_0(n-1-s_0)u_i = f_i \text{ on } M, \tag{5.3}$$

$i = 1, 2$, where $f_i \in L^{2,\alpha}(M)$ for some non-negative $\alpha > (n-1)/2 - \operatorname{Re} s_0$. Then $u_i \in L^{2,-\alpha}(M)$ and the following relation holds

$$\langle u_1, f_2 \rangle = \langle u_2, f_1 \rangle. \tag{5.4}$$

PROOF. Suppose first that $s_0 \notin \Lambda(P)$. Applying Theorem 4.1 (uniqueness statement in (ii)) it follows that $u_i = G(s_0)f_i$ which in turn implies that $u_i \in L^{2,-\alpha}(M)$ for $i = 1, 2$. Thus in this case the relation (5.4) is nothing else but the relation (5.1) proved before.

Next suppose that $s_0 \in \Lambda(P)$. Invoke Theorem 4.2 and choose a function $W = ce^{-\rho}$, c a large positive constant, such that $s_0 \notin \Lambda(P_W)$, $P_W = -\Delta + W$. Note now that in view of (5.3) and our assumptions the function u_i is also a s_0-outgoing solution of the (respective) differential equation

$$P_W u_i - s_0(n-1-s_0)u_i = g_i \text{ on } M, \tag{5.5}$$

where

$$g_i = f_i + (W - V)u_i \in L^{2,\alpha}(M), \tag{5.5'}$$

$i = 1, 2$. Hence, by the special case of the theorem proved above (applied to solutions of (5.5)) it follows that $u_i \in L^{2,-\alpha}(M)$ and that

$$\langle u_1, g_2 \rangle = \langle u_2, g_1 \rangle. \tag{5.6}$$

Substituting for g_i in (5.6) from (5.5′) we obtain (5.4) proving the lemma. □

The following theorem gives a necessary and sufficient condition for s_0 to be a pole of $G(s)$.

THEOREM 5.2. *Consider the differential equation*

$$Pu - s_0(n-1-s_0)u = 0 \text{ on } M. \tag{5.7}$$

The following holds.

(i) *$s_0 \in \Lambda(P)$ iff there exist a non-trivial s_0-outgoing solution of* (5.7).

(ii) *The vector space of s_0-outgoing solutions of* (5.7) *is finite dimensional.*

PROOF. Fix a point s_0 and let $D(s_0)$ be an open disc in the s-plane centered at s_0 such that $G(s)$ is analytic in $D(s_0)$ except possibly for a pole at s_0. Fix also a non-negative number α such that $\alpha > (n-1)/2 - \operatorname{Re} s$ for all $s \in D(s_0)$. Next, invoke Theorem 4.2 and choose a function W on M of the form: $W = ce^{-\rho}$, c a constant, such that $G_W(s)$ (the resolvent of $-\Delta + W$) is analytic in $D(s_0)$. Here $G(s)$ and $G_W(s)$ are considered as functions in $D(s_0)$ with values in $B(L^{2,\alpha}(M), L^{2,-\alpha}(M))$. Proceeding as in the proof of Theorem 4.1, it follows from the resolvent equation that

$$G(s)(I + T(s)) = G_W(s) \text{ in } D(s_0) \setminus \{s_0\}, \tag{5.8}$$

where

$$T(s) = (V - W)G_W(s). \tag{5.8'}$$

I is the identity operator in $B(L^{2,\alpha}(M), L^{2,\alpha}(M))$ and $T(s)$ is considered as an analytic function in $D(s_0)$ with values in $B(L^{2,\alpha}(M), L^{2,\alpha}(M))$. As was remarked already before $T(s)$ is a compact operator in $B(L^{2,\alpha}(M), L^{2,\alpha}(M))$ for each s in $D(s_0)$.

Next observe that $(I + T(s))^{-1}$ exists in $B(L^{2,\alpha}(M), L^{2,\alpha}(M))$ for any $s \in D(s_0) \setminus \{s_0\}$. Indeed, if this were not the case for some such s then it would follow from the Fredholm theory that the equation: $v + T(s)v = 0$ admits a non-trivial solution $v_0 \in L^{2,\alpha}(M)$. Applying (5.8) it would follow that $G_W(s)v_0 = 0$ which is impossible.

The analytic Fredholm theory and the last remark imply that $(I + T(s))^{-1}$ exists as an analytic function in $D(s_0)$ (with values in $B(L^{2,\alpha}(M), L^{2,\alpha}(M))$) except possibly for a pole at s_0. It follows now from (5.8) that

$$G(s) = G_W(s)(I + T(s))^{-1} \tag{5.9}$$

for $s \in D(s_0) \setminus \{s_0\}$.

From (5.9) it follows that s_0 is a pole of $G(s)$ iff s_0 is a pole of $(I + T(s))^{-1}$ (note that the map $f \to G_W(s)f$ in one to one!). This and the Fredholm theorem yield the following

LEMMA. *s_0 is a pole of $G(s)$ iff the equation*

$$v + T(s_0)v = 0, \qquad v \in L^{2,\alpha}(M), \tag{5.10}$$

admits a non-trivial solution.

Denote by $\mathcal{K}(s_0)$ the vector space of solutions of (5.10) and let $\mathcal{L}(s_0)$ denote the vector space of s_0-outgoing solutions of (5.7) (note that by Fredholm's theorem $\mathcal{K}(s_0)$ is finite dimensional). We shall show that the map

$$\tau \colon \mathcal{K}(s_0) \ni v \to G_W(s_0)v \tag{5.11}$$

is an isomorphism between $\mathcal{K}(s_0)$ and $\mathcal{L}(s_0)$. This and the above lemma will establish the theorem.

It is clear that τ is a one to one map. Now, let $v \in \mathcal{K}(s_0)$ and set $u = G_W(s_0)v$. Hence u is a s_0-outgoing solution of the equation

$$-\Delta u + Wu - s_0(n - 1 - s_0)u = v. \tag{5.12}$$

Since

$$v = -T(s_0)v = (W - V)G_W(s_0)v = (W - V)u, \tag{5.13}$$

it follows from (5.12) and (5.13) that $u \in \mathcal{L}(s_0)$. Next let $u \in \mathcal{L}(s_0)$ and set $v = (W - V)u$. Then u is s_0-outgoing solution of (5.12). Invoking Theorem 4.1 for P_W it follows that $u = G_W(s_0)v$ and that $v \in \mathcal{K}(s_0)$ (since $v = (W - V)G_W(s_0)v = -T(s_0)v$). The above arguments show that $\mathcal{K}(s_0)$ and $\mathcal{L}(s_0)$ are isomorphic vector spaces. The proof of Theorem 5.2 is now complete. □

We now introduce the notion of resonances and resonance functions.

DEFINITION. A pole s_0 of $G(s)$ is said to be a resonance (of P) if equation (5.7) admits a s_0-outgoing solution u such that $u \notin L^2(M)$. Any such solution u is called a resonance function.

It follows from Theorem 3.3 that any s_0-outgoing solution u of equation (5.7) has an asymptotic expansion of the form

$$u = c_j \rho^{n-1-s_0} + O(\rho^{-\infty}) \tag{5.14}$$

in Z_j where c_j is some constant, $j = 1, \dots, m$. Using (5.14) and the self-adjointness of $\widetilde{P}$ it is easy to verify the following assertions.

(i) All resonances of P are contained in the half-plane $\operatorname{Re} s \leq (n-1)/2$.

(ii) All non-real poles of $G(s)$ in the half-plane $\operatorname{Re} s < (n-1)/2$ are resonances of P.

(iii) If s_0 is a real pole of $G(s)$ such that $\operatorname{Re} s_0 < (n-1)/2$ then s_0 is a resonance except when all s_0-outgoing solutions of (5.7) decay rapidly at the cusps. This exceptional case arises iff $s_0(n-1-s_0) \in \sigma_{\text{dis}}(\widetilde{P})$ with all corresponding eigenfunctions decaying rapidly at the cusps.

Later on we shall see that there are no resonances on the line $\operatorname{Re} s = (n-1)/2$ except possibly the point $s_0 = (n-1)/2$

Let s_0 be a pole of $G(s)$. We shall denote by $\Pi(s_0)$ the residue of $G(s)$ at s_0. It follows from Theorem 4.1 that $\Pi(s_0)$ is a finite rank operator in $B(L^{2,\alpha}, L^{2,-\alpha})$ where α is any non-negative number such that $\alpha > (n-1)/2 - \operatorname{Re} s_0$. $\Pi(s_0)$ is a symmetric operator in the sense that

$$\langle \Pi(s_0)\varphi, \psi\rangle = \langle \varphi, \Pi(s_0)\psi\rangle \tag{5.15}$$

for all φ, $\psi \in L^{2,\alpha}(M)$. This follows form (5.1) and the formula

$$\Pi(s_0) = \frac{1}{2\pi i}\int_{|s-s_0|=\varepsilon} G(s)\, ds \tag{5.16}$$

valid for all $\varepsilon > 0$ sufficiently small.

Suppose now that s_0 is a pole of the first order of $G(s)$. Let $G_W(s)$ and $T(s)$ be defined as in the proof of Theorem 5.2. Recall that for s in some disc $D(s_0)$ centered at s_0 we have

$$G(s) = G_W(s)\big(I + T(s)\big)^{-1} \tag{5.9}$$

where $G_W(s)$ is an analytic function in $D(s_0)$ with values in $B(L^{2,\alpha}, L^{2,-\alpha})$ and $\big(I + T(s)\big)^{-1}$ is a meromorphic function in $D(s_0)$ with values in $B(L^{2,\alpha}, L^{2,-\alpha})$. By choosing $D(s_0)$ small enough we may assume that α is any given non-negative number such that $\alpha > (n-1)/2 - \operatorname{Re} s_0$. Since s_0 is a pole of the first order of $G(s)$ it follows from (5.9) that s_0 is also a pole of the first order of $\big(I + T(s)\big)^{-1}$. Applying the residue theorem it follows from (5.9) that

$$\Pi(s_0) = G_W(s_0)Q(s_0) \tag{5.17}$$

where $Q(s_0)$ is the residue of $\big(I + T(s)\big)^{-1}$ at s_0. Clearly $Q(s_0) \in B(L^{2,\alpha}, L^{2,\alpha})$. Formula (5.17) will be useful in the sequel.

THEOREM 5.3. *Let s_0 be a pole of the first order of $G(s)$. The range of $\Pi(s_0)$ admits the following characterization:*

$$\operatorname{Ran}\Pi(s_0) = \mathcal{L}(s_0) \tag{5.18}$$

where $\mathcal{L}(s_0)$ denotes the vector space of s_0-outgoing solutions of equation (5.7).

PROOF. Let $f \in L^{2,\alpha}(M)$ where α is any non-negative number such that $\alpha > (n-1)/2 - \operatorname{Re} s_0$. Set

$$u = \Pi(s_0)f. \tag{5.19}$$

We shall show that $u \in \mathcal{L}(s_0)$. To this end observe that the following relation holds

$$\begin{aligned}\langle G(s)f, (P - s_0(n-1-s_0))\varphi\rangle &= \langle f, \varphi\rangle \\ &+ (s(n-1-s) - s_0(n-1-s_0))\langle G(s)f, \varphi\rangle\end{aligned} \tag{5.20}$$

for all $\varphi \in C_0^\infty(M)$ and all s in a sufficiently small neighborhood of s_0, $s \neq s_0$. Note also that the r.h.s of (5.20) is analytic in a neighborhood of s_0. Hence, integrating both sides of (5.20) along a circle $|s - s_0| = \varepsilon$ where ε is taken sufficiently small, using (5.16) and (5.19), it follows that

$$\big\langle u, (P - s_0(n-1-s_0))\varphi\big\rangle = 0$$

for all $\varphi \in C_0^\infty(M)$. This proves that u is a solution of (5.7). Moreover, from (5.17) and (5.19) it follows that $u = G_W(s_0)f_1$ where $f_1 = Q(s_0)f$, $f_1 \in L^{2,\alpha}(M)$. This shows that u is also s_0-outgoing. Thus $u \in \mathcal{L}(s_0)$ proving that $\operatorname{Ran}\Pi(s_0) \subset \mathcal{L}(s_0)$.

To prove the converse inclusion we consider $\operatorname{Ran}\Pi(s_0)$ and $\mathcal{L}(s_0)$ as finite dimensional subspaces of $L^{2,-\alpha}(M)$ (Theorem 5.2). Note that the Hilbert space $L^{2,\alpha}(M)$ is the dual of $L^{2,-\alpha}(M)$ under the duality relation $\langle\cdot,\cdot\rangle$. We shall show that if f is any function in $L^{2,\alpha}(M)$ such that

$$\langle u, f\rangle = 0 \tag{5.21}$$

for all $u \in \operatorname{Ran}\Pi(s_0)$, then

$$\langle v, f\rangle = 0 \tag{5.22}$$

for all $v \in \mathcal{L}(s_0)$. By a well known result this will imply that $\mathcal{L}(s_0) \subset \operatorname{Ran}\Pi(s_0)$.

Observe that condition (5.21) is equivalent to the condition

$$\Pi(s_0)f = 0 \tag{5.21$'$}$$

since (5.21) can be written in the form:

$$0 = \langle \Pi(s_0)h, f\rangle = \langle h, \Pi(s_0)f\rangle$$

for all $h \in L^{2,\alpha}(M)$.

Hence, suppose that (5.21$'$) holds for some $f \in L^{2,\alpha}(M)$. We shall show that (5.22) must hold. To this end consider the vector valued function $G(s)f$ which is a meromorphic function of s with values in $L^{2,-\alpha}(M)$. Since s_0 is a pole of the first order of $G(s)$ it follows from (5.21$'$) that s_0 is a removable singularity of $G(s)f$. Set

$$u_0 = \lim_{s\to s_0} G(s)f \tag{5.23}$$

where the limit is taken in $L^{2,-\alpha}(M)$. We claim that u_0 verifies the differential equation

$$Pu_0 - s_0(n-1-s_0)u_0 = f \text{ on } M \tag{5.24}$$

and that, moreover, u_0 is s_0-outgoing. To see this invoke once more Theorem 4.2 and choose a function $W = ce^{-\rho}$ such that $G_W(s)$ is an analytic function of s in a neighborhood of s_0. By the resolvent equation we find that

$$G_W(s)f + G_W(s)(W-V)G(s)f = G(s)f \tag{5.25}$$

for s in a neighborhood of s_0, $s \neq s_0$. Letting $s \to s_0$ it follows from (5.23) and (5.25) that

$$u_0 = G_W(s)(f + (W-V)u_0). \tag{5.26}$$

From (5.26) it follows readily that u_0 is a solution of (5.24) and that u_0 is s_0-outgoing. Now, let $v \in \mathcal{L}(s_0)$. Applying Lemma 5.1 to the functions u_0 and v, noting that the functions are s_0-outgoing solutions of equation (5.24) and equation (5.7), respectively, we conclude that (5.22) holds. This completes the proof. □

The following proposition is an easy corollary of Theorem 5.3.

PROPOSITION 5.4. *Let s_0 be a pole of the first order of $G(s)$. The residue operator $\Pi(s_0)$ has the following representation*

$$\Pi(s_0) = \sum_{i,j=1}^{\ell} c_{ij}\langle\cdot, \psi_i\rangle\psi_j \tag{5.27}$$

where ℓ = dimension $\Pi(s_0)$ and $\psi_1, \dots, \psi_\ell$ are any ℓ linearly independent s_0-outgoing solutions of (5.7). Here $(c_{ij})_{i,j=1}^{\ell}$ is some non-singular symmetric matrix of complex constants.

We now turn our attention to the special poles of $G(s)$ which are located on the line $\operatorname{Re} s = (n-1)/2$. Observe that all such poles are of the first order except possibly the point $s_0 = (n-1)/2$ which may be a pole of the first or of the second order. This is a consequence of the following elementary resolvent norm estimates:

(i) $\|G(s_0+\varepsilon)\| = O(\varepsilon^{-1})$ as $\varepsilon \downarrow 0$ if $\operatorname{Re} s_0 = (n-1)/2$, $s_0 \neq (n-1)/2$.

(ii) $\|G((n-1)/2+\varepsilon)\| = O(\varepsilon^{-2})$ as $\varepsilon \downarrow 0$.

Here $\|\cdot\|$ stands for the operator norm in $B(L^2(M), L^2(M))$.

PROPOSITION 5.5. *Let s_0 be a point on the line $\operatorname{Re} s = (n-1)/2$. Set $\lambda_0 = s_0(n-1-s_0)$ (note that $\lambda_0 \in \sigma_{\text{ess}}(\widetilde{P})$). Let $\mathcal{E}_{\{\lambda\}}$ be the L^2 orthogonal projection on the λ_0-eigenspace of $\widetilde{P}$ ($\mathcal{E}_{\{\lambda_0\}}$ denotes the zero projection if λ_0 is not an eigenvalue of $\widetilde{P}$). Finally let $\Pi_i(s_0)$, $i = 1, 2$, be the coefficients of $(s-s_0)^{-i}$ in the Laurent expansion of $G(s)$ about s_0. (By Theorem 4.1 $\Pi_i(s_0) \in B(L^{2,\alpha}(M), L^{2,-\alpha}(M))$ for any $\alpha > 0$.) The following holds:*

(i) *If $s \neq (n-1)/2$ then*

$$\mathcal{E}_{\{\lambda_0\}}f = (2s_0 - n + 1)\Pi_1(s_0)f \tag{5.28}$$

for any $f \in L^{2,\alpha}(M)$, $\alpha > 0$.

(ii) *If $s_0 = (n-1)/2$ then*

$$\mathcal{E}_{\{\lambda_0\}}f = \Pi_2(s_0)f \tag{5.29}$$

for any $f \in L^{2,\alpha}(M)$, $\alpha > 0$.

PROOF. The following is a well known formula derived from the spectral theorem:

$$\mathcal{E}_{\{\lambda_0\}} f = \lim_{s \to s_0} \big(s_0(n-1-s_0) - s(n-1-s)\big) G(s) f \tag{5.30}$$

for $f \in L^2(M)$. Here s tends to s_0 from the right along the line $s = s_0 + t$, $t > 0$, and the limit is the strong limit in $L^2(M)$.

Now, let $f, h \in L^{2,\alpha}(M)$ for some $\alpha > 0$. It follows from Theorem 4.1 that the function $\langle G(s)f, h\rangle$ admits a meromorphic continuation to the half-plane $\operatorname{Re} s > (n-1)/2 - \alpha$. Using (5.30), we obtain

$$\langle \mathcal{E}_{\{\lambda_0\}} f, h\rangle = \lim_{s \to s_0} \big(s_0(n-1-s_0) - s(n-1-s)\big) \langle G(s) f, h\rangle \tag{5.31}$$

Formulas (5.28) and (5.29) follow from (5.31) by a straightforward computation. □

COROLLARIES.

(i) *It follows from* (5.28) *and Theorem* 5.3 *that there are no resonances on the line* $\operatorname{Re} s = (n-1)/2$ *except possibly the point* $s_0 = (n-1)/2$.

(ii) *It follows from* (5.28) *that a point* s_0 *with* $\operatorname{Re} s_0 = (n-1)/2$, $s_0 \neq (n-1)/2$, *is a pole of* $G(s)$ *iff* $s_0(n-1-s_0)$ *is an eigenvalue of* $\widetilde{P}$.

(iii) *It follows form* (5.29) *that* $\lambda_0 = \big((n-1)/2\big)^2$ *(the bottom of* $\sigma_{\mathrm{ess}}(\widetilde{P})$*) is an eigenvalue of* $\widetilde{P}$ *iff* $s_0 = (n-1)/2$ *is a pole of the second order of* $G(s)$.

(iv) *Since the* Π_i *are finite rank operators it follows from* (5.28)-(5.29) *that all eigenvalues of* $\widetilde{P}$ *embedded in the continuous spectrum have a finite multiplicity.*

(v) *Since* $\sigma_{\mathrm{dis}}(\widetilde{P}) = \{\lambda = s(n-1-s) : s > (n-1)/2, s \in \Lambda(\widetilde{P})\}$ *is a finite set, it follows from* (ii) *and* (iv) *that* $\sigma_p(\widetilde{P})$ *consists of an increasing sequence of eigenvalues of finite multiplicity having no finite limit point.*

This completes the proof of Theorem 2.1.

6. The Generalized Eigenfunctions

DEFINITION. A function u on M will be called a generalized eigenfunction of P, with eigenvalue λ, if

(i) $u \in L^{2,\alpha}(M)$ for some α.

(ii) u satisfies the differential equation

$$Pu = \lambda u \text{ on } M. \tag{6.1}$$

It follows from Theorem 3.1 that a generalized eigenfunction $u(z)$ with eigenvalue $\lambda = s(n-1-s)$ has the following asymptotic expansion as $\rho(z) \to \infty$ in the cusp Z_j, $j = 1, \ldots, m$:

$$u = c_j \rho^s + b_j \rho^{n-1-s} + O(\rho^{-\infty}) \tag{6.2}$$

for $s \neq (n-1)/2$,

$$u = \rho^{(n-1)/2}(c_j \log \rho + b_j) + O(\rho^{-\infty}) \tag{6.2$'$}$$

for $s = (n-1)/2$ where c_j, b_j, are constants.

Note that L^2 eigenfunctions and resonance functions are generalized eigenfunctions with eigenvalue $\lambda = s(n-1-s)$ which are s (or $n-1-s$) outgoing solutions of (6.1). Note also that if u is an L^2 eigenfunction of P with eigenvalue λ embedded in the continuous spectrum (so that $\operatorname{Re} s = (n-1)/2$) then the asymptotic expansions (6.2) and (6.2′) can hold only if $c_j = b_j = 0$ for all j. Thus such an eigenfunction verifies the rapid decay condition:

$$u - O(\rho^{-\infty}) \tag{6.3}$$

in all cusps.

THEOREM 6.1. *Let s be a complex number, $s \notin \Lambda(P)$ and $s \neq (n-1)/2$. Given complex numbers $c_1^0, \dots, c_m^0$, there exists a unique generalized eigenfunction u of P with eigenvalue $\lambda = s(n-1-s)$ such that u has the asymptotic expansion*:

$$u = c_j^0 \rho^s + b_j \rho^{n-1-s} + O(\rho^{-\infty}) \tag{6.4}$$

at the cusp Z_j, $j = 1, \dots, m$ (b_j some constants).

PROOF. We first prove uniqueness. One has to show that if u is a generalized eigenfunction possessing an asymptotic expansion (6.4) with constants $c_j^0 = 0$ for $j = 1, \dots, m$, then u is a null function. This, however, is an immediate corollary of Theorem 4.1 since such u is a s_0-outgoing solution of equation (6.1) and by assumption $s_0 \notin \Lambda(P)$.

The uniqueness result just proved implies that for a fixed s (restricted as above) the set of generalized eigenfunctions with eigenvalue $\lambda = s(n-1-s)$ is a finite dimensional vector space of dimension $\leq m$. To complete the proof of the theorem we show that this space contains a basis of m generalized eigenfunctions $E_i(z;s)$, $i = 1, \dots, m$, having the asymptotic expansion:

$$E_i(z;s) = \delta_{ij}\rho(z)^s + b_{ij}\rho(z)^{n-1-s} + O(\rho(z)^{-\infty}) \tag{6.5}$$

as $\rho(z) \to \infty$ in the cusp Z_j, $j = 1, \dots, m$. Here δ_{ij}, $i, j = 1, \dots, m$ is the Kronecker delta.

To define E_i pick a function $\chi_i \in C^\infty(M)$ with $\operatorname{supp}\chi_i \subset Z_i$, such that $\chi_i(z) = 1$ in Z_i outside some compact. Introduce the function

$$f_i(z;s) = \big(P - s(n-1-s)\big)\big(\chi_i(z)\rho(z)^s\big) \tag{6.6}$$

and note that f_i decays rapidly at the cusps. Now, define

$$E_i(z;s) = \chi_i(z)\rho(z)^s - \big(G(s)f_i(\cdot;s)\big)(z). \tag{6.7}$$

It is readily seen that $E_i(z;s)$ is a generalized eigenfunction of P with the desired properties. In particular it possesses the asymptotic expansion (6.5) since $\chi_i\rho^s = \delta_{ij}\rho^s$ outside some compact while the function $G(s)f_i$ is s-outgoing (see Theorems 4.1 and 3.3). $\square$

We shall refer to the functions $E_i(z;s)$, $i = 1, \dots, m$, as the *canonical* generalized eigenfunctions of P. Clearly $E_i(z;s)$ is a meromorphic function in s with poles contained in the set $\Lambda(P)$. The following question arises: Under what conditions a pole of $G(s)$ is a removable singularity for one of the functions $E_i(\cdot;s)$. In the following we resolve the question for poles of the first order. In this connection recall that if $s_0 \in \Lambda(P)$ then there exists a s_0-outgoing non-trivial generalized eigenfunction of P with eigenvalue $\lambda_0 = s_0(n-1-s_0)$ (Theorem 5.2).

THEOREM 6.2. *Let s_0 be a pole of the first order of $G(s)$, $s_0 \neq (n-1)/2$. A necessary and sufficient condition for s_0 to be a removable singularity of $E_i(\cdot; s)$ (some i) is that all s_0-outgoing generalized eigenfunctions $\psi(z)$ of P with eigenvalue $\lambda_0 = s_0(n-1-s_0)$ decay rapidly as $\rho(z) \to \infty$ in the cusp Z_i.*

PROOF. Observing that the function $f_i(z,s)$ (defined by (6.6)) is analytic in the s-plane and decays rapidly in z at all cusps it follows readily from (6.7) that s_0 is a removable singularity of $E_i(\cdot; s)$ iff

$$\Pi(s_0) f_i(\cdot; s_0) = 0. \tag{6.8}$$

Here $\Pi(s_0)$ is the residue of $G(s)$ at s_0 and we have used the fact that s_0 is a pole of the first order. Using the symmetry of $\Pi(s_0)$ on $L^{2,\infty}(M)$ condition (6.8) can be written in the equivalent form:

$$\langle \Pi(s_0)\varphi, f_i(\cdot; s_0)\rangle = 0 \tag{6.8$'$}$$

$\forall \varphi \in L^{2,\infty}(M)$. Recalling that $\operatorname{Ran}\Pi(s_0) = \mathcal{L}(s_0)$ (Theorem 5.3) condition (6.8$'$) can be written in the form:

$$\langle \psi, f_i(\cdot; s_0)\rangle = 0 \tag{6.9}$$

$\forall \psi \in \mathcal{L}(s_0)$ (vector space of s_0-outgoing generalized eigenfunctions with eigenvalue $s_0(n-1-s_0)$). Thus inserting for f_i in (6.9) its expression from (6.6) we arrive at the following condition: s_0 is a removable singularity of $E_i(\cdot; s)$ iff

$$\int_M \psi(P - s_0(n-1-s_0)) v_i \, d\mu = 0 \tag{6.10}$$

for all $\psi \in \mathcal{L}(s_0)$ where

$$v_i(z) = \chi_i(z)\rho(z)^{s_0}. \tag{6.11}$$

Now, by Theorem 3.3 a function ψ is in $\mathcal{L}(s_0)$ has the following asymptotic expansion (in canonical coordinates), in the cusp Z_j:

$$\psi(z) = c_j(\psi) y^{n-1-s_0} + O(y^{-\infty}) \tag{6.12}$$

as $y \to \infty$ where $c_j(\psi)$ is some constant. We also have, outside some compact:

$$v_i(z) = \delta_{ij} y^{s_0}. \tag{6.13}$$

Using (6.12) and (6.13), and the corresponding expansions for $\partial\psi/\partial y$ and $\partial v_i/\partial y$, one obtains via integration by parts (Green's formula properly applied) that

$$\int_M \psi(P - s_0(n-1-s_0)) v_i \, d\mu = c_i(\psi)(2s_0 - n + 1). \tag{6.14}$$

We omit the details of the computation which is straightforward and standard.

Finally, combining (6.10) and (6.14) we conclude that s_0 is a removable singularity of $E_i(\cdot; s)$ iff $c_i(\psi) = 0$ for all $\psi \in \mathcal{L}(s_0)$. Taking note of (6.12) this proves the theorem. □

We shall use Theorem 6.2 to prove the following

THEOREM 6.3. *The poles of $G(s)$ situated on the line* $\operatorname{Re} s = (n-1)/2$, *except possibly a pole at* $s_0 = (n-1)/2$, *are removable singularities of the functions* $E_i(\cdot; s)$, $i = 1, \ldots, m$.

REMARK. We shall see later on that the point $(n-1)/2$ is also a removable singularity of the $E_i(\cdot; s)$.

PROOF OF THEOREM 6.3. Let s_0 be a pole of $G(s)$ with $\operatorname{Re} s_0 = (n-1)/2$, $s_0 \neq (n-1)/2$. Thus s_0 is a pole of the first order. From Theorem 5.3 and Proposition 5.5 it follows that $\mathcal{L}(s_0)$ consists of L^2 eigenfunctions for P with eigenvalue λ_0 embedded in $\sigma_{\text{ess}}(\widetilde{P})$. Such eigenfunctions decay rapidly at all cusps (see (6.3)). Hence by Theorem 6.2 s_0 is a removable singularity for $E_i(\cdot; s)$, $i = 1, \ldots, m$. □

REMARK. Let $s_0 \in \Lambda(P)$, s_0 real and $> (n-1)/2$. Then s_0 is a pole of the first order and $s_0(n-1-s_0) \in \sigma_{\text{dis}}(\widetilde{P})$. Applying Theorem 5.3 it follows that also in this case: $\mathcal{L}(s_0) =$ the eigenspace $\mathcal{E}_{\{\lambda_0\}}$, $\lambda_0 = s_0(n-1-s_0)$. Thus it follows from Theorem 6.2 that s_0 is a removable singularity of $E_i(\cdot; s)$ (some i) iff all L^2 eigenfucitons of P with eigenvalue λ_0 decay rapidly in the cusp Z_i.

We shall now derive some useful relations between the resolvent and the canonical generalized eigenfunctions. In this connection we shall need the following lemma.

LEMMA 6.4. *Let u be a s-outgoing solution of the differential equation*

$$Pu - s(n-1-s)u = f \tag{6.15}$$

where $f \in L^{2,\infty}(M)$ *and* $s \notin \Lambda(P)$, $s \neq (n-1)/2$. *The following asymptotic expansion holds in the cusp* Z_j $(j = 1, \ldots, m)$:

$$u = c_j \rho^{n-1-s} + O_{L^2}(\rho^{-\infty}) \tag{6.16}$$

where c_j is a constant given by

$$c_j = (\omega_j(2s-n+1))^{-1} \langle E_j(\cdot; s), f \rangle. \tag{6.17}$$

Here $\omega_j = \operatorname{vol}(X_j)$ *in the measure* $d\mu_j'$ *(see §2).*

PROOF (SKETCH). That u has the asymptotic expansion (6.1) we know already from Theorem 3.3. The additional information in the lemma is that the constants c_j in (6.16) have the "explicit" expression (6.17). Now to establish formula (6.17) one applies Green's formula to the functions $u(z)$ and $E_j(z; s)$, taking note of the appropriate differential equations satisfied by these functions, using the asymptotic expansions (6.16) and (6.5) and the corresponding asymptotic expansions of $\partial u/\partial y$ and $\partial E_j/\partial y$. (In this computation one has to use a standard approximation argument since M is not compact.) We omit the details of the computation. □

THEOREM 6.5. *Suppose that neither s nor $n-1-s$ is a pole of $G(s)$ and that* $s \neq (n-1)/2$. *Let* $f \in L^{2,\infty}(M)$. *The following formula holds:*

$$G(s)f - G(n-1-s)f = (2s-n+1)^{-1} \sum_{j=1}^{m} \omega_j^{-1} \langle f, E_j(\cdot; n-1-s) \rangle E_j(\cdot; s). \tag{6.18}$$

PROOF. Consider the function $v = G(s)f - G(n-1-s)f$. Since $G(s)f$ is a s-outgoing solution of (6.15) and $G(n-1-s)f$ is a $n-1-s$-outgoing solution of the same equation, it follows from Lemma 6.4 that v has the following asymptotic expansion at the cusp Z_j:

$$v = \big(\omega_j(2s-n+1)\big)^{-1}(\langle E_j(\cdot;s), f\rangle\rho^{n-1-s} + \langle E_j(\cdot;n-1-s), f\rangle\rho^{s}) + O_{L^2}(\rho^{-\infty}). \tag{6.19}$$

Also, v is a generalized eigenfunction of P with eigenvalue $s(n-1-s)$. Thus it can be written in the form:

$$v = \sum_{j=1}^{m} \gamma_j E_j(\cdot, s) \tag{6.20}$$

where the γ_j are certain constants. Comparing the asymptotic expansions at the cusp Z_j of both sides of (6.20), using (6.17) and (6.5), we find that

$$\gamma_j = ((2s-n+1)\omega_j)^{-1}\langle f, E_j(\cdot;n-1-s)\rangle. \tag{6.21}$$

Inserting γ_j from (6.21) in (6.20) the theorem follows. □

Suppose now that $\operatorname{Re} s = (n-1)/2$, $s \neq (n-1)/2$. Since $n-1-s = \bar{s}$ it follows readily from the reality of P that

$$\begin{gathered} E_j(z;n-1-s) = \overline{E_j(z;s)}, \\ \overline{G(s)f} = G(\bar{s})\bar{f}. \end{gathered} \tag{6.22}$$

Using (6.22) it follows from (6.18) by a straightforward computation that the following formula holds

$$\operatorname{Im}\langle G(s)f, \bar{f}\rangle = -(4t)^{-1}\sum_{j=1}^{m}\omega_j^{-1}|\langle E_j(\cdot;s), \bar{f}\rangle|^2 \tag{6.23}$$

for $s = (n-1)/2 + it$, $t \in \mathbb{R}\setminus\{0\}$, $s \notin \Lambda(P)$ and $f \in L^{2,\infty}(M)$.

We note that (6.23) is the basis for the generalized eigenfunction expansion theorem associated with the operator $\widetilde{P}$. An easy corollary of (6.23) is the following proposition which complements Theorem 6.3.

PROPOSITION. *Suppose that $(n-1)/2 \in \Lambda(P)$. Then the point $(n-1)/2$ is a removable singularity of $E_j(\cdot;s)$ for $j = 1, \dots, m$.*

PROOF. It was observed before that if $(n-1)/2$ is a pole of $G(s)$ then it is at most of order 2. Thus setting $s_t = (n-1)/2 + it$, $t \in \mathbb{R}\setminus\{0\}$, we have:

$$\langle G(s_t)f, f\rangle = O(t^{-2}) \tag{6.24}$$

as $t \to 0$ for any $f \in L^{2,\infty}(M)$. From (6.23) and (6.24) it follows that

$$\langle E_j(\cdot;s_t), f\rangle = O(t^{-1/2})$$

as $t \to 0$ for any $f \in L^{2,\infty}(M)$ and all j. This implies that E_j cannot have a pole at the point $(n-1)/2$. □

References

1. S. Agmon, *Lectures on exponential decay of solutions of second-order elliptic equations*, Mathematical Notes, vol. 29, Princeton University Press, 1982.
2. ——, *Bounds on exponential decay of eigenfunctions,*, Schrödinger Operators (G. Graffi, ed.), Lecture Notes, vol. 1159, Springer, 1985.
3. Y. Colin de Verdière,, *Pseudo Laplacians* I, Ann. Inst. Fourier (Grenoble) **32** (1982), 275–286; II **33** (1983), 87–113.
4. J. Elstrodt, *Die Resolvente zum Eigenwertproblem der automorphen Formen in der hyperbolischen Ebene* I, Math Ann. **203** (1973), 295–330; II, Math. Z. **132** (1973), 99–134; III, Math. Ann. **208** (1974), 99–132.
5. L. D. Faddeev, *Expansion in eigenfunctions of the Laplace operator in the fundamental domain of a discrete group on the Lobachevski plane*, Trudy Moscov. Mat. Obsch. **17** (1967), 323–350; English transl., Trans. Moscow Math. Soc. **17** (1967), 357–386.
6. D. A. Hejhal, *The Selberg trace formula for* $PSL(2;\mathbb{R})$, Lecture Notes, vol. 1001, Springer, 1983.
7. P. Lax and R. S. Phillips, *Scattering theory for automorphic functions*, Ann. of Math. Stud., vol. 87, Princeton University Press,, Princeton, 1976.
8. H. Maass, *Über eine neue Art von nichtanalytischen automorphen Funktionen und die Bestimmung Dirichletscher Reihen durch Funktionalgelichungen*, Math. Ann. **121** (1949), 141–183.
9. W. Müller, *Spectral theory for Riemannian manifolds with cusps and a related trace formula*, Math. Nachr. **111** (1983), 197–288.
10. S. J. Patterson, *The Laplacian operator on a Riemann surface* I, Compositio Math. **31** (1975), 83–107; II, Compositio Math. **32** (1976), 71–112; III, Compositio Math. **33** (1976), 227–259.
11. R. Phillips and P. Sarnak, *On cusp forms for cofinite subgroups of* $PSL(2,\mathbb{R})$, Inven. Math. **80** (1985), 339–364.
12. ——, *Perturbation theory for the Laplacian on automorphic forms*, J. Amer. Math. Soc. **5** (1992), 1–32.
13. W. Roelcke, *Über die Wellengleichung bei Grenzkreisgruppen erster Art*, S. B. Heidelberger Akad. Wiss. Math. -Nat. Kl. (1956), 159–267.
14. ——, *Das Eigenwertproblem der automorphen Formen in der hyprebolischen Ebene* I, Math. Ann. **167** (1966), 292–337; II **168** (1967), 261–324.
15. A. Selberg, *Harmonic analysis and discontinuous groups in weakly symmetric Riemannian spaces with applications to Dirichlet series*, J. Indian Math. Soc. **20** (1956), 47–87.
16. ——, *Harmonic analysis*, Collected works, Springer-Verlag,, Berlin and New York, 1989.
17. S. Steinberg, *Meromorphic families of compact operators*, Arch. Rat. Mech. Anal. **31** (1968), 372–379.
18. A. B. Venkov, Trudy Mat. Inst. Steklov; English transl., *Spectral theory of automorphic functions*, Proc. Steklov Inst. of Math. **153** (1981).
19. S. A. Wolpert, *The spectrum of a Riemann surface with a cusp*, Prospects in Complex Geometry (J. Noguchi and T. Ohsawa, eds.), Lecture Notes, vol. 1468, Springer, 1991.

Appendix A. A Proof of Theorem 3.2 and Related Estimates

We start by recalling that if $w(x_j)$ is a function in $H^1(X_j)$ such that

$$\int_{X_j} w(x_j)d\mu_j' = 0,$$

then the following (Poincaré) inequality holds

$$\int_{X_j} |\nabla_j' w|^2 \, d\mu_j' \geq \gamma_j \int_{X_j} |w|^2 \, d\mu_j' \tag{A.1}$$

where $\gamma_j > 0$ is the second eigenvalue of the Laplace-Beltrami operator Δ_j' on the compact manifold X_j.

Next, let $u \in H_0^1(M)$ (class of H_{loc}^1 functions with compact support in M). Suppose that in every cusp Z_j the function $u(x_j, y)$ verifies the zero-mean condition (3.3) for all $y \geq a > a_j$. Then, using (A.1), it follows that

$$\text{(A.2)} \quad \int_{Z'_j} |\nabla u|^2 \, d\mu = \int_{Z'_j} y^2 \left(\left|\frac{\partial u}{\partial y}\right|^2 + |\nabla'_j u|^2 \right) d\mu \geq \gamma_j \int_{Z'_j} y^2 |u|^2 \, d\mu, \qquad \text{for } j = 1, \ldots, m,$$

where $Z'_j = \{(x_j, y) \in Z_j : y \geq a\}$.

REMARK. $|\nabla u(z)|$ and $|\nabla'_j w(x_j)|$ denote the norms of the vectors $\nabla u(z)$ and $\nabla'_j w(x_j)$ induced by the metrics g and h_j, respectively.

Using (A.2) one proves easily

PROPOSITION A.1. *Let u be a function in $H_0^1(M)$ which verifies the zero-mean condition* (3.3) *in all cusps Z_j. The following inequality holds*:

$$\text{(A.3)} \qquad \int_M |\nabla u|^2 \, d\mu \geq \int_M \lambda(z) |u|^2 \, d\mu$$

where $\lambda(z)$ is some positive continuous function (*independent of u*) *such that*

$$\text{(A.4)} \qquad \lambda(z) = \gamma_j y^2$$

in the cusp Z_j for $y > a + 1$, $j = 1, \ldots, m$.

In the following $\lambda(z)$ denotes some fixed positive continuous function, satisfying (A.4), for which Proposition A.1 holds. (The number $a > \max_j a_j$ will also be fixed.)

We shall consider solutions of the differential equation

$$\text{(A.5)} \qquad -\Delta u + qu = f.$$

We first assume that $q(z)$ is a locally bounded complex valued function on M satisfying the following condition:

$$\text{(A.6)} \qquad \operatorname{Re} q(z) \geq -\frac{1}{4}\lambda(z) + Q(z)$$

where $Q(z)$ is some non-negative locally bounded function on M. Under this condition we have:

THEOREM A.2. *Let u be a solution* (*in $H_{\text{loc}}^2(M)$*) *of equation* (A.5) *with $f \in L^{2,\ell-1}(M)$ for some ℓ. Suppose that u verifies the zero-mean condition* (3.3) *in the cusps* (*for $y > a$*) *and that $u \in L^{2,\alpha}(M)$ for some α. The following inequality holds*:

$$\text{(A.7)} \qquad (\|u\|_{\ell+1} + \|uQ^{1/2}\|_\ell + \|\nabla u\|_\ell) \leq C_\ell \|f\|_{\ell-1}$$

where C_ℓ is a constant depending only on ℓ, a, the function ρ and M. In particular C_ℓ does not *depend on Q.*

Before proving the theorem we show that Theorem 3.2 is an easy corollary of Theorem A.2.

PROOF OF THEOREM 3.2. Let u satisfy the conditions of Theorem 3.2. Thus u is a solution of (A.5) with q bounded and $f \in L^{2,\infty}(M)$. It is easy to see that there exists a non-negative continuous function χ with compact support such that

$\operatorname{Re} q(z) + \chi(z) \geq -\lambda(z)/4$ on M. Setting $q_1 = q + \chi$ it follows that q_1 verifies condition (A.6) (with $Q \equiv 0$). Considering u as a solution of the equation:

$$-\Delta u + q_1 u = f_1, \tag{A.8}$$

where $f_1 = f + \chi u$, it is readily seen that u verifies all the conditions of Theorem A.2. Applying the theorem, noting that $f_1 \in L^{2,\infty}(M)$, it follows that $(|u| + |\nabla u|) \in L^{2,\infty}(M)$. □

We turn to the

PROOF OF THEOREM A.2. Let $\psi(z)$ be a real C^1 function with compact support in M such that ψ depends only on the variable y in the cusp Z_j, $j = 1, \dots, m$. We shall use the identity

$$\operatorname{Re}[\nabla(\psi^2 \bar{u}) \cdot \nabla u] = |\nabla(\psi u)|^2 - |u|^2 |\nabla \psi|^2, \tag{A.9}$$

see [1, p. 20]. With u the function in the theorem, it follows by integration of (A.9) that

$$-\operatorname{Re} \int_M \psi^2 \bar{u} \Delta u \, d\mu = \int_M (|\nabla(\psi u)|^2 - |u|^2 |\nabla \psi|^2) \, d\mu. \tag{A.10}$$

It follows from (A.10), using (A.5) and (A.6), that the following inequality holds:

$$\int_M \left(|\nabla(\psi u)|^2 + \left(Q - \frac{1}{4}\lambda \right) |\psi u|^2 - |\nabla \psi|^2 |u|^2 \right) d\mu \leq \int_M |\psi|^2 |u| \, |f| \, d\mu. \tag{A.11}$$

Now, $\psi u \in H_0^1(M)$. Also, ψu verifies the zero-mean condition at all cusps. Thus, applying Proposition A.1 to ψu, combining (A.3) with (A.11), it follows that

$$\int_M \left(\frac{1}{4} |\nabla(\psi u)|^2 + Q|\psi u|^2 + \frac{1}{2}\lambda |\psi u|^2 - |\nabla \psi|^2 |u|^2 \right) d\mu \leq \int_M |\psi|^2 |u| \, |f| \, d\mu. \tag{A.12}$$

Before we proceed we observe that a simple approximation argument shows that (A.12) holds for a wider class of ψ functions. Namely, (A.12) holds for a real function $\psi \in C^1(M)$ such that ψ depends only on the variable y at the cusps and such that (abusing notation) $\psi(y)$ and $\psi'(y)$ decay exponentially at the cusps as $y \to \infty$. (Here and elsewhere one uses the assumption that $u \in L^{2,\alpha}(M)$ for some α.) We may even allow functions ψ of the above type such that $\psi'(y)$ has a finite number of simple discontinuities at the cusp.

Next, let $\ell \in \mathbb{R}$, ℓ as in the theorem. Fix a number $N_0 > \max(a_1, \dots, a_m)$, N_0 chosen sufficiently large so that the following inequality holds:

$$\lambda(z) > 4|\ell|^2 \tag{A.13}$$

at all points $z = (x_j, y) \in Z_j$ such that $y \geq N_0$, $j = 1, \dots, m$. Let $\psi_\ell(z)$ be a function, which coincides with $\rho(z)^\ell$ outside a compact, defined as follows:

$$\psi_\ell(z) = y^\ell \tag{A.14}$$

at all points z in Z_j with coordinate $y \geq N_0$, $j = 1, \dots, m$, and

$$\psi_\ell(z) = N_0^\ell \text{ at all other points.} \tag{A.14$'$}$$

$\psi_\ell(z)$ is a positive Lipschitz function. At points z in Z_j with $y \geq N_0$, we have

$$|\nabla\psi_\ell(z)|^2 = \left|y\frac{\partial}{\partial y}\psi_\ell(z)\right|^2 = \ell^2 y^{2\ell} < \frac{1}{4}\lambda(z)\psi_\ell(z)^2$$

(in view of (A.13) and (A.14)) while at other points: $|\nabla\psi_\ell(z)| = 0$. Thus

$$|\nabla\psi_\ell(z)|^2 < \frac{1}{4}\lambda(z)\psi_\ell(z)^2 \tag{A.15}$$

everywhere on M.

We now approximate the function ψ_ℓ by functions $\psi_{\ell,N}$ which decay exponentially at the cusps. These are defined as follows. For $N > N_0$ (N_0 as above) set

$$\psi_{\ell,N}(z) = c_N e^{-\delta y} \tag{A.16}$$

for $y \geq N$ in the cusp Z_j, $j = 1, \ldots, m$, and set

$$\psi_{\ell,N}(z) = \psi_\ell(z) \text{ at all other points.} \tag{A.16$'$}$$

Choose the number $\delta > 0$ sufficiently small so that

$$(\delta y)^2 < \frac{1}{4}\lambda(z) \text{ in all cusps,} \tag{A.17}$$

and then choose the constants c_N so that $\psi_{\ell,N}(z)$ becomes a continuous function on M. It follows readily from (A.15), (A.16)-(A.16$'$) and (A.17) that $\psi_{\ell,N}(z)$ is a positive Lipschitz function such that

$$|\nabla\psi_{\ell,N}(z)|^2 < \frac{1}{4}\lambda(z)\psi_{\ell,N}(z)^2 \tag{A.18}$$

everywhere on M.

We now apply the inequality (A.12) with $\psi = \psi_{\ell,N}$ and use (A.18). (As was observed before this is a valid application of (A.12) in view of the exponential decay of $\psi_{\ell,N}$ and its y-derivative.) We find that

$$\begin{aligned}\int_M \left(\frac{1}{4}|\nabla(\psi_{\ell,N}u)|^2 + Q|\psi_{\ell,N}u|^2 + \frac{1}{4}\lambda|\psi_{\ell,N}u|^2\right) d\mu \\ \leq \int_M |f|\,|u|\psi_{\ell,N}^2\, d\mu \\ \leq \left(\int_M \lambda^{-1}|\psi_{\ell,N}f|^2\, d\mu\right)^{1/2}\left(\int_M \lambda|\psi_{\ell,N}u|^2\, d\mu\right)^{1/2},\end{aligned} \tag{A.19}$$

by the Cauchy-Schwarz inequality. From (A.19) it follows in particular that

$$\left(\int_M \lambda|\psi_{\ell,N}u|^2\, d\mu\right)^{1/2} \leq 4\left(\int_M \lambda^{-1}|\psi_{\ell,N}f|^2\, du\right)^{1/2}. \tag{A.20}$$

Thus, using (A.20) to bound the last term in (A.19), we get

$$\begin{aligned}\int_M \left(\frac{1}{4}|\nabla(\psi_{\ell,N}u)|^2 + Q|\psi_{\ell,N}u|^2 + \frac{1}{4}\lambda|\psi_{\ell,N}u|^2\right) d\mu \\ \leq 4\int_M \lambda^{-1}|\psi_{\ell,N}f|^2 d\mu.\end{aligned} \tag{A.21}$$

Now, note that

$$\begin{aligned}|\nabla(\psi_{\ell,N}u)|^2 &\geq (\psi_{\ell,N}|\nabla u| - |\nabla\psi_{\ell,N}|\,|u|)^2\\ &\geq \frac{1}{2}\left(\psi_{\ell,N}|\nabla u|\right)^2 - (|\nabla\psi_{\ell,N}||u|)^2\\ &\geq \frac{1}{2}\left(\psi_{\ell,N}|\nabla u|\right)^2 - \frac{1}{4}\lambda|\psi_{\ell,N}u|^2, \qquad \text{(A.22)}\end{aligned}$$

where the last inequality follows from (A.18). Using (A.22) to bound from below the integral of the first term in (A.21), using also the estimate (A.20), it follows readily from (A.21) that

$$\text{(A.23)}\quad \int_M \left(\frac{1}{8}\left(\psi_{\ell,N}|\nabla u|\right)^2 + Q|\psi_{\ell,N}u|^2 + \frac{1}{4}\lambda|\psi_{\ell,N}u|^2\right) d\mu \leq 5\int_M \lambda^{-1}|\psi_{\ell,N}f|^2 d\mu.$$

Observing that $\psi_{\ell,N}(z) \leq \psi_\ell(z)$ for all z and that $\psi_{\ell,N}(z) = \psi_\ell(z)$ on every compact if N is sufficiently large, it follows from (A.23) upon letting $N \to \infty$ that

$$\text{(A.23}'\text{)}\quad \int_M \left(\frac{1}{8}\left(\psi_\ell|\nabla u|\right)^2 + Q|\psi_\ell u|^2 + \frac{1}{4}|\psi_\ell u|^2\right) d\mu \leq 5\int_M \lambda^{-1}|\psi_\ell f|^2 d\mu.$$

Taking note of the explicit from at the cusps of the functions ψ_ℓ and λ (see (A.14) and (A.4)) it follows from (A.23′) that the inequality (A.7) holds. □

We shall use the full strength of Theorem A.2 in the second appendix. We shall also need the following

LEMMA A.3. *Let u be a solution (in $H^2_{\text{loc}}(M)$) of equation* (A.5) *where q is a locally bounded function such that $\operatorname{Re} q \geq -k$, $k \geq 0$ a constant. Let Ω and Ω_0 be two bounded open sets in M such that $\overline{\Omega} \subset \Omega_0$. The following inequality holds*:

$$\text{(A.24)}\quad \int_\Omega |\nabla u|^2\, d\mu \leq \int_{\Omega_0} |f|^2\, d\mu + (C+k)\int_{\Omega_0} |u|^2\, d\mu$$

where C is a constant which depends only on Ω and Ω_0.

PROOF. Let ψ be a real function in $C^1(M)$ such that $\operatorname{supp}\psi \subset \Omega_0$, $0 \leq \psi \leq 1$, $\psi = 1$ on Ω. Applying formula (A.10) and using equation (A.5) we find that

$$\int_M |\nabla(\psi u)|^2\, d\mu = \operatorname{Re}\int_M \psi^2\bar{u}f\, d\mu - \operatorname{Re}\int_M \psi^2|u|^2 q\, d\mu + \int_M |\nabla\psi|^2|u|^2 d\mu$$

from which it follows that

$$\begin{aligned}\int_\Omega |\nabla u|^2\, d\mu &\leq \int_{\Omega_0} |u||f|\, d\mu + k\int_{\Omega_0} |u|^2\, d\mu + \int_{\Omega_0} |\nabla\psi|^2|u|^2\, d\mu\\ &\leq \int_{\Omega_0} |f|^2\, d\mu + (k+C)\int_{\Omega_0} |u|^2\, d\mu\end{aligned}$$

where $C = 1/4 + \sup_{\Omega_0}|\nabla\psi|^2$. □

Appendix B. On the Proof of Theorem 4.2

In this appendix we give some indications concerning the proof of Theorem 4.2. The crux of the proof are certain weighted L^2 estimates for s-outgoing solutions $u(z)$ of the differential equation

$$-\Delta u + ce^{-\rho}u - s(n-1-s)u = f \tag{B.1}$$

on M where c is a large positive constant. These estimates are described in

THEOREM B.1. *Let R and ε be given numbers, $R > 1$ and $0 < \varepsilon < 1$. There exists a positive constant c_0 depending only on R, ε, the weight function ρ and the manifold M, such that the following proposition holds: Let u be a s-outgoing solution of equation* (B.1) *with $c \geq c_0$ and $f \in L^{2,\alpha}(M)$ where*

$$\begin{gathered}\frac{1}{2}(n-1) - \operatorname{Re} s + \varepsilon \leq \alpha \leq 2R, \\ \left|s - \frac{1}{2}(n-1)\right| \leq R.\end{gathered} \tag{B.2}$$

Then u verifies the inequality

$$\|\rho^{n-1-s}\,|\nabla(\rho^{s+1-n}u)|\,\|_\alpha + \|e^{-\rho/2}u\|_{\alpha+1/2} \leq K\|f\|_\alpha \tag{B.3}$$

where K is a constant depending only on R, ε, the function ρ and M.

REMARK. Note that at points $z = (x_j, y) \in Z_j$:

$$\rho^{n-1-s}|\nabla(\rho^{s+1-n}u)| = \left|y\frac{\partial u}{\partial y} + (s-n+1)u\right|. \tag{B.4}$$

Using (B.4) it is easy to see that if $\alpha > 0$ then (B.3) and (B.2) imply that

$$\|u\|_{-\alpha} \leq K_0\|f\|_\alpha.$$

The derivation of Theorem 4.2 from Theorem B.1 follows a standard argument. We shall not give here details of this part of the proof but only observe that the proof of Theorem 4.2 is achieved by showing that equation (B.1) admits a unique s-outgoing solution u for any $f \in L^{2,\alpha}(M)$ if c, s and α verify the conditions of Theorem B.1. This is proved by the *method of continuity* (in the parameter s), using the *a priori* estimates (B.3).

A basic step in the proof of Theorem B.1 is the derivation of the estimate (B.3) for solutions $u(z)$ of (B.1) which are supported in the cusps and which depend only on the y coordinate of z. Our problem of estimating solutions of (B.1) becomes in this case a one dimensional problem which involves solutions $u(y)$ of the ordinary differential equation:

$$y^2u'' + (2-n)yu' + s(n-1-s)u - ce^{-y}u = f, \tag{B.5}$$

on the half-line $\mathbb{R}_+$. The relevant weighted L^2 estimates for solutions of (B.5) are given in the following lemma.

LEMMA B.2. *Let $u(y)$ be a solution of equation* (B.5) *on $\mathbb{R}_+$ where n is a real number, s a complex number and c a non-negative number. Suppose that $u \in H^2_{\rm loc}(\mathbb{R}_+)$, $f \in L^2_{\rm loc}(\mathbb{R}_+)$ and that*

$$\int_0^\infty y^{2\ell}|f|^2\,dy < \infty \tag{B.6}$$

for some $\ell > -\operatorname{Re} s - 1/2$. Suppose that $u \equiv 0$ in a neighborhood of the origin and also that $u \equiv 0$ for $0 \le y \le 8\ell + 4n$ if $\ell > -n/2$. Finally, assume that

$$\int_0^\infty y^{2\ell_0}|yu' + (s-n+1)u|^2\,dy < \infty \tag{B.7}$$

for some $\ell_0 > -\operatorname{Re} s - 1/2$, then the following inequality holds:

$$\left(\ell + \operatorname{Re} s + \frac{1}{2}\right)\int_0^\infty y^{2\ell}|yu' + (s-n+1)u|^2\,dy + \frac{c}{4}\int_0^\infty y^{2\ell+1}e^{-y}|u|^2\,dy \le \left(\ell + \operatorname{Re} s + \frac{1}{2}\right)^{-1}\int_0^\infty y^{2\ell}|f|^2\,dy. \tag{B.8}$$

An immediate corollary of Lemma B.2 is the following

PROPOSITION B.3. *Let u be an s-outgoing solution of equation* (B.1) *with* $c \ge 0$ *and $f \in L^{2,\alpha}(M)$ for some $\alpha > (n-1)/2 - \operatorname{Re} s$. Suppose that*

$$\operatorname{supp} u \subset \bigcup_{j=1}^m Z_j.$$

Suppose also that, in every cusp Z_j, $u(z)$ vanishes at all points $z = (x_j, y) \in Z_j$ such that $a_j \le y \le \max(a_j, 8\alpha) + 1$. Suppose finally that at the cusps the function $u(z)$ depends only on the y coordinate of z. Then the following inequality holds:

$$\tau\|\rho^{n-1-s}\,|\nabla(\rho^{s+1-n}u)|\,\|_\alpha^2 + \frac{c}{4}\|e^{-\rho/2}u\|_{\alpha+1/2}^2 \le \tau^{-1}\|f\|_\alpha^2 \tag{B.9}$$

where

$$\tau = \alpha + \operatorname{Re} s - \frac{1}{2}(n-1). \tag{B.9$'$}$$

PROOF. Define functions $u_j(y)$ and $f_j(y)$ on $\mathbb{R}_+$, $j = 1, \dots, m$, as follows. Set

$$u_j(y) = u(x_j, y) \quad \text{and} \quad f_j(y) = f(x_j, y) \tag{B.10}$$

for $(x_j, y) \in Z_j$, $y \ge a_j$, and set $u_j(y) = f_j(y) \equiv 0$ for $0 \le y < a_j$.

It follows from our assumptions on u that $u_j(y)$ is a well defined function in $H^2_{\rm loc}(\mathbb{R}_+)$ which satisfies equation (B.5) with $n = \dim M$ and $f = f_j(y)$. Moreover, it is easily checked that f_j and u_j satisfy the conditons of Lemma B.2 with $\ell = \alpha - n/2$. (Observe that condition (B.6) is satisfied by f_j since $f(z) \in L^{2,\alpha}(M)$ while condition (B.7) holds for u_j since $u(z)$ is an s-outgoing solution of (B.1), (see (3.21)). Applying Lemma B.2 to u_j we find that

$$\tau\int_0^\infty |y|^{2\alpha-n}|yu_j' + (s-n+1)u_j|^2\,dy + \frac{c}{4}\int_0^\infty y^{2\alpha-n+1}e^{-y}|u_j|^2\,dy \le \tau^{-1}\int_0^\infty |y|^{2\alpha-n}|f_j|^2 dy, \tag{B.11}$$

τ given by (B.9′). Multiplying both sides of (B.11) by $\omega_j = \mathrm{vol}(X_j)$ and summing over j, taking note of (B.10), (B.4), (2.10) and (2.3), we obtain the inequality (B.9). □

Deferring the proof of Lemma B.2, we turn to the

PROOF OF THEOREM B.1. Let u be an s-outgoing solution of equation (B.1) with $c \geq 0$ and $f \in L^{2,\alpha}(M)$. Assume that α and s verify the bounds (B.2) where ε and R are fixed numbers: $0 < \varepsilon < 1 < R$. Define functions $u_j(z)$ and $f_j(z)$ on Z_j, $j = 1, \dots, m$, as follows: for $z = (x_j, y)$ set

$$\begin{aligned} u_j(z) &= \omega_j^{-1} \int_{X_j} u(\cdot, y)\,d\mu_j', \\ f_j(z) &= \omega_j^{-1} \int_{X_j} f(\cdot, y)\,d\mu_j', \end{aligned} \tag{B.12}$$

$\omega_j = \mathrm{vol}(X_j)$. It is clear that u_j and f_j depend only on the coordinate y in Z_j and that

$$-\Delta u_j + ce^{-\rho} u_j - s(n-1-s)u_j = f_j \text{ on } Z_j. \tag{B.13}$$

Next, pick a function $\zeta \in C^\infty(\mathbb{R}_+)$ verifying $0 \leq \zeta \leq 1$ on $\mathbb{R}_+$, such that

$$\zeta(y) = \begin{cases} 0 & \text{for } 0 \leq y \leq y_0 - 1, \\ 1 & \text{for } y \geq y_0, \end{cases} \tag{B.14}$$

where $y_0 = \sum_{j=1}^m a_j + 16R + 1$. Now, define functions $\tilde{u}$ and $\tilde{f}$ on M as follows, set

$$\begin{aligned} \tilde{u}(z) &= \zeta(y) u_j(z), \\ \tilde{f}(z) &= \zeta(y) f_j(z) \end{aligned} \tag{B.15}$$

if $z = (x_j, y) \in Z_j$, $j = 1, \dots, m$, and set $\tilde{u}(z) = \tilde{f}(z) = 0$ at all other points z.

Clearly, $\tilde{u}$ and $\tilde{f}$ depend only on y in each cusp. Moreover, it follows from (B.13), (B.14) and (B.15) that $\tilde{u}$ is a solution (in $H^2_{\mathrm{loc}}(M)$) of the differential equation

$$-\Delta \tilde{u} + ce^{-\rho} \tilde{u} - s(n-1-s)\tilde{u} = \tilde{f} + \chi \tag{B.16}$$

where the function χ is given in the cusp Z_j by

$$\chi = \left((n-2)y\zeta' - y^2 z''\right)u_j - 2y^2 \zeta' \frac{\partial u_j}{\partial y}, \tag{B.17}$$

$\chi(z) = 0$ outside the cusps.

It follows from (B.17) that the support of χ is contained in the closure of the bounded set:

$$\Omega = \bigcup_j \{z = (x_j, y) \in Z_j : y_0 - 1 < y < y_0\}. \tag{B.18}$$

Hence it follows from (B.17), (B.12), (B.14) and (B.18) that

$$\|\chi\|_\alpha \leq K_1(\|u\|_{0,\Omega} + \|\,|\nabla u|\,\|_{0,\Omega}) \tag{B.19}$$

where here and in the following $K_1, K_2, \dots, K_8$, denote constants which depend only on R, ε, the function ρ and the manifold M. (In particular the K_i do not depend on the parameter c in (B.1).) We are using the notation

$$\|w\|_{0,\Omega} := \left(\int_\Omega |w|^2 \, d\mu\right)^{1/2}$$

for $w \in L^2_{\mathrm{loc}}(M)$.

Also observe that from (B.15), (B.12) and (B.14) it follows that

$$\|\tilde{f}\|_\alpha \le \|f\|_\alpha. \tag{B.20}$$

Now, it is easily checked that the function $\tilde{u}$ verifies all the conditions of Proposition B.3. (In particular $\tilde{u}$ is an s-outgoing solution of (B.16) since u is s-outgoing.) Hence applying the inequality (B.9) to $\tilde{u}$, with f replaced by $\tilde{f} + \chi$, using (B.20), (B.19) and elementary inequalities, we find that

$$\begin{aligned} \|\rho^{n-1-s}\,|\nabla(\rho^{s+1-n}\tilde{u})|\,\|_\alpha + \sqrt{c}\|e^{-\rho/2}\tilde{u}\|_{\alpha+1/2} \\ \le K_2(\|f\|_\alpha + \|u\|_{0,\Omega} + \|\,|\nabla u|\,\|_{0,\Omega}). \end{aligned} \tag{B.21}$$

Next,consider the function

$$v = u - \tilde{u} \tag{B.22}$$

which (by (B.1) and (B.16)) satifies the differential equation

$$-\Delta v + ce^{-\rho}v - s(n-1-s)v = f - \tilde{f} - \chi. \tag{B.23}$$

It follows from the definition of $\tilde{u}$ and (B.22) that v verifies the zero-mean condition (3.3) in each cusp Z_j for $y \ge y_0$. We also have that $v \in L^{2,\gamma}(M)$ for some γ (since v is an *s-outgoing* solution the result is implied by (3.21)). We wish to apply to v Theorem A.2. To this end we need to choose the parameter c in (B.1) sufficiently large. In this connection let c_1 be the smallest non-negative constant c such that

$$ce^{-\rho(z)} - \operatorname{Re} s(n-1-s) \ge -\frac{1}{4}\lambda(z) \tag{B.24}$$

for all $z \in M$ and all s verifying (B.2) where $\lambda(z)$ is the function introduced in Proposition A.1 with $a = y_0$. The constant c_1 exists since $\lambda(z) \to \infty$ as $\rho(z) \to \infty$ in the cusps. (We note that c_1 and other constants c_i to be introduced in the sequel have the same dependence as the constants K_i.) From now on assume that the parameter c in equations (B.1) and (B.23) is $\ge c_1$ and set:

$$c = c_1 + (3b)^2, \tag{B.25}$$

$b \ge 0$. It follows readily that the function $q(z) = ce^{-\rho(z)} - s(n-1-s)$ verifies condition (A.6) with $Q(z) = (3b)^2 e^{-\rho(z)}$ for all s verifying (B.2). This and the remarks made before show that v verifies the conditions of Theorem A.2 with $\ell = \alpha + 1$ and $Q(z)$ as above. Applying (A.7), using (B.19) and (B.20), we find that

$$\|v\|_{\alpha+2} + 3b\|e^{-\rho/2}v\|_{\alpha+1} + \|\,|\nabla v|\,\|_{\alpha+1} \le K_3(\|f\|_\alpha + \|u\|_{0,\Omega} + \|\,|\nabla u|\,\|_{0,\Omega}),$$

which implies the (weaker) estimate

$$\|v\|_{\alpha+2} + 3b\|e^{-\rho/2}v\|_{\alpha+1/2} + \|\,|\nabla v|\,\|_\alpha \le K_4(\|f\|_\alpha + \|u\|_{0,\Omega} + \|\,|\nabla u|\,\|_{0,\Omega}). \tag{B.26}$$

Next, observe that from the inequality:

$$\begin{aligned}|\rho^{n-1-s}\nabla(\rho^{s+1-n}v(z)| &= |\nabla v(z) + (s+1-n)v(z)\rho(z)^{-1}\nabla\rho(z)| \\ &\leq |\nabla v(z)| + K_5|v(z)|,\end{aligned}$$

it follows that

$$\|\rho^{n-1-s}\,|\nabla(\rho^{s+1-n}v)|\,\|_\alpha \leq \|\,|\nabla v|\,\|_\alpha + K_6\|v\|_\alpha \tag{B.27}$$

for all s verifying (B.2). Combining (B.27) with (B.26) we get

$$\begin{aligned}\|v\|_{\alpha+2} + 3b\|e^{-\rho/2}v\|_{\alpha+1/2} + \|\rho^{n-1-s}\,|\nabla(\rho^{s+1-n}v)|\,\|_\alpha \\ \leq K_4(\|f\|_\alpha + \|u\|_{0,\Omega} + \|\,|\nabla u|\,\|_{0,\Omega}) + K_6\|v\|_\alpha\end{aligned} \tag{B.28}$$

for any $b \geq 0$. Now, let c_2 be the smallest non-negative constant such that $\rho(z)^4 + c_2^2 e^{-\rho(z)}\rho(z) \geq K_6^2$ on M. Noting that

$$\|v\|_{\alpha+2} + c_2\|e^{-\rho/2}v\|_{\alpha+1/2} \geq K_6\|v\|_\alpha, \tag{B.29}$$

it follows from (B.28) and (B.29) that for $b \geq c_2$:

$$\begin{aligned}2b\|e^{-\rho/2}v\|_{\alpha+1/2} + \|\rho^{n-1-s}|\nabla(\rho^{s+1-n}v)|\,\|_\alpha \\ \leq K_4(\|f\|_\alpha + \|u\|_{0,\Omega} + \|\,|\nabla u|\,\|_{0,\Omega}).\end{aligned} \tag{B.30}$$

We are now in a position to derive the estimates for $u = \tilde{u} + v$, u the given s-outgoing solution of (B.1). Writing the parameter c in equation (B.1) in the form (B.25), it follows readily, combining (B.30), (B.22) and (B.21), that the following inequality holds for $b \geq c_2$:

$$\begin{aligned}2b\|e^{-\rho/2}u\|_{\alpha+1/2} + \|\rho^{n-1-s}|\nabla(\rho^{s+1-n}u)|\,\|_\alpha \\ \leq K_7(\|f\|_\alpha + \|u\|_{0,\Omega} + \|\,|\nabla u|\,\|_{0,\Omega}).\end{aligned} \tag{B.31}$$

Next, we shall apply Lemma A.3 to the function u. In this connection note that u is a solution of equation (A.5) with $q = ce^{-\rho} - s(n-1-s)$. We have:

$$\begin{aligned}\operatorname{Re} q &\geq -\operatorname{Re} s(n-1-s) \\ &= \operatorname{Re}\left(s - \frac{n-1}{2}\right)^2 - \left(\frac{n-1}{2}\right)^2 \\ &\geq -R^2 - \left(\frac{n-1}{2}\right)^2\end{aligned}$$

in view of (B.2). Hence applying Lemma A.3 to u we get

$$\|\,|\nabla u|\,\|_{0,\Omega} \leq \|f\|_{0,\Omega_0} + \left(C + R + \frac{n-1}{2}\right)\|u\|_{0,\Omega_0} \tag{B.32}$$

where Ω_0 is an open bounded set such that $\Omega_0 \supset \overline{\Omega}$ and C is a constant depending only on Ω and Ω_0. Thus, combining (B.31) with (B.32), we get

$$2b\|e^{-\rho/2}u\|_{\alpha+1/2} + \|\rho^{n-1-s}|\nabla(\rho^{s+1-n}u)|\,\|_\alpha \leq K_8(\|f\|_\alpha + \|u\|_{0,\Omega_0}) \tag{B.33}$$

for $b \geq c_2$.

Finally, let c_3 be a positive constant such that $c_3^2\rho^{2\alpha+1}e^{-\rho} \geq K_8^2$ on Ω_0. We have:

$$c_3\|e^{-\rho/2}u\|_{\alpha+1/2} \geq K_8\|u\|_{0,\Omega_0}. \tag{B.34}$$

Hence, combining (B.33) with (B.34) we find that

$$\|e^{-\rho/2}u\|_{\alpha+1/2} + \|\rho^{n-1-s}|\nabla(\rho^{s+1-n}u)|\,\|_\alpha \le K_8\|f\|_\alpha$$

for any $b \ge c_4$ where $c_4 = \max(c_2, c_3, 1)$. This yields Thereom B.1 with $c_0 = c_1 + (3c_4)^2$. □

We conclude with the

PROOF OF LEMMA B.2. It will be convenient to consider instead of u (the solution of (B.5)) the function:

$$v(y) = u(y)y^{1+s-n} \tag{B.35}$$

which is a solution of the differential equation

$$y^2v'' + (n-2s)yv' - ce^{-y}v = g \text{ on } \mathbb{R}_+ \tag{B.36}$$

where

$$g(y) = f(y)y^{1+s-n}. \tag{B.36$'$}$$

By our assumptions: $v \in H^2_{\text{loc}}(\mathbb{R}^+)$, $g \in L^2_{\text{loc}}(\mathbb{R}_+)$, v and g vanish in some neighborhood of the origin. The two growth conditions imposed in the lemma take the following forms. Condition (B.6) implies that

$$\int_0^\infty y^{2\alpha-2}|g(y)|^2\,dy < \infty \tag{B.37}$$

where

$$\alpha = \ell + n - \operatorname{Re} s > n - \frac{1}{2} - 2\operatorname{Re} s, \tag{B.37$'$}$$

while condition (B.7) takes the form:

$$\int_0^\infty y^{2\alpha_0}|v'(y)|^2\,dy < \infty \tag{B.38}$$

for some number $\alpha_0 > n - 1/2 - 2\operatorname{Re} s$. In view of (B.37′) we shall assume with no loss of generality that $\alpha_0 \le \alpha$.

Finally note that if $\alpha > 1/2$, then

$$v \equiv 0 \text{ for } 0 \le y \le 4\alpha - 2. \tag{B.39}$$

This follows from our assumption that $u \equiv 0$ for $y \le 8\ell + 4n$. (Indeed, by (B.37′) we have: $8\ell + 4n = 4\alpha + 4(\alpha - n + 2\operatorname{Re} s) > 4\alpha - 2$.)

We proceed to establish the inequality (B.8). To this end we introduce the family of auxiliary functions $\psi_N(y)$ defined for every $N \ge 1$ as follows:

$$\begin{aligned} \psi_N(y) &= y^{2\alpha-1} \text{ for } 0 < y \le N, \\ &= N^{2(\alpha-\alpha_0)}y^{2\alpha_0-1} \text{ for } y > N. \end{aligned} \tag{B.40}$$

Note that ψ_N is a piecewise continuously differentiable function such that for any $k \in \mathbb{R}$:

$$\left(y^k\psi_N(y)\right)' = y^{k-1}\psi_N(y)(J(y)+k) \tag{B.41}$$

for $y > 0$, $y \neq N$, where

$$J(y) = 2\alpha - 1 \text{ for } 0 < y \leq N, \qquad \text{(B.42)}$$
$$= 2\alpha_0 - 1 \text{ for } y > N.$$

Note also that from (B.42) and our assumptions on α_0 it follows that

$$2(n - 1 - 2\operatorname{Re} s + \delta) \leq J(y) \leq 2\alpha - 1 \qquad \text{(B.42}')$$

for some $\delta > 0$.

Now, with v defined by (B.35), we multiply both sides of equation (B.36) by $\overline{v'(y)}\psi_N(y)$, integrate from 0 to t and then take real parts ($N \geq 1$ and $t > 0$ fixed). We find that

$$\frac{1}{2}\int_0^t y^2\psi_N \frac{d}{dy}|v'|^2\,dy + \operatorname{Re}(n-2s)\int_0^t y\psi_N|v'|^2\,dy - \frac{c}{2}\int_0^t e^{-y}\psi_N\frac{d}{dy}|v|^2\,dy = \operatorname{Re}\int_0^t \psi_N(y)g(y)\overline{v'(y)}\,dy. \qquad \text{(B.43)}$$

Integrating by parts, using (B.41), it follows from (B.43) that

$$\frac{1}{2}t^2\psi_N(t)|v'(t)|^2 - \frac{1}{2}\int_0^t y(2+J)\psi_N|v'|^2\,dy + \operatorname{Re}(n-2s)\int_0^t y\psi_N|v'|^2\,dy - \frac{c}{2}e^{-t}\psi_N(t)|v(t)|^2 + \frac{c}{2}\int_0^t (y^{-1}J - 1)e^{-y}\psi_N|v|^2 dy = \operatorname{Re}\int_0^t \psi_N g\overline{v'} dy. \qquad \text{(B.44)}$$

Rewriting (B.44), we obtain the following inequality:

$$\begin{aligned}\int_0^t (1 - n + 2\operatorname{Re} s + \tfrac{1}{2}J(y))y\psi_N|v'|^2\,dy &+ \frac{c}{2}\int_0^t (1 - y^{-1}J(y))e^{-y}\psi_N|v|^2\,dy \\ &= \frac{1}{2}t^2\psi_N(t)|v'(t)|^2 - \frac{c}{2}e^{-t}\psi_N(t)|v(t)|^2 - \operatorname{Re}\int_0^t \psi_N g\overline{v'}\,dy \\ &\leq \frac{1}{2}t^2\psi_N(t)|v'|(t)^2 + \int_0^t \psi_N|g||v'|dy.\end{aligned} \qquad \text{(B.45)}$$

Observe that by (B.42) and (B.39):

$$1 - y^{-1}J(y) \geq 1 - (2\alpha - 1)y^{-1} \geq \frac{1}{2} \text{ on } \operatorname{supp} v. \qquad \text{(B.46)}$$

We also have

$$\int_0^t \psi_N|g||v'|\,dy \leq \frac{\delta}{2}\int_0^t y\psi_N|v'|^2\,dy + \frac{1}{2\delta}\int_0^t y^{2\alpha-2}|g|^2\,dy \qquad \text{(B.47)}$$

for any $\delta > 0$ (by an elementary inequality, since $\psi_N \leq y^{2\alpha-1}$ on $\mathbb{R}_+$). Picking in (B.47) a δ such that the first inequality (B.42′) holds, combining (B.45) with

(B.42′), (B.46) and (B.47), we find that

$$\text{(B.48)}\quad \frac{\delta}{2}\int_0^t y\psi_N|v'|^2\,dy + \frac{c}{4}\int_0^t e^{-y}\psi_N|v|^2\,dy \le \frac{1}{2}t^2\psi_N(t)|v'(t)|^2 + \frac{1}{2\delta}\int_0^t y^{2\alpha-2}|g|^2\,dy$$

Now, it follows from (B.38) that there exists a sequence $t_j \to \infty$ such that

$$\text{(B.49)}\qquad t_j^{2\alpha_0+1}|v'(t_j)|^2 \to 0 \text{ as } t_j \to \infty.$$

Hence, letting $t \to \infty$ through the sequence $\{t_j\}$, it follows from (B.48) and (B.49) (since $\psi_N(t_j) = t_j^{2\alpha_0-1}$ for $t_j > N$) that

$$\text{(B.50)}\qquad \frac{\delta}{2}\int_0^\infty y\psi_N|v'|^2\,dy + \frac{c}{4}\int_0^\infty e^{-y}\psi_N|v|^2\,dy \le \frac{1}{2\delta}\int_0^\infty y^{2\alpha-2}|g|^2\,dy.$$

Finally, letting $N \to \infty$ in (B.50), we get:

$$\text{(B.51)}\qquad \frac{\delta}{2}\int_0^\infty y^{2\alpha}|v'|^2\,dy + \frac{c}{4}\int_0^\infty y^{2\alpha-1}e^{-y}|v|^2\,dy \le \frac{1}{2\delta}\int_0^\infty y^{2\alpha-2}|g|^2\,dy.$$

To improve the constants in (B.51) we return to the inequality (B.45). Assume that $0 < t < N$. Taking note of (B.40) and (B.42), using (B.46), it follows from (B.45) that

$$\begin{aligned}\text{(B.52)}\quad \gamma\int_0^t y^{2\alpha}|v'|^2\,dy &+ \frac{c}{4}\int_0^t y^{2\alpha-1}e^{-y}|v|^2\,dy \\ &\le \frac{1}{2}t^{2\alpha+1}|v'(t)|^2 + \int_0^t y^{2\alpha-1}|g|\,|v'|\,dy \\ &\le \frac{1}{2}t^{2\alpha+1}|v'(t)|^2 + \left(\int_0^t y^{2\alpha}|v'|^2dy\right)^{1/2}\left(\int_0^t y^{2\alpha-2}|g|^2\,dy\right)^{1/2}\end{aligned}$$

where $\gamma = \alpha + 1/2 - n + 2\operatorname{Re} s$. Now, since

$$\text{(B.53)}\qquad \int_0^\infty y^{2\alpha}|v'|^2\,dy < \infty \quad \text{(by (B.51))},$$

it follows that there exists a sequence $t_j \to \infty$ such that $t_j^{2\alpha+1}|v'(t_j)|^2 \to 0$ as $t_j \to \infty$. Letting $t \to \infty$ through the sequence $\{t_j\}$, taking note of (B.53), we find by some simple manipulations that

$$\text{(B.54)}\qquad \gamma\int_0^\infty y^{2\alpha}|v'|^2dy + \frac{c}{4}\int_0^\infty y^{2\alpha-1}e^{-y}|v|^2\,dy \le \frac{1}{\gamma}\int_0^\infty y^{2\alpha-2}|g|^2\,dy.$$

Inserting in (B.54) for v its expression in u from (B.35), for g its expression from (B.36′) and for α its expression in ℓ from (B.37′), noting that $\gamma = \ell + \operatorname{Re} s + 1/2$, we obtain (B.8). □

INSTITUTE OF MATHEMATICS, THE HEBREW UNIVERSITY OF JERUSALEM, GIVAT RAM, 91904 JERUSALEM, ISRAEL

E-mail address: agmon@cs.huij.ac.il

Centre de Recherches Mathématiques
CRM Proceedings and Lecture Notes
Volume 8, 1995

The General Theory of N-Body Quantum Systems

W. Hunziker and I. M. Sigal

1. Basic Dynamics

Introduction. In this section we discuss two fundamental properties of Schrödinger operators

$$H = p^2 + V(x) \text{ on } \mathcal{H} = L^2(X),$$

where X is a Euclidean space, $x \in X$, and $p^2 = -\Delta$. The first is Kato's celebrated theorem, which states that

$$H = H^*, \quad H > -\infty,$$

for a large class of potentials, including N-body Coulomb systems. This was the first proof that atoms fit into the abstract Hilbert space scheme of quantum mechanics. In particular, H generates a unitary group

$$e^{-iHt} \colon \psi_0 \to \psi_t$$

describing the time evolution of any intitial state ψ_0, and it has a spectral representation

$$H = \int \lambda \, dE_\lambda,$$

which in turn defines the energy distribution

$$d(\psi, E_\lambda \psi)$$

for any state ψ (i.e. any unit vector $\psi \in \mathcal{H}$) as a probability measure on the spectrum $\sigma(H)$. The result that H is bounded from below explains the stability of atoms. It is in fact obtained in the stronger form

$$p^2 < \alpha H + \beta \tag{1.1}$$

for some constants α, β. This bound on the kinetic energy in terms of the conserved total energy plays a fundamental role. Here we use it to characterize bound states and continuum states by their space-time behavior (Ruelle's theorem).

In classical mechanics, (1.1) only holds if the potential $V(x)$ is bounded from below. Then the inequalitites $|x| \leq R$ and $H \leq E$ define a finite volume in phase space. By Liouville's theorem, the net flux into (or out of) any cylinder $|x| \leq R$ is

---

1991 *Mathematics Subject Classification.* Primary: 35J10, 81V70; Secondary: 35Q40, 81Q15.

These notes contain parts of a review which is currently being written by the two authors. The presentation is self-contained, but in a preliminary form. In particular, references and bibliographical comments are omitted. An expanded version will appear elsewhere.

therefore zero. This means that the orbits for which x_t is bounded only in one time-direction form a set of measure zero. Almost all orbits thus fall into two classes: either x_t is bounded for all times, or it is unbounded in both time directions.

Ruelle's theorem is the exact quantum analogue. The finite phase volume is reflected in the compactness of the set of states satisfying $\langle |x| \rangle \leq R$ and $\langle H \rangle \leq E$, and the unitarity of $\exp(-iHt)$ replaces Liouville's Theorem. As a result, the Hilbert space $\mathcal{H}$ splits into two orthogonal subspaces invariant under $\exp(-iHt)$:

$$\mathcal{H} = \mathcal{H}_B \oplus \mathcal{H}_C.$$

$\mathcal{H}_B$ is the subspace of bound states, spanned by the eigenvectors of H. The orbits ψ_t in $\mathcal{H}_B$ are characterized by the condition that x remains with probability $1-\varepsilon$ in some finite ball $|x| < R(\varepsilon)$ for all times. $\mathcal{H}_C$ is the subspace of continuum states, for which x is with probability 1 outside any finite ball in the time average, taken over $-\infty < t < 0$, or over $0 < t < +\infty$.

This general result sets the stage for the further analysis of N-body systems, where we will eventually arrive at much sharper statements concerning the localization of bound states and the asymptotic behavior of continuum states as $t \to \pm\infty$.

1.1. Selfadjointness. We make no effort to work with the most general class of potentials. Kato potentials are convenient and describe many important physical systems.

DEFINITION. A *Kato potential* on X is a real function $V \in L^2_{\text{loc}}(X)$ which, as a multiplication operator on $L^2(X)$, satisfies:

$$\|V\psi\| \leq \alpha \|p^2\psi\| + \beta(\alpha)\|\psi\| \tag{1.2}$$

for any $\alpha > 0$ and all $\psi \in C_0^\infty(X)$.

THEOREM 1.1 (KATO'S THEOREM). *If V is a Kato potential on X, then $H = p^2 + V$ is selfadjoint on $L^2(X)$ with domain $D(p^2)$, and bounded from below. Moreover, p^2 is H-bounded, with a bound*:

$$\|p^2\psi\|^2 \leq (1-\alpha)^{-1}\big(\|H\psi\| + \beta(\alpha)\|\psi\|\big) \ \textit{if}\ \alpha < 1. \tag{1.3}$$

PROOF. V is a closed operator on its natural domain. Since $C_0^\infty(X)$ is a core of p^2, (1.2) extends to all $\psi \in D(p^2)$. Thus H is defined as a symmetric operator on $D(p^2)$, where

$$z - H = \big[1 - V(z-p^2)^{-1}\big](z - p^2), \tag{1.4}$$

for $\text{Re}(z) < 0$. From (1.2) we find $\big\|V(z-p^2)^{-1}\big\| \leq \alpha + \beta(\alpha)\big|\text{Re}(z)\big|^{-1} < 1$, if we choose $a < 1$ and $\big|\text{Re}(z)\big|$ sufficiently large. Then (1.4) shows that $\text{ran}(z-H) = L^2(X)$ and that $z - H$ has a bounded inverse. Thus the resolvent set $\rho(H)$ of H contains a left halfplane, which proves the first part of the theorem. The fact that p^2 is bounded relative to H (and vice versa) follows already from the closed graph theorem. The explicit bound (1.3) is easily derived from (1.2). □

We give a summary on Kato potentials at the end of this section.

1.2. Bound States and Continuum States.

1.2.1. *Compactness.* The Hamiltonians constructed in Theorem 1.1 have a "local compactness property": the operator

$$f(x)(z-H)^{-1} \text{ is compact} \tag{1.5}$$

for any $z \in \rho(H)$, if f is a bounded (L^∞) function on X vanishing as $|x| \to \infty$. In fact, if f and g are both of this type, then

$$f(x)g(p) \text{ is compact.} \tag{1.6}$$

Here $p = (p_1, \ldots, p_n)$ are the momentum operators $p_k = -i\partial/\partial x_k$ corresponding to cartesian coordinates $x = (x_1, \ldots, x_n)$ in X. (1.6) follows by observing that $f(x)g(p)$ is a norm-limit of Hilbert-Schmidt operators, obtained by restricting f and g to compact supports. From (1.3) we see that $(1+p^2)(z-H)^{-1}$ is bounded. Therefore the product $f(x)(1+p^2)^{-1}(1+p^2)(z-H)^{-1}$ is compact.

THEOREM 1.2 (RUELLE'S THEOREM). *Suppose that $H = H^*$ on $L^2(X)$ has the local compactness property (1.5). Let $\mathcal{H}_B$ be the subspace spanned by all eigenvectors of H, and $\mathcal{H}_C = \mathcal{H}_B^\perp$. Then*

$$\phi \in \mathcal{H}_B \iff \lim_{R\to\infty} \left\|(1-F_R)e^{-iHt}\phi\right\| = 0, \text{ uniformly in } 0 \le t < \infty, \tag{1.7}$$

$$\psi \in \mathcal{H}_C \iff \lim_{t\to\infty} t^{-1} \int_0^t ds \left\|F_R e^{-iHs}\psi\right\|^2 = 0, \text{ for any } R < \infty, \tag{1.8}$$

where $F_R(x)$ is the characteristic function of the ball $|x| < R$.

Remarks. Replacing H by $-H$, we obtain the analogous theorem for negative times. We also note that two states ϕ and ψ with the space-time characteristic (1.7) and (1.8) are orthogonal:

$$(\phi, \psi) = 0. \tag{1.9}$$

In fact, (1.8) implies $F_R e^{-iHt}\psi \to 0$ for some sequence $t \to \infty$. Thus we can make

$$(\phi,\psi) = (e^{-iHt}\phi, F_R e^{-iHt}\psi) + \left((1-F_R)e^{-iHt}\phi, e^{-iHt}\psi\right)$$

arbitrary small by first choosing R and then t large enough.

PROOF OF THEOREM 1.2. Let $H = H^*$ on a Hilbert space $\mathcal{H}$. Then

$$\lim_{t\to\infty} t^{-1} \int_0^t ds\, e^{-iHs}\psi = 0$$

for all ψ if zero is not an eigenvalue of H. This follows for $\psi = H\phi$ by explicit integration, and these ψ are dense by hypothesis. If H has no eigenvalues, then zero is not an eigenvalue of $H \otimes 1 - 1 \otimes H$ on $\mathcal{H} \otimes \mathcal{H}$, and we obtain

$$\begin{aligned} 0 &= \lim_{t\to\infty} t^{-1} \int_0^t ds\, (\phi\otimes\psi, e^{-iHs}\psi \otimes e^{iHs}\phi) \\ &= \lim_{t\to\infty} t^{-1} \int_0^t ds \left|(\phi, e^{-iHs}\psi)\right|^2 \end{aligned} \tag{1.10}$$

for all $\phi, \psi \in \mathcal{H}$. Now let $H = H^*$ be arbitrary and suppose that $K(i+H)^{-1}$ is compact. Then we claim that

$$\lim_{t\to\infty} t^{-1} \int_0^t ds\, \|Ke^{-iHs}\psi\|^2 = 0 \tag{1.11}$$

for all $\psi \in \mathcal{H}_C$. Since it suffices to prove this for $\psi = (i+H)^{-1}\phi$, we may assume that K itself is compact. Then K is a norm limit of finite rank operators, which leaves us with proving (1.11) for K's of rank one: $Ke^{-iHs}\psi = (\phi, e^{-iHs}\psi)\eta$. For $\psi \in \mathcal{H}_C$ we can replace ϕ by its component in $\mathcal{H}_C$. Then (1.11) follows from (1.10) since H has no eigenvectors in $\mathcal{H}_C$. In the context of Theorem 1.2 we see that $\psi \in \mathcal{H}_C$ implies (1.8). On the other hand, (1.7) is evident for eigenvectors ϕ of H and thus for all $\phi \in \mathcal{H}_B$. The converse statements now follow from (1.9). □

1.3. Kato Potentials.

LEMMA 1.3. (i) *A real function $V \in L^p(X)$ is a Kato potential on X if $2 \le p$ and $\dim(X) < 2p \le \infty$.*

(ii) *Let $X = X_1 \oplus X_2$ be an orthogonal decomposition of X with adapted coordinates $x = (x_1, x_2)$, and suppose that $V = V(x_1)$. Then V is a Kato potential on X_1 if and only if it is a Kato potential on X.*

(iii) *The Kato potentials on X form a real vector space.*

(iv) *If V is a Kato potential on X which vanishes at infinity, then V is H-compact.*

EXAMPLE. $V(x) = 1/|x|$ is a Kato potential on R^3, since the singular part in $|x| < 1$ is in $L^2(R^3)$, while the tail in $|x| > 1$ is in $L^\infty(R^3)$. As a consequence of (ii), the total Coulomb potential of a system of N charged particles in R^3 is a Kato potential on R^{3N}.

PROOF OF LEMMA 1.3. (i) is proved in [M. Reed and B. Simon, *Methods of modern mathematical physics* II, Academic Press 1975, Theorem X.20.]. To prove (ii), let V be a Kato potential on X and apply (1.3) to product vectors $\psi = u \otimes v$, i.e. $\psi(x_1, x_2) = u(x_1)v(x_2)$. This gives

$$\alpha\|u\|\,\|v\|^{-1}\|p_2^2 v\| \ge \|Vu\| - \alpha\|p_1^2 u\| - \beta(\alpha)\|u\|,$$

with norms in the respective L^2-spaces. Since 0 is in the spectrum of p_2^2, the l.h.s. is arbitrary small for suitable v, so that the r.h.s must be ≤ 0. Therefore V is a Kato potential on X_1. Conversely, let V be a Kato potential on X_1. We use that for positive a, b, c: $a \le b + c \Longrightarrow a^2 \le 2(b^2 + c^2) \Longrightarrow a \le \sqrt{2}(b + c)$. Viewing $\psi \in L^2(X)$ as a function of x_2 taking values in $L^2(X_1)$, we obtain:

$$\|V\psi(x_2)\|^2 < 2\alpha^2\|p_1^2\psi(x_2)\|^2 + 2\beta(\alpha)^2\|\psi(x_2)\|^2$$

with norms in $L^2(X_1)$. Integrating over x_2, and using that $\|p_1^2\psi\| < \|p^2\psi\|$, we find:

$$\|V\psi\| \le \sqrt{2}\big(\alpha\|p^2\psi\| + \beta(\alpha)\|\psi\|\big),$$

which shows that V is a Kato potential on X. (iv) We show that $K = V(1+p^2)^{-1}$ is compact by using the same approximation argument as in the proof of (1.6). Since $V(x) \to 0$ as $|x| \to \infty$, we can restrict V to some ball $|x| < R$ with arbitrary small error in K, i.e. we may assume $V \in L^2(X)$. Similarly, we can cut off the tail of the function $(1+p^2)^{-1}$ in some region $|p| > R$. This tail is bounded by $f_R(p) = 2(R^2+p^2)^{-1}$, and its contribution to K by:

$$\alpha\|p^2 f_R(p)\| + \beta(\alpha)\|f_R(p)\| < 2(\alpha + \beta(\alpha)R^{-2}),$$

which is arbitrary small for small α and large R. As a result, K is again approximated in norm by Hilbert-Schmidt operators and therefore compact. □

2. N-Body Systems

Introduction. The standard N-body system consists of N particles moving in R^d, interacting by 2-body forces which vanish in the limit of large separation. The fact that such a system can break apart into separated, non-interacting subsystems of the same type is reflected in the behavior of the potential $V(x)$ as $x \to \infty$ in configuration space. We first describe the asymptotic situations in purely geometric terms, independent of the interactions. The resulting structure is summarized in abstract form, and used as a basis to introduce a more general class of systems. We keep the name "N-Body Systems" for this class, although the notion of particles moving in some physical space R^d will have disappeared. All further results are formulated and proven in this general setting. A first example is Zhislin's theorem, which characterizes the essential spectrum of N-body Hamiltonians.

2.1. Configurations. The standard system consists of N particles, labelled by $k = 1, \dots, N$, with masses $m_k > 0$, and with positions $x_k \in R^d$ relative to the center-of-mass (CM). Its configuration space is:

$$X = \{x = (x_1, \dots, x_N) \mid x_k \in R^d; \Sigma m_k x_k = 0\}.$$

We make X a euclidean space by defining the scalar product:

$$x \cdot y = \sum_k m_k (x_k, y_k)_{R^d}. \tag{2.1}$$

Sometimes we also write $x \cdot x = x^2 = |x|^2$. The classical kinetic energy is:

$$T = \frac{1}{2}\dot{x} \cdot \dot{x}, \tag{2.2}$$

where $\dot{x}$ denotes the velocities, and the Schrödinger Hamiltonian takes the form:

$$H = -\frac{1}{2}\Delta + V(x) \text{ on } L^2(X), \tag{2.3}$$

where Δ is the Laplacian corresponding to the metric (2.1). In the standard case,

$$V(x) = \sum_{i<k} V_{ik}(x_i - x_k); \quad V_{ik} \colon R^d \to R,$$

is a sum of 2-body potentials which vanish at infinity. For $N > 2$ the behavior of the *total* potential $V(x)$ is quite different: it is possible to let $x \to \infty$ along directions where $x_i = x_k$ for some pair (or pairs) of particles, i.e. in some subspace:

$$a = \{x \mid x_i = x_k \text{ if } i \sim k\}, \tag{2.4}$$

where $\sim$ is the equivalence relation in the set $\{1, \dots, N\}$ which defines a. The corresponding equivalence classes (the subsets of particles whose positions coincide for $x \in a$) are called the "clusters" of a. Suppose that $x \to \infty$ parallel to a: $x \to$

$x+sy$, $y \in a$, $s \to \infty$. Then the pair potentials between equivalent particles (acting inside the clusters) are not affected, while the others (acting between different clusters) may or may not go to zero — depending on the separation of the clusters. For an unambiguous description of the asymptotic situations we we introduce the set L of all subspaces (2.4), including

$$\begin{aligned} a =& \{0\}: \ i \sim k \text{ for all } i, k; \\ a =& \ X: i \sim k \text{ only if } i = k. \end{aligned}$$

L is closed under intersections: if $a, b \in L$, then $a \cap b$ is given by the union of the coincidence conditions $x_i = x_k$ defining a and b. Therefore L is a lattice with respect to the partial ordering $a \subset b$ for subsets of X. This is the basic geometric structure of (generalized) N-body systems:

DEFINITION. An *N-body configuration space* is a pair (X, L): X is a finite dimensional euclidean space; L is a finite lattice of subspaces of X, closed under intersections, with $\{0\} \in L$ and $X \in L$.

Hereafter we will work exclusively in this general setting, using standard N-body systems only as a guideline and for illustrations. In general, the letters $a, b, c, \ldots$ will denote elements of L. We write:

$$\begin{aligned} a \leq b \text{ for } &: a \subset b; \\ a < b \text{ for } &: a \subset b, \quad a \neq b. \end{aligned}$$

For each $a \in L$ there is an orthogonal decomposition:

$$X = a \oplus a^{\perp}, \tag{2.5}$$

and a corresponding decomposition of $x \in X$:

$$x = x_a + x^a; \quad x_a \in a; \quad x^a \in a^{\perp}. \tag{2.6}$$

As an example we consider the standard system with $N = 3$ and with a defined by $x_1 = x_2$, corresponding to the clusters $\{12\}\{3\}$. Then

$$x_a = (\xi, \xi, x_3); \quad x^a = (x_1 - \xi, x_2 - \xi, 0),$$

where $\xi = (m_1 x_1 + m_2 x_2)/(m_1 + m_2)$ is the CM of the cluster $\{12\}$. We see that x_a describes the configuration of the CM's of the clusters, and x^a the internal configurations of all clusters relative to their own CM's. We also note the relations:

$$\begin{aligned} x \cdot x &= x_a \cdot x_a + x^a \cdot x^a; \\ x_a \cdot x_a &= (m_1 + m_2)|\xi|^2 + m_3 |x_3|^2; \\ x^a \cdot x^a &= m_1 m_2 (m_1 + m_2)^{-1} |x_1 - x_2|^2, \end{aligned} \tag{2.7}$$

which express the familiar decomposition properties of the kinetic energy into CM-part and internal part. Returning to the general theory, we define the *intercluster distance*:

$$|x|_a \equiv \min_{c \cap a < a} |x^c| \tag{2.8}$$

for any $a > \{0\}$, and we note that

$$|x|_b \leq |x|_a \text{ if } b \geq a. \tag{2.9}$$

This follows since $b \geq a$ and $c \cap a < a$ imply $c \cap b < b$. The name *intercluster distance* is taken from the standard case, where

$$|x|_a = \min_{i \not\sim k} (m_i m_k)^{1/2} (m_i + m_k)^{-1/2} |x_i - x_k|, \tag{2.10}$$

with $\sim$ denoting the equivalence defining a. (2.10) is derived as follows. In the standard case, any $c \in L$ is an intersection of *elementary* subspaces $e \in L$, each given by a single coincidence $x_i = x_k$. If $c \cap a < a$, it follows that $e \cap a < a$ for some elementary $e \geq c$. Since $e \geq c$ implies $|x^e| \leq |x^c|$, we see that it suffices to take the minimum (2.8) over elementary c's. If c is given by $x_i = x_k$, then $c \cap a < a$ is the same as $i \not\sim k$. (2.10) now follows from from the explicit form of (x^c, x^c), taken from (2.7).

With the notion of intercluster distance we can obtain full control over the asymptotic situations as $x \to \infty$. For each x there is a minimal element $m(x)$ of L containing x:

$$m(x) \equiv \bigcap_{a \ni x} a \in L. \tag{2.11}$$

In the standard case, $m(x)$ is given by the equivalence: $i \sim k$ if and only if $x_i = x_k$. Since subspaces are closed, the function $m \colon X \mapsto L$ is lower semi-continuous:

$$m(y) \geq m(x) \tag{2.12}$$

for all y in some neighbourhood of x. The level sets of this function are:

$$a^* \equiv \{x \mid m(x) = a\} = a \setminus \bigcup_{b<a} b. \tag{2.13}$$

Since the b's here are lower dimensional subspaces of a, we see that $a^* \neq \emptyset$ for all $a \in L$.

LEMMA 2.1.

$$a^* \cap b^* = \emptyset \textit{ if } a \neq b;$$

$$\bigcup_{a>\{0\}} a^* = X \setminus \{0\}; \tag{2.14}$$

$$\textit{if } a > \{0\}, \textit{ then } x \in a^* \iff x \in a; \quad |x|_a > 0. \tag{2.15}$$

PROOF. As level sets, the a^* form a disjoint covering of X, and (2.14) follows from $\{0\}^* = \{0\}$. If $a > \{0\}$, then $x \in a$ and $x \notin a^*$ are equivalent to $x \in b < a$, which in turn is equivalent to $x \in a$ and $|x|_a = 0$. □

The covering (2.14) provides a complete classification of the asymptotic situations. Each $y \in X$, $y \neq 0$, belongs to some unique a^*. In the language of the standard N-body system, $y \in a^*$ means exactly this: if $x \to \infty$ parallel to y, then the internal configurations of the clusters of a remain fixed, while the intercluster distance $|x|_a$ diverges proportional to $|x|$.

2.2. N-Body Potentials. An N-body system is described by its configuration space (X, L) and by a potential V, which is always assumed to satisfy the following minimal conditions:

A.1 V is a Kato potential on X.

A.2 For any $a \in L$, $a > \{0\}$, there is a decomposition

$$V(x) = V^a(x^a) + I_a(x); \quad |I_a(x)| \leq f(|x|_a) \tag{2.16}$$

for sufficiently large $|x|_a$, where $f(s) \to 0$ as $s \to \infty$.

For $a = \{0\}$ we define $V^a = V$, $I_a = 0$. For $a = X$, V^a is a constant, and we normalize V by setting $V^a = 0$. In the standard case, A.1 and A.2 are satisfied if all pair potentials V_{ik} are Kato potentials on R^d vanishing at infinity; I_a is the sum of all pair potentials acting between different clusters of a.

LEMMA 2.2. *The decomposition* (2.16) *is unique, and V^a is again a Kato potential on X.*

PROOF. Let $y \in a^*$. Then $|x + sy|_a \to \infty$ as $s \to \infty$, so that

$$V^a(x^a) = \lim_{s \to \infty} V(x + sy).$$

Stated differently:

$$V^a \psi = \lim_{s \to \infty} T_{-s} V T_s \psi \tag{2.17}$$

for any $\psi \in C_0^\infty(X)$, where T_s is the unitary translation operator

$$(T_s \psi)(x) = \psi(x - sy). \tag{2.18}$$

Since T_s commutes with Δ, the inequality (1.2) for V also holds for $T_{-s} V T_s$ and is inherited by V^a in the limit (2.17). □

2.3. Subsystems. Each $a \in L$ defines a *subsystem* a, described by the potential V^a on the configuration space (X^a, L^a):

$$X^a = a^\perp; \quad L^a = \{b \cap a^\perp \mid b \in L; b \geq a\}. \tag{2.19}$$

In the standard case, this subsystem is formed by the non-interacting clusters of a, each one with its CM fixed at the origin. In general, we remark that the lattice L^a is isomorphic to the sublattice $\{b \mid b \geq a\}$ of L, since $b \to b \cap a^\perp$ has the inverse $b = a \oplus (b \cap a^\perp)$. It is in fact convenient to use $b \geq a$ to denote the elements of L^a. Using this isomorphism we find that

$$|x|_b^a \equiv \min_{c \cap b < b; c \geq a} |x^c| \tag{2.20}$$

is the intercluster distance $|x|_b$ in the subsystem a, defined for all $b > a$.

LEMMA 2.3. *Subsystems are N-body systems satisfying* A.1 *and* A.2.

PROOF. Since V^a is a Kato potential on X depending only on x^a, it is also a Kato potential on X^a (Lemma 1.3). To prove A.2 we start from the decomposition

$$V^a = V^b + I_b - I_a \equiv V^b + I_b^a. \tag{2.21}$$

Using (2.9) we conclude that

$$|I_b^a(x)| \leq f(|x|_b) \tag{2.22}$$

for some function f vanishing at infinity. Next, we determine the relation between $|x|_b$ and $|x|_b^a$:

$$|x|_b = \min_{c\cap b<b} |x^c| = \min\Big(\min_{c\cap b<b;c\geq a} |x^c|, \min_{c\cap b<b;c\cap a<a} |x^c|\Big) \tag{2.23}$$
$$= \min\big(|x|_b^a, |x|_a\big),$$

since $c \cap a < a$ already implies $c \cap b < b$. By (2.20), $|x|_b^a = |x+y|_b^a$ for all $y \in a$, while $|x+y|_a$ becomes arbitrary large for suitable $y \in a^*$. Hence

$$|x|_b^a = \max_{y\in a} |x+y|_b. \tag{2.24}$$

By (2.21), I_b^a is translation invariant along a, like $|x|_b^a$. Therefore (2.22) and (2.24) imply $\big|I_b^a(x)\big| \leq f\big(|x|_b^a\big)$. □

In this way, L becomes a lattice of subsystems of the same general type. This allows proofs by "induction in subsystems". To establish some property P for all N-body systems, we first verify P for the trivial subsystem:

$$a = X: \quad X^a = \{0\}; \quad V^a = 0.$$

Then we prove P for the full system:

$$a = \{0\}: \quad X^a = X; \quad V^a = V,$$

assuming that P is already established for all proper subsystems $a > \{0\}$.

2.4. The Essential Spectrum. Let H be the N-body Hamiltonian (2.3) with a potential $V(x)$ satisfying A.1 and A.2. Let $\sigma(H)$ denote the spectrum of H, and $\sigma_{\text{disc}}(H)$ the set of all discrete eigenvalues of H (eigenvalues of finite multiplicity, which are isolated from the rest of the spectrum). We define the essential spectrum by:

$$\sigma_{\text{ess}}(H) \equiv \sigma(H) \setminus \sigma_{\text{disc}}(H).$$

For any $a \in L$ we introduce:

$$H_a = H - I_a = -\frac{1}{2}\Delta + V^a \text{ on } L^2(X). \tag{2.25}$$

Corresponding to the decomposition (2.5) we write:

$$L^2(X) = L^2(a) \otimes L^2(a^\perp),$$

which simply means that we regard $\psi \in L^2(X)$ as a function of the two variables (x_a, x^a). With respect to this factorization we have the representations:

$$\begin{aligned} \Delta &= \Delta_a \otimes 1 + 1 \otimes \Delta^a; \\ H_a &= -\frac{1}{2}\Delta_a \otimes 1 + 1 \otimes H^a; \\ H^a &= -\frac{1}{2}\Delta^a + V^a. \end{aligned} \tag{2.26}$$

Here Δ_a and Δ^a are the Laplacians with respect to x_a, x^a, and H^a is the Hamiltonian of the subsystem a, acting on the Hilbert space $L^2(a^\perp)$. If $a > \{0\}$, then $\sigma(-\Delta_a) = [0,\infty)$, so that:

$$\sigma(H_a) = [0,\infty) + \sigma(H^a) = [\Sigma_a, \infty); \quad \Sigma_a = \min\big(\sigma(H^a)\big). \tag{2.27}$$

In the standard case, H_a describes the motion of a system of non-interacting clusters, and Σ_a is the lowest possible internal energy of all the clusters.

LEMMA 2.4. *If $a < b$ then $\sigma(H_b) \subset \sigma(H_a)$; i.e.*

$$\Sigma_a \leq \Sigma_b \tag{2.28}$$

PROOF. Using (2.21) we write $H_a = H_b + I_b^a$. Let T_s be the translation operator (2.18) for some $y \in b^*$. Since T_s commutes with H_b,

$$\|(\lambda - H_a)T_s\psi\| \leq \|(\lambda - H_b)\psi\| + \|I_b^a T_s\psi\|.$$

Now let $\lambda \in \sigma(H_b)$. Then the first term on the r.h.s. is arbitrary small for some suitable state $\psi \in C_0^\infty(X)$, while the second term always vanishes in the limit $s \to \infty$ as a consequence of (2.22). Therefore, the l.h.s. is arbitrary small for some state $T_s\psi$, which proves that $\lambda \in \sigma(H_a)$. $\square$

In the special case $a = \{0\}$ it follows that $\sigma(H)$ contains the continuous parts $[\Sigma_b, \infty)$ for all $b > \{0\}$. Therefore:

$$\sigma_{\text{ess}}(H) \supset [\Sigma, \infty); \quad \Sigma \equiv \min_{b > \{0\}} \Sigma_b. \tag{2.29}$$

THEOREM 2.5 (ZHISLIN'S THEOREM).

$$\sigma_{\text{ess}}(H) = [\Sigma, \infty).$$

PROOF. We will use a partition of unity on X, i.e. a finite family $\{j_a\}$ of real C^∞-functions on X, satisfying

$$\sum_a j_a^2 = 1. \tag{2.30}$$

Such a partition allows to split H into symmetric *localized parts* $j_a H j_a$, modulo a *localization error* (double commutator):

$$\begin{aligned} H &= \sum_a j_a H j_a + \frac{1}{2}[j_a, [j_a, H]] \\ &= \sum_a j_a H j_a - \frac{1}{2}\sum_a |\nabla j_a|^2. \end{aligned} \tag{2.31}$$

For the problem at hand, a runs over all $a \in L$ with $a > \{0\}$, and the j_a are constructed as follows. By Lemma 2.1, the sets

$$S_a \equiv \{x \mid |x| = 1; |x|_a > 0\}$$

form an open covering of the unit sphere $S \subset X$. For any such covering there exists a partition of unity with $\text{supp}(j_a) \subset S_a$. Since $\text{supp}(j_a)$ is compact, we have $|x|_a > \varepsilon$ for some $\varepsilon > 0$, all a and all $x \in \text{supp}(j_a)$. Now we extend the functions j_a from S to X. For $|x| > 1$ we define $j_a(x) = j_a(x/|x|)$, while for $|x| < 1$ we choose any smooth extension obeying (2.30). The resulting partition of unity on X then has the special properties:

$$j_a(\lambda x) = j_a(x) \text{ for } |x| \geq 1, \lambda \geq 1; \tag{2.32}$$

$$|x|_a \geq \varepsilon|x| \text{ for } |x| \geq 1, x \in \text{supp}(j_a). \tag{2.33}$$

(2.32) implies that $\nabla j_a = O(1/|x|)$ as $|x| \to \infty$. The last term in (2.31) is therefore H-compact by (1.5). In the first term, we write $H = H_a + I_a$ and we note that, by (2.33), $j_a I_a j_a$ is a Kato potential which vanishes at infinity, and is therefore also H-compact (Lemma 1.4). As a result,

$$H = \sum_a j_a H_a j_a + K, \tag{2.34}$$

where K is H-compact, so that

$$\sigma_{\text{ess}}(H) = \sigma_{\text{ess}}\left(\sum_a j_a H_a j_a\right)$$

by a theorem of Weyl. Now, since $H_a \geq \Sigma$ for all a,

$$\left(\psi, \sum_a j_a H_a j_a \psi\right) = \sum_a (j_a\psi, H_a j_a \psi) \geq \Sigma\left(\psi, \sum_a j_a^2 \psi\right) = \Sigma(\psi,\psi)$$

for all $\psi \in D(\Delta)$. We conclude that

$$\sigma_{\text{ess}}(H) \subset \sigma\left(\sum_a j_a H_a j_a\right) \subset [\Sigma, \infty),$$

which is the converse of (2.29). □

3. Exponential Bounds

Introduction. In this section we first discuss the exponential decay of bound state wave functions $\psi(x)$ as $x \to \infty$ in configuration space. Let $H\psi = E\psi$, $E < \Sigma$. In the standard 2-body case (where $\Sigma = 0$), this implies that $\psi(x)$ decays exponentially like $\exp(-f(x))$; $f(x) = |x_1 - x_2|\sqrt{-2mE}$, m being the reduced mass of the pair $\{1,2\}$. For general N-body systems, $f(x)$ will be a positive function, homogeneous of degree 1, which is implicitly determined by the energy E and by the thresholds Σ_a. Although the general form of this bound f is not known, useful weaker bounds can be given explicitely. All these bounds are expressed in terms of the euclidean metric of X, which describes their mass-dependence in the standard case.

3.1. Exponential Bounds for Eigenfunctions.

THEOREM 3.1 (DEIFT, HUNZIKER, SIMON, VOCK). *Suppose that H is an N-body Hamiltonian satisfying* A.1 *and* A.2. *Let $H\psi = E\psi$, $E < \Sigma$. If $f(x)$ is homogeneous of degree* 1, *and has the Lipschitz properties*

$$|f(x) - f(y)| \leq \lambda_a |x-y|; \quad \lambda_a \equiv \sqrt{2(\Sigma_a - E)}; \tag{3.1}$$

for all $a \in L$, $a > \{0\}$, and all $x, y \in a$, then f is an exponential bound for ψ in the sense that

$$e^{\alpha f}\psi \in L^\infty(X) \text{ for any } \alpha < 1. \tag{3.2}$$

In particular, the pointwise supremum $\bar{f}$ of all these exponential bounds f is an exponential bound.

The maximal bound $\bar{f}$ allowed by (3.1) is uniquely determined by the energy E and the thresholds Σ_a. Non-optimal bounds obtained from Theorem 3.1 are also useful, especially the isotropic bound

$$f(x) = |x|\sqrt{2(\Sigma - E)} \tag{3.3}$$

due to O'Connor. In general, the maximal bound $\bar{f}$ can be highly anisotropic, with range in:

$$|x|\sqrt{2(\Sigma - E)} \leq \bar{f}(x) \leq |x|\sqrt{-2E}.$$

The general explicit form of $\bar{f}$ is not known. Agmon has given a geometric characterization of $\bar{f}$ in terms of the metric:

$$ds^2 = 2\Big(\sum_{m(x)} - E\Big)\, dx^2 \text{ on } X, \tag{3.4}$$

where $m(x)$ is defined by (2.11). A path $p \subset X$, given by a function $x(t)$ on $0 \leq t \leq 1$, has the Agmon length:

$$s(p) = \int_0^1 dt\, \lambda_{m(x(t))}\big|\dot{x}(t)\big|. \tag{3.5}$$

This is well defined for square-integrable $\dot{x} = dx/dt$, since the function $x \to \lambda_{m(x)}$ is lower semicontinuous, due to (2.12) and (2.28). The Agmon distance $d(x,y)$ between x and y is the infimum of $s(p)$, taken over all paths p joining x, y. We refer to [Carmona R. and Simon B. *Pointwise bounds on eigenfunctions and wave packets in N-body systems. V. Lower Bounds and Path Inegrals*, Commun. Math. Phys. **80**, (1981) 59–98], for a proof that this infimum a minimum, and for a discussion of Agmon geodesics.

THEOREM 3.2 (AGMON'S THEOREM). *The exponential bound $\bar{f}$ given in Theorem* 3.1 *is* $\bar{f}(x) = d(0,x) \equiv \rho(x)$.

PROOF. Evidently $\rho(x)$ is homogeneous of degree 1. By the geodesic triangle inequality,

$$\big|\rho(x) - \rho(y)\big| \leq d(x,y) \leq \lambda_a|x-y|$$

for $x, y \in a$. This proves $\rho \leq \bar{f}$. To show the converse we choose a path p from 0 to x with $s(p) < \rho(x) + \varepsilon$. Approximating $\dot{x}(t)$ by a step function in L^2-sense, we see that p may be taken as a polygon of straight lines $p_1, \ldots, p_n$. For each p_k we define:

$$a_k = \bigcap_{a \supset p_k} a \in L.$$

Then $\lambda_{m(x)} = \lambda_{a_k}$ for all $x \in p_k$, with the possible exception of a single point (a straight line $p \not\subset a$ can intersect a only in one point). Therefore $s(p_k) = \lambda_{a_k}|p_k|$, where $|p_k|$ is the euclidean length of p_k. On the other hand (3.1) implies $|\Delta\bar{f}_k| \leq \lambda_{a_k}|p_k|$ for the increment $\Delta\bar{f}_k$ of $\bar{f}$ along p_k. It follows that

$$\bar{f}(x) = \sum_k \Delta\bar{f}_k \leq \sum_k \lambda_{a_k}|p_k| = s(p) \leq \rho(x) + \varepsilon. \quad \square$$

We now begin with the derivation of Theorem 3.1. Instead of (3.2) we will only prove the L^2-bound:

$$e^{\alpha f}\psi \in L^2(X) \text{ for any } \alpha < 1. \tag{3.6}$$

Since f is uniformly Lipschitz, the L^∞-bound (3.2) then follows by a general argument given in [P. Deift, W. Hunziker, B. Simon, and E. Vock, Commun. Math. Phys. **64** (1978) 1–34]. The basic tool for estimating exponential tails of eigenfunctions is simple:

LEMMA 3.3. *Suppose that $H\psi = E\psi$. Let $J, f \in C^2(X)$ be nonnegative, with bounded derivatives, and let* $\operatorname{supp}(\nabla J)$ *be compact. If*

$$J\left(H - \frac{1}{2}|\nabla f|^2 - E\right)J \geq \delta J^2 \tag{3.7}$$

for some $\delta > 0$, then

$$\|e^f J\psi\| \leq \delta^{-1}\|e^f[H, J]\psi\|. \tag{3.8}$$

Remark. The hypothesis allows $f(x) \to \infty$ as $x \to \infty$. The bound (3.8) is finite, since f is bounded on $\operatorname{supp}(\nabla J)$. Typically, J will be a smoothed characteristic function of some set $|x| > R$. Then (3.8) implies $\exp(f)\psi \in L^2$.

PROOF. Suppose first that f is bounded, and let

$$H_f \equiv e^f H e^{-f} = H - \frac{1}{2}|\nabla f|^2 + \frac{i}{2}(\nabla f \cdot p + p \cdot \nabla f).$$

Then $(H_f - E)u = 0$ for $u = e^f\psi$, so that (3.7) implies

$$\delta\|Ju\|^2 \leq \operatorname{Re}\big(Ju, (H_f - E)Ju\big) \leq \|Ju\|\,\big\|[H_f, J]u\big\|,$$

which proves (3.8). If f is unbounded, we replace it by $f_\varepsilon = f(1 + \varepsilon f)^{-1}$, $\varepsilon > 0$. Since $|\nabla f_\varepsilon| \leq |\nabla f|$, (3.8) holds for f_ε uniformly in ε, and extends to f in the limit $\varepsilon \to 0$. $\square$

Our next step is to prove a smooth version of Theorem 3.1:

LEMMA 3.4. *Theorem* 3.1 *holds if the Lipschitz condition* (3.1) *is replaced by the the stronger hypothesis*:

$$f \in C^2(X\backslash\{0\});\quad \big|\nabla f(x)\big|^2 \leq 2(\Sigma_{m(x)} - E) \tag{3.9}$$

for all $x \in S =$ unit sphere of X.

PROOF. Let $g = \alpha f$, $\alpha < 1$. Then

$$|\nabla g|^2 \leq 2(\Sigma_{m(x)} - E - \varepsilon) \tag{3.10}$$

for some $\varepsilon > 0$ and all $x \in S$. Each $y \in S$ has a neighbourhood $S_y \subset S$ given by

$$S_y = \Big\{x \in S \mid 2|x|_{m(y)} > |y|_{m(y)};\, \big|\nabla g(x)\big|^2 < \big|\nabla g(y)\big|^2 + \varepsilon\Big\}, \tag{3.11}$$

where we have used the definitions (2.8) and (2.11). Proceeding as in the proof of Theorem 2.1, we pick a finite covering $\{S_y\}$ of S, and then construct a partition of unity $\{j_y\}$ on X with the properties:

$$S \cap \operatorname{supp}(j_y) \subset S_y;\quad j_y(x) = j_y\big(x/|x|\big) \text{ for } |x| > 1.$$

From (2.18) we obtain:

$$H - \frac{1}{2}|\nabla g|^2 - E = \sum_y j_y \left(H - \frac{1}{2}|\nabla g|^2 - E\right) j_y - \frac{1}{2}\sum_y |\nabla j_y|^2.$$

By construction, $|\nabla j_y(x)| \to 0$ and $I_{m(y)} j_y(x) \to 0$ as $x \to \infty$. Let J be a real, smooth, bounded function supported in $\{|x| > R\}$. Using (3.10), (3.11) and $H_a \geq \Sigma_a$, we find

$$J\left(H - \frac{1}{2}|\nabla g|^2 - E\right) J \geq \left(\frac{\varepsilon}{2} - o(R)\right) J^2.$$

Taking R sufficiently large, we conclude from Lemma 3.3 that $e^g\psi \in L^2(X)$. □

To prove Theorem 3.1 we want to approximate f by a smooth exponential bound. However, since the Lipschitz conditions (3.1) on f become stronger on lower dimensional subspaces, we cannot apply smoothing by convolution directly. As an intermediate approximation step we use:

LEMMA 3.5. *Let f satisfy the hypothesis of Theorem 2.2, and let $\delta > 0$. Then there exists $\varepsilon > 0$ and a function g on X with the properties:*

$$\begin{aligned} &g(\lambda x) = \lambda g(x); \quad |g(x) - f(x)| \leq \delta|x|; \quad \textit{and} \\ &|g(x) - g(y)| \leq |x - y|\lambda_a(1 + \delta) \end{aligned} \tag{3.12}$$

for all $a > \{0\}$ and all x, y in the cone $|x^a| < \varepsilon|x|$.

The proof of this Lemma requires some basic N-body geometry and will be given at the end of this section.

PROOF OF THEOREM 2.2. Given $\delta > 0$, we first construct g satisfying (3.12). Then we set

$$h(x) = \int_X dy\, \rho(y) g(x - |x|y),$$

where ρ is a suitable C_0^∞-approximation of the δ-function. By construction, $h(\lambda x) = \lambda h(x)$ and $h \in C^\infty(X \setminus \{0\})$. It also follows from (3.12) that

$$|h(x) - f(x)| \leq 2\delta|x|; \quad |\nabla h(x)| \leq \lambda_a(1 + 2\delta) \text{ for } x \in a,$$

if $\operatorname{supp}(\rho)$ is taken sufficiently small. Now $(1 - 2\delta)h$ is an exponential bound by Lemma 3.4, and so is

$$(1 - \mu)h + (\mu - 2\delta)\alpha|x|$$

for $2\delta < \mu < 1$, where $\alpha|x|$ stands for the isotropic bound (3.3). Thus $(1 - \mu)f \leq (1 - \mu)h + 2\delta|x|$ is seen to be an exponential bound by choosing $2\delta < \alpha(\mu - 2\delta)$. Since μ is arbitary small, f is an exponential bound. □

PROOF OF LEMMA 3.5. Let $a \to |a|$ be some monotone integer function on the lattice L, e.g.

$$|a| \equiv \dim(a). \tag{3.13}$$

For any $\varepsilon \in (0, 1)$ we define two systems of cones along the subspaces $a \in L$:

$$U_a \equiv \{x \mid |x^a| < 2\varepsilon^{|a|}|x|\}; \quad V_a \equiv \{x \mid |x^a| < \varepsilon^{|a|}|x|\}. \tag{3.14}$$

In particular, $U_{\{0\}} = V_{\{0\}} = \emptyset$; $U_X = V_X = X \setminus \{0\}$. We show that, for ε sufficiently small:

$$U_a \cap U_b \subset V_{a \cap b} \text{ if } a \cap b < a, b. \tag{3.15}$$

There is a constant M depending only on the lattice L, such that

$$|x^{a\cap b}| \leq M \max\big(|x^a|, |x^b|\big) \tag{3.16}$$

for all a, b, x (both sides are seminorms on X, and $x_a = x_b = 0$ implies $x \in a \cap b$ so that $|x^{a\cap b}| = 0$). It follows that (3.15) is satisfied if $2M\varepsilon < 1$. For each $a > \{0\}$ we now define a transformation $T_a\colon f \to \tilde{f}$ in the space of homogeneous functions of order 1 on X:

$$(T_a f)(x) = \tilde{f}(x) = f(S_a x); \quad S_a x = x_a + j\big(|x^a| \,/\, \varepsilon^{|a|}|x|\big)x^a,$$

where j is a smooth function with $0 \leq j \leq 1$; $j(r) = 0$ for $0 \leq r \leq 1$; $j(r) = 1$ for $2 \leq r < \infty$. $\tilde{f}$ is a mild rearrangement of f inside U_a with the following relevant properties :

(i) If f is Lipschitz on X with constant L, then $\big|\tilde{f}(x) - f(x)\big| \leq \varepsilon L|x|$, since $|S_a x - x| < \varepsilon|x|$.

(ii) $\tilde{f}(x) = f(x_a)$ on V_a. Therefore, if f is Lipschitz with constant λ_a on a, $\tilde{f}$ is Lipschitz with constant λ_a on V_a.

(iii) Let $b \leq a$. If $f(x) = f(x_b)$ on V_b, then $\tilde{f}(x) = f(x)$ on V_b. This follows since x_b is a function of x_a only and since $S_a V_b \subset V_b$.

(iv) If f is Lipschitz on $b > a$ with constant λ_b, then $\tilde{f}$ is Lipschitz on b with constant $\lambda_b\big(1 + O(\varepsilon)\big)$.

To prove (iv) we note that $S_a b \subset b$, so it suffices to consider the case $b = X$. Writing out $\nabla\tilde{f}$ in terms of ∇f the estimate is straightforward.

Now suppose that f obeys (3.1). We set $f_0 = f$ and transform it successively into

$$f_k = A_k f_{k-1}; \quad A_k = \prod_{|a|=k} T_a; \quad k = 1, \ldots, |X|, \tag{3.17}$$

where missing factors are treated as 1. We claim that the order of the factors T_a in (3.17) is irrelevant, and that f_k has the properties:

(3.18) $\big|f_k(x) - f(x)\big| \leq \varepsilon(k-1)\big(1 + O(\varepsilon)\big)^{k-1}\sqrt{-2E}|x|$;

(3.19) On each V_a with $|a| \leq k$, f_k depends only on x_a and is Lipschitz with constant $\lambda_a\big(1 + O(\varepsilon)\big)^{k-1}$;

(3.20) $f_k(x)$ is Lipschitz on all $b \in L$ with constant $\lambda_b\big(1 + O(\varepsilon)\big)^{k-1}$.

The proof is by induction in k. Consider 2 operators T_a, T_b with $|a| = |b| = k+1$, $a \neq b$. By (3.19) and (iii), $T_a f_k = f_k$ on $V_{a\cap b}$, and therefore on $U_a \cap U_b$ by (3.15). Thus T_a and T_b affect f_k only in disjoint regions, and we see from (i) and (ii) that (3.18) and (3.19) hold for $k+1$. (3.20) now follows from (3.19) if $|b| \leq k+1$. Let $|b| > k+1$ and $b \not\supset a$. Then $T_a f_k = f_k$ on b since $U_a \cap b \subset V_{a\cap b}$. If $b \supset a$, then, by (iv), T_a raises the Lipschitz constant of f_k on b by a factor $\big(1 + O(\varepsilon)\big)$. Since different T_a act in disjoint sets, we find that (3.20) holds for $k+1$.

The function $g = f_{|X|}$ satisfies (3.12) if ε is taken sufficiently small. This proves Lemma 3.5 and concludes the proof of Theorem 3.1. □

4. Mourre's Inequality and the Structure of the Continuum

Introduction. Zhislin's theorem characterizes the continuous spectrum of H as a set, but it says nothing about the spectral properties of H in the energy range $[\Sigma, \infty)$. The key to this problem is Mourre's inequality, which can be understood, to some extent, in the following heuristic way. For trajectories $\psi_t = e^{-iHt}\psi$ in $\mathcal{H}_C$ we expect

$$\langle x^2 \rangle_t = (\psi_t, x^2 \psi_t) \approx \theta t^2, \quad (t \to \infty) \tag{4.1}$$

for some constant $\theta \geq 0$. This constant should be equal to the second time-derivative of $\langle x^2/2 \rangle_t$ *in the time-mean*, or for states ψ with *sufficiently sharp energy distribution*. Writing this out for

$$\frac{d^2}{dt^2} \left\langle \frac{x^2}{2} \right\rangle_t = \langle i[H, A] \rangle_t;$$

$$A = i\left[H, \frac{x^2}{2}\right] = \frac{1}{2}(p \cdot x + x \cdot p); \tag{4.2}$$

$$i[H, A] = p^2 - \big(x \cdot \nabla V(x)\big), \tag{4.3}$$

we are led to a special case of the Mourre inequality: if E is not an eigenvalue of H, then

$$B_\Delta(H) \equiv E_\Delta(H) i[H, A] E_\Delta(H) \geq \big(\theta(E) - \varepsilon\big) E_\Delta(H) \tag{4.4}$$

for some $\theta(E) \geq 0$ and any $\varepsilon > 0$, where E_Δ is the spectral projection of H for a sufficiently small open interval $\Delta \ni E$. The Mourre constant $\theta(E)$ is directly related to the effect of thresholds. A *threshold of* H is an eigenvalue λ^a of H^a for some $a > \{0\}$, corresponding to a stationary state $\psi^a \in L^2(X^a)$ of the subsystem a. In the energy range above λ^a, we expect the existence of scattering trajectories ψ_t with a behavior like

$$\psi_t \approx e^{-i(1/2p_a^2 + \lambda^a)t} \psi_a \otimes \psi^a, \quad (t \to \infty), \tag{4.5}$$

where $\psi_a \in L^2(X_a)$ is arbitrary. For such trajectories, with energy distribution sharply localized at $E > \lambda^a$, we will have:

$$\langle x^2 \rangle_t \approx \langle x_a^2 \rangle_t \approx \langle p_a^2 \rangle t^2 \approx 2(E - \lambda^a) t^2, \quad (t \to \infty). \tag{4.6}$$

For fixed E, all thresholds $\leq E$ will contribute scattering states with energies localized near E. For $E > \Sigma$, *the Mourre constant* $\theta(E)$ *is in fact twice the distance of* E *from the highest threshold* $\leq E$. (4.4) also holds for $E < \Sigma$ if we define $\theta(E) = 0$. This is trivial if $E \notin \sigma(H)$. If E is an eigenvalue of H, then $\langle x^2 \rangle_t$ is constant for the corresponding bound states. In this case $E_\Delta(H)$ reduces to an eigenprojection for small Δ, and $B_\Delta(H) = 0$ follows from the *virial theorem:*

$$(\psi, i[H, A]\phi) = 0 \tag{4.7}$$

if ψ, ϕ are eigenvectors of H with the same eigenvalue E. (4.7) is formally obvious for any $A = A^*$, since $\big(\psi, i[H, A]\phi\big) = iE\big[(\psi, A\phi) - i(A\psi, \phi)\big] = 0$.

All these are special cases of a more general Mourre inequality: (4.4) *holds for any real* E*, up to an error term which is a compact operator.* From this powerful inequality we will derive a number of basic results concerning the structure of the continuous spectrum of H:

(i) Eigenvalues away from thresholds have finite multiplicities and can accumulate only at thresholds, and only from below. Since thresholds are eigenvalues of subsystems, it follows that the set of thresholds is closed and countable.

(ii) Eigenfunctions for non-threshold eigenvalues decay at least like

$$\exp\bigl(-|x|\sqrt{2(\lambda - E)}\bigr),$$

where λ is the lowest threshold $> E$.

(iii) H has no eigenvalues (and thus no thresholds) > 0.

Later, we will exploit Mourre's inequality to analyze the large time behavior of continuum states, thereby confirming the heuristic arguments used in this introduction.

4.1. The Virial Theorem and Mourre's Inequality. We define $i[H,A]$ by the explicit form (4.3) as a Schrödinger operator on $D(p^2)$, assuming that the *virial* $x \cdot \nabla V(x)$ also satisfies the conditions A.1 and A.1 imposed on the potential $V(x)$. Then the virial theorem (4.7) can be proven by using some regularization of A, e.g.

$$A \to A_\varepsilon = \frac{1}{2}\bigl(p \cdot xe^{-\varepsilon x^2} + e^{-\varepsilon x^2} x \cdot p\bigr); \quad (\varepsilon > 0). \tag{4.8}$$

A_ε is bounded relative to p^2, and $i[H, A_\varepsilon]$ is defined by an expression similar to (4.3). The formal argument given for (4.7) is correct for A_ε, since $\psi, \phi \in D(A_\varepsilon)$. Thus $\bigl(\psi, i[H, A_\varepsilon]\phi\bigr) = 0$, and (4.7) follows in the limit $\varepsilon \to 0$.

DEFINITION. A *threshold of H* is an eigenvalue of H^a for some $a > \{0\}$. $\tau(H)$ is the set of all thresholds of H. The *Mourre constant* $\theta(E)$ is defined for any real E by:

$$\theta(E) = \begin{cases} 0 & \text{for } E < \Sigma; \\ \inf_{\lambda \in \tau(H); \lambda \le E} 2(E - \lambda) & \text{for } E \ge \Sigma. \end{cases} \tag{4.9}$$

THEOREM 4.1 (MOURRE'S THEOREM). *Suppose that both the potential $V(x)$ and its virial $\bigl(x \cdot \nabla V(x)\bigr)$ satisfy* A.1 *and* A.2. *Let $i[H,A]$ be defined by* (4.3), *and let $E_\Delta(H)$ be the spectral projection of H for an interval Δ. Then*:

(i) *Given $E \in R$ and $\varepsilon > 0$, there exists an open interval $\Delta \ni E$ and a compact operator K such that*

$$B_\Delta(H) \equiv E_\Delta(H) i[H,A] E_\Delta(H) \ge \bigl(\theta(E) - \varepsilon\bigr) E_\Delta(H) + K. \tag{4.10}$$

(ii) *Non-threshold eigenvalues of H have finite multiplicities and can accumulate only at thresholds. Therefore $\tau(H)$ is closed and countable.*

Notation. We will use the abbreviation "$\Delta \ni E$ " to say that Δ is an open interval containing E, and "$\Delta \to \{E\}$" for a sequence of such intervals with length $|\Delta| \to 0$.

COROLLARY 1. (4.4) *follows from* (4.10) *if E is not an eigenvalue of H.*

PROOF. Multiplying (4.10) from both sides by $E_\Delta(H)$ we see that K may be replaced by $E_\Delta K E_\Delta$. Then we let $\Delta \to \{E\}$ while keeping K fixed. Since E is not an eigenvalue of H, $E_\Delta(H) \xrightarrow{s} 0$ and therefore $\bigl\|K E_\Delta(H)\bigr\| \to 0$. □

PROOF OF THEOREM 4.1. We proceed by induction in subsystems, assuming that Theorem 4.1 holds for all H^a with $a > \{0\}$ in the following form. $\tau(H^a)$ consists of the eigenvalues of all H^b with $b > a$. $\theta^a(E)$ is defined with respect to $\tau(H^a)$, and (4.10) reads:

$$B_\Delta(H^a) \geq \big(\theta^a(E) - \varepsilon\big)E_\Delta(H^a) + K \text{ on } L^2(X^a), \tag{4.11}$$

where $B_\Delta(H^a) = E_\Delta(H^a)i[H^a, A^a]E_\Delta(H^a)$; $i[H^a, A^a] = -\Delta^a - \big(x^a \cdot \nabla V^a(x^a)\big)$. This induction hypothesis is trivially satisfied for $a = X$. □

LEMMA 4.2. *Part* (ii) *of Theorem* 4.1 *follows from part* (i).

PROOF. By part (ii) of the induction hypothesis, $\tau(H)$ is closed and countable. Let $E_n \to E$ be an infinite sequence of eigenvalues of H with orthonormal eigenvectors ψ_n. From Theorem 2.1 we know that $E \geq \Sigma$, so that $\theta(E) = 0$ implies $E \in \tau(H)$. By (4.7) and (4.11),

$$0 \geq \big(\theta(E) - \varepsilon\big) + (\psi_n, K\psi_n)$$

for any $\varepsilon > 0$ and large n. Since $\psi_n \xrightarrow{w} 0$ we have $\|K\psi_n\| \to 0$ and therefore $\theta(E) = 0$, i.e. $E \in \tau(H)$. □

LEMMA 4.3.

$$B_\Delta(H^a) \geq \big(\theta(E) - \varepsilon\big)E_\Delta(H^a) \tag{4.12}$$

for any $E \in R$, *any* $\varepsilon > 0$ *and some* $\Delta \ni E$.

PROOF. If E is not an eigenvalue of H^a we have

$$B_\Delta(H^a) \geq \big(\theta^a(E) - \varepsilon\big)E_\Delta(H^a)$$

by the induction hypothesis and the Corollary to Theorem 4.1, and (4.12) follows since $\theta^a(E) \geq \theta(E)$. Now let E be an eigenvalue of H^a with eigenprojection P. Then we have to prove (4.12) with $\theta(E) = 0$. Since $\dim(P) = \infty$ is not excluded, we represent P as a strong limit of finite rank projections $P_n \leq P$. By the virial theorem:

$$B_\Delta = (1 - P_n)B_\Delta(1 - P_n) + P_nB_\Delta(1 - P) + (1 - P)B_\Delta P_n.$$

Using (4.11) to estimate the first term on the r.h.s, we obtain:

$$\begin{aligned} B_\Delta &\geq -\varepsilon E_\Delta + (1 - P_n)K(1 - P_n) - 2\big\|P_nB_\Delta E_\Delta(1 - P)\big\| \\ &\geq -\varepsilon E_\Delta - \big\|K(P - P_n)\big\| - \big\|KE_\Delta(1 - P)\big\| - 2\big\|P_nE_\Delta(1 - P)\big\|. \end{aligned}$$

Since K and P_nB_Δ are compact, and since $E_\Delta(1 - P) \xrightarrow{s} 0$ as $\Delta \to \{E\}$, we can first fix n and then Δ to make the last three norms arbitrary small. □

LEMMA 4.4. *Given a compact* $I \subset R$ *and* $\varepsilon > 0$, *there exists* $\delta > 0$ *such that*

$$B_\Delta(H^a) \geq \big(\theta(E + \varepsilon) - 2\varepsilon\big)E_\Delta(H^a) \tag{4.13}$$

for any $E \in I$ *and any* $\Delta \ni E$ *of length* $|\Delta| < \delta$.

PROOF. Suppose this is false. Then (4.13) is violated for some sequence $E_n \to E$ in I and corresponding $\Delta_n \ni E_n$ with $|\Delta_n| \to 0$. From (4.12) we obtain:

$$B_\Delta(H^a) \geq \big(\theta(E) - \varepsilon/2\big)E_\Delta(H^a) \tag{4.14}$$

for some $\Delta \ni E$. Let n be so large that $|E_n - E| < \varepsilon/2$ and $\Delta_n \subset \Delta$. Since $\theta(E+x) \leq \theta(E) + x$ for any $x \geq 0$, we then have:

$$\theta(E) \geq \theta(E_n + \varepsilon) - \varepsilon + E - E_n \geq \theta(E_n + \varepsilon) - \frac{3\varepsilon}{2}.$$

From this, $\Delta_n \subset \Delta$ and (4.14) we deduce that

$$B_{\Delta_n}(H^a) \geq \big(\theta(E_n + \varepsilon) - 2\varepsilon\big)E_{\Delta_n}(H^a),$$

which contradicts our assumption. □

LEMMA 4.5.

$$B_\Delta(H_a) \geq \big(\theta(E+\varepsilon) - 2\varepsilon\big)E_\Delta(H_a) \tag{4.15}$$

for any $E \in R$, any $\varepsilon > 0$ and some $\Delta \ni E$.

PROOF. We represent the functions $\psi(x) = \psi(x_a, x^a)$ by their partial Fourier transforms with respect to x_a, i.e. by functions $\psi_F(k)$ on $L^2(X_a)$ taking values in $L^2(X^a)$. In this representation:

$$(\psi, \phi) = \int_{X_a} dk\, \big(\psi_F(k), \phi_F(k)\big)_{L^2(X^a)};$$

$$(H_a\psi)_F(k) = (k^2/2 + H^a)\psi_F(k);$$

$$\big(E_\Delta(H_a)\psi\big)_F(k) = E_{\Delta - k^2/2}(H^a)\psi_F(k);$$

$$\big(i[H_a, A]\psi\big)_F(k) = \big(k^2 + i[H^a, A^a]\big)\psi_F(k).$$

For $\psi = E_\Delta(H_a)\psi$ we thus obtain:

$$\big(\psi, B_\Delta(H_a)\psi\big) = \int_{X_a} dk \Big(\psi_F(k), \big(k^2 + B_{\Delta - k^2/2}(H^a)\big)\psi_F(k)\Big)_{L^2(X^a)}.$$

Since H^a is bounded from below, the integrand has compact support: $\psi_F(k) = E_{\Delta - k^2/2}(H^a)\psi_F(k) = 0$ for large k^2. By (4.13) it thus has a lower bound:

$$\big(k^2 + \theta(E - k^2/2 + \varepsilon) - 2\varepsilon\big)\big\|\psi_F(k)\big\|^2 \geq \big(\theta(E+\varepsilon) - 2\varepsilon\big)\big\|\psi_F(k)\big\|^2,$$

which proves (4.15). □

PROOF OF (4.10). Let $f \in C_0^\infty(\Delta)$ be real with $f = 1$ on some $\Delta_1 \ni E$. Applying the localization formula (2.31) to the Schrödinger operator $i[H, A]$, we obtain in analogy to (2.34):

$$f(H)i[H, A]f(H) = \sum_a f(H)j_a i[H_a, A]j_a f(H) + \text{ compact}.$$

We will prove below that

$$L \equiv f(H)j_a - j_a f(H_a) \text{ is compact} \tag{4.16}$$

for any $f \in C_0^\infty(R)$. Using this and (4.15) we arrive at

$$f(H)i[H, A]f(H) \geq \big(\theta(E+\varepsilon) - 2\varepsilon\big)f(H)^2 + \text{ compact}.$$

Multiplying both sides with $E_{\Delta_1}(H)$ we obtain

$$B_{\Delta_1}(H) \geq \big(\theta(E+\varepsilon) - 2\varepsilon\big)E_{\Delta_1}(H) + \text{ compact.}$$

This is equivalent to (4.10), since $\theta(E+\varepsilon) = \theta(E) + \varepsilon$ for small ε if $E \notin \tau(H)$. □

PROOF OF (4.16). Let $\hat{f}$ be the Fourier transform of f and $R_a = (i + H_a)^{-1}$. Then

$$\begin{aligned} LR_a &= \int_{-\infty}^{+\infty} dt\, \hat{f}(t) e^{-itH}(j_a - e^{itH} j_a e^{-itH_a}) R_a \\ &= i \int_{-\infty}^{+\infty} \hat{f}(t) \int_0^t ds\, e^{i(s-t)H} K e^{-isH_a}, \end{aligned}$$

where $K = (Hj_a - j_a H_a)R_a = \big(1/2[p^2, j_a] - j_a I_a\big)R_a$ is compact by the results of chapter 2. Since $\|LR_a\| \leq \text{const.}\, \|K\|$, it suffices to prove compactness of LR_a for K's of rank one: $K\psi = (u, \psi)v$. Then the integrand is the operator

$$\psi \to \hat{f}(t)(e^{isH_a}u, \psi)e^{-i(t-s)H}v,$$

which is norm continuous in s, t. Therefore LR_a is compact. Now we set $f(x) = (i + x)g(x)$. Then $g(H)j_a - j_a g(H_a) = LR_a + g(H)\big(1/2[p^2, j_a]R_a + j_a I_a R_a\big)$ is compact. Since g is still arbitrary in $C_0^\infty(R)$ we have proven (4.16). This also concludes the proof of Theorem 4.1. □

4.2. Exponential Bounds for Eigenfunctions.

THEOREM 4.6 (THE FIRST FROESE-HERBST THEOREM). *Under the hypothesis of Theorem* 4.1, *let* $H\psi = E\psi$ *and*

$$\alpha = \sup\{\beta \in R \mid e^{\beta r}\psi \in L^2(X)\}, \quad \big(r = |x|\big).$$

Then $E + 1/2\alpha^2$ *is either infinite or a threshold of* H.

PROOF. The proof is indirect: we assume $0 \leq \alpha < \infty$ and $E + 1/2\alpha^2 \notin \tau(H)$, which will lead to a contradiction. We construct a sequence of smooth, bounded functions $F_n(x)$ which approximate $F(x) = \alpha r$ from above in successively larger regions:

$$F_n(x) = \alpha_n r j(r)\big(1 - j(\varepsilon_n r)\big). \tag{4.17}$$

Here $j(r)$ is a smoothed characteristic function of $\{r > 1\}$, α_n is some sequence with $\alpha_n \searrow \alpha$, and $\varepsilon_n \searrow 0$ is chosen such that:

$$\|e^{F_n}\psi\| \to \infty. \tag{4.18}$$

We define:

$$\psi_n = e^{F_n}\psi \|e^{F_n}\psi\|^{-1}; \quad \langle \ldots \rangle_n = (\psi_n, \ldots, \psi_n). \quad \square \tag{4.19}$$

LEMMA 4.7. (i) $\psi_n \xrightarrow{w} 0$;

(ii) $p^2\psi_n \xrightarrow{w} 0$;

(iii) $\|G(x)\psi_n\| \to 0$, *and* $\|G(x)(1 + p^2)^{1/2}\psi_n\| \to 0$ *for any* $G \in L^\infty(X)$ *which vanishes at* ∞.

PROOF. (i) follows from (4.18); (iii) follows from (ii) and from the compactness property (1.6). To prove (ii) it remains to show that $\|p^2\psi\|$ is bounded. We set:

$$H_n = e^{F_n} H e^{-F_n} = H - \frac{1}{2}|\nabla F_n|^2 + i\gamma_n; \tag{4.20}$$

$$\gamma_n = \frac{1}{2}(\nabla F_n \cdot p + p \cdot \nabla F_n). \tag{4.21}$$

Then $H_n \psi_n = E\psi_n$, and

$$E = \mathrm{Re}\langle H_n \rangle_n = \langle H \rangle_n - \frac{1}{2}\big\langle |\nabla F_n|^2 \big\rangle_n.$$

Since $\nabla F_n(x)$ is uniformly bounded in (n, x), we obtain successive bounds for $\langle H \rangle_n$, $\langle p^2 \rangle_n$, $\|\gamma_n \psi_n\|$, $\|H\psi_n\|$ and $\|p^2 \psi_n\|$. □

Remark. In the following estimates we will need higher derivatives of $F_n(x)$. They are of the form:

$$F_{n,k_1,\ldots,k_s}(x) = \alpha_n \left(\frac{\partial^s r}{\partial x_{k_1} \ldots \partial x_{k_s}} \right) j(r)\big(1 - j(\varepsilon_n r)\big) \tag{4.22}$$

plus terms of fixed compact support (involving derivatives of $j(r)$), plus terms of order ε_n (involving derivatives of $j(\varepsilon_n r)$). By Lemma 4.7 (iii), and since $\varepsilon_n \to 0$, these terms will give no contributions in the limit $n \to \infty$ to the expectation values $\langle \ldots \rangle_n$ estimated below. The same is true for the leading term (4.22) if $s > 1$, since it is of order r^{1-s} as $r \to \infty$.

LEMMA 4.8.

$$\lim_{n\to\infty} \left\| \left(H - \frac{1}{2}\alpha^2 - E \right) \psi_n \right\| = 0.$$

PROOF.

$$\begin{aligned} 0 &= \big\langle (H_n^* - E)(H_n - E) \big\rangle_n \\ &= \left\langle \left(H - \frac{1}{2}|\nabla F_n|^2 - E \right)^2 \right\rangle_n + \langle \gamma_n^2 \rangle_n + \left\langle i \left[H - \frac{1}{2}|\nabla F_n|^2, \gamma_n \right] \right\rangle_n . \end{aligned} \tag{4.23}$$

We show that the third term vanishes as $n \to \infty$. By (4.22) we need only consider

$$\big\langle i[V, \gamma_n] \big\rangle_n = -\langle \nabla V \cdot \nabla F_n \rangle_n,$$

since the other commutators involve higher derivatives of F_n. By (4.22),

$$\lim_{n\to\infty} \langle \nabla V \cdot \nabla F_n \rangle_n = \lim_{n\to\infty} \big\langle (x \cdot \nabla V) r^{-1} j(r) \big\rangle_n = 0,$$

since $x \cdot \nabla V(x)$ is p^2-bounded and $r^{-1}j(r) \to 0$ as $r \to \infty$. Because the first two terms in (4.23) are positive, we now have:

$$\lim_{n\to\infty} \left\| \left(H - \frac{1}{2}|\nabla F_n|^2 - E \right) \psi_n \right\| = 0.$$

Therefore it suffices to show that $\big\| \big(\alpha_n - |\nabla F_n|\big)\psi_n \big\| \to 0$. By Lemma 4.7, the contribution from any bounded region to this norm vanishes as $n \to \infty$, so we need only show that:

$$\alpha_n I_n \to 0 \text{ for } I_n = \big\| j(\varepsilon_n r) \psi_n \big\|.$$

This is trivial if $\alpha = 0$. Otherwise, we split I_n into contributions from the regions $j(\varepsilon_n r) < \delta$ and $j(\varepsilon_n r) > \delta$, obtaining:

$$I_n \leq \delta + \left\| e^{\alpha_n(1-\delta)r}\psi) \right\| \, \| e^{F_n}\psi \|^{-1}.$$

This bound becomes arbitrary small by first fixing δ small and then n sufficiently large, since eventually $\alpha_n(1-\delta) < \alpha$. □

LEMMA 4.9.

$$\liminf_{n\to\infty} \langle i[H, A] \rangle_n > 0.$$

PROOF. Let $B = i[H, A]$. Since $E + 1/2\alpha^2 \notin \tau(H)$, we have a Mourre inequality:

$$\langle E_\Delta(H) B E_\Delta(H) \rangle_n \geq \theta \langle E_\Delta(H) \rangle_n + \langle K \rangle_n \tag{4.24}$$

with $\theta > 0$ and compact K, for some interval $\Delta = (E + \alpha^2/2 - \varepsilon, E + \alpha^2/2 + \varepsilon)$. By Lemma 4.7, $\langle K \rangle_n \to 0$ for $n \to \infty$. Setting $\overline{E}_\Delta = 1 - E_\Delta$ we also obtain

$$\| \overline{E}_\Delta \psi_n \| \leq \varepsilon^{-1} \left\| \left(H - \frac{1}{2}\alpha^2 - E \right) \psi_n \right\| \to 0 \tag{4.25}$$

from Lemma 4.8. Now:

$$\langle B \rangle_n \geq \langle E_\Delta B E_\Delta \rangle_n - \| \overline{E}_\Delta \psi_n \| \left(2\| B E_\Delta \| + \| B \overline{E}_\Delta \psi_n \| \right).$$

By (4.25) the second term vanishes as $n \to \infty$, since $H\psi_n$ is bounded. Therefore:

$$\liminf_{n\to\infty} \langle B \rangle_n \geq \liminf_{n\to\infty} \langle E_\Delta B E_\Delta \rangle_n \geq \liminf_{n\to\infty} \theta \langle E_\Delta \rangle_n = \theta > 0. \quad \square$$

The next statement is in contradiction to Lemma 4.9 and thus completes the proof of Theorem 4.6:

LEMMA 4.10.

$$\limsup_{n\to\infty} \langle i[H, A] \rangle_n \leq 0.$$

PROOF. From the identity $0 = 2\,\mathrm{Im}\big(A\psi_n, (H_n - E)\psi_n\big)$ we obtain:

$$\langle i[H, A] \rangle_n = \frac{1}{2} \left\langle i\left[|\nabla F_n|^2, A \right] \right\rangle_n - 2\,\mathrm{Re}\langle \gamma_n A \rangle_n.$$

The only contribution involving only first order derivatives of F_n is:

$$-\langle p_k (x_k F_{n,l} + F_{n,k} x_l) p_l \rangle_n.$$

By (4.22) this is equal to

$$-2\alpha_n \left\langle (p \cdot x) \frac{j(r)}{r} (x \cdot p) \right\rangle_n \leq 0,$$

modulo terms vanishing as $n \to \infty$. □

4.3. Accumulation of Eigenvalues.

THEOREM 4.11 (PERRY'S THEOREM). *Under the hypothesis of Theorem* 4.1, *eigenvalues can accumulate at thresholds only from below.*

PROOF. Proceeding inductively, we assume that thresholds can accumulate at a given threshold E only from below: there is an interval $(E \ldots E + 1/2\alpha^2](\alpha > 0)$ containing no thresholds. Now suppose there is an infinite sequence of eigenvalues $E_n \searrow E$ in this interval, with corresponding orthonormal eigenfunctions $\phi_n \xrightarrow{w} 0$. This leads to a contradiction. The proof is a straight copy of the proof of Theorem 4.6. The function

$$F(x) = \alpha r j(r)$$

corresponds to the functions $F_n(x)$ of (4.17) for $\alpha_n = \alpha$ and $\varepsilon_n = 0$. By Theorem 4.6,

$$e^F \phi_n \in L^2(X).$$

Therefore:

$$\psi_n = e^F \phi_n \| e^F \phi_n \|^{-1} \xrightarrow{w} 0,$$

and the rest of the proof (Lemma 4.7 to Lemma 4.10) goes through with minimal changes (as an eigenvalue, E is replaced by E_n). □

4.4. Absence of Positive Eigenvalues.

THEOREM 4.12 (THE SECOND FROESE-HERBST THEOREM). *Under the hypothesis of Theorem* 4.1, *H has no eigenvalue $E > 0$.*

PROOF. Since thresholds arise from eigenvalues of subsystems, we can proceed by induction: assuming that H has no positive threshold, we prove that H has no positive eigenvalue. The proof is again indirect: we derive a contradiction from $H\psi = E\psi$; $E > 0$. By Theorem 4.6,

$$e^{\alpha r}\psi \in L^2(X) \text{ for any } \alpha > 0.$$

We first fix ρ such that

$$\int_{r<\rho} dx |\psi|^2 \leq \int_{r>2\rho} dx |\psi|^2, \tag{4.26}$$

then we choose a C^∞-function $F(r) \leq r$, with $F'(r) \geq 0$, and $F(r) = r$ for $r > \rho$, and we define:

$$\psi_\alpha = e^{\alpha F}\psi \| e^{\alpha F}\psi \|^{-1}; \quad \langle \ldots \rangle_\alpha = (\psi_\alpha, \ldots, \psi_\alpha).$$

By (4.26):

$$\int_{r<\rho} dx |\psi_\alpha|^2 \leq e^{-2\alpha\rho}. \quad \square \tag{4.27}$$

LEMMA 4.13.

$$\langle H \rangle_\alpha \equiv (\psi_\alpha, H\psi_\alpha) \geq E + \frac{\alpha^2}{2} - c_1 \alpha^2 e^{-2\alpha\rho} \tag{4.28}$$

for some constant c_1 and all $\alpha > 0$.

PROOF. We define:

$$H_\alpha = e^{\alpha F} H e^{-\alpha F} = H - \frac{\alpha^2}{2}|\nabla F|^2 + i\alpha\gamma; \tag{4.29}$$

$$\gamma = \frac{1}{2}(\nabla F \cdot p + p \cdot \nabla F). \tag{4.30}$$

Then $H_\alpha \psi_\alpha = E\psi_\alpha$ and

$$\langle H\rangle_\alpha = E + \frac{\alpha^2}{2}\langle |\nabla F|^2\rangle_\alpha.$$

Since $|\nabla F|^2 = 1$ for $r > \rho$, we obtain from (4.27):

$$\left|\langle |\nabla F|^2\rangle_\alpha - 1\right| \leq c_1 e^{-2\alpha\rho}. \quad \square$$

LEMMA 4.14.

$$\langle i[H,A]\rangle_\alpha = \langle p^2\rangle_\alpha - \langle x \cdot \nabla V(x)\rangle_\alpha \leq c_2\alpha^2 e^{-2\alpha\rho} + \alpha c_3. \tag{4.31}$$

for some constants c_2, c_3 *and all* $\alpha > 0$.

PROOF. From the identity $0 = 2\,\mathrm{Im}\big(A\psi_\alpha, (H_\alpha - E)\psi_\alpha\big)$ we obtain:

$$\langle i[H,A]\rangle_\alpha = \frac{\alpha^2}{2}\left\langle i\big[|\nabla F|^2, A\big]\right\rangle_\alpha - 2\alpha\,\mathrm{Re}\langle\gamma A\rangle_\alpha.$$

Since $|\nabla F|^2 = 1$ for $r > \rho$, the first term is bounded by $c_2\alpha^2 exp(-2\alpha\rho)$. For the second term we compute:

$$2\,\mathrm{Re}(\gamma A) = p_k(x_k F_{,l} + F_{,k}\, x_l)p_l - \frac{d}{2}F_{,ll} - \frac{1}{2}x_k F_{,llk}\,,$$

where $d = \dim(X)$. The first term is positive since $F_{,l} = x_l r^{-1}F'(r)$, $F'(r) \geq 0$, and the remaining two terms are bounded. $\square$

4.4.1. *Completion of the Proof of Theorem* 4.12. Subtracting (4.28) from (4.31) we arrive at

$$\frac{1}{2}\langle p^2\rangle_\alpha - \langle V\rangle_\alpha - \langle x \cdot \nabla V(x)\rangle_\alpha \leq -E - \frac{1}{2}\alpha^2 + (c_1 + c_2)\alpha^2 e^{-2\alpha\rho} + \alpha c_3.$$

This is a contradiction: the left hand side is bounded below, while the right hand side goes to $-\infty$ for $\alpha \to \infty$.

5. Local Time Decay of Continuum States

Introduction. Ruelle's theorem describing the escape of continuum states from bounded regions in configuration space can be substantially sharpened by using Mourre's inequality. The resulting *local decay estimate* goes back to Mourre, and was first proven for general N-body systems by Perry, Sigal and Simon. The proof rests on the additional hypothesis that the second virial

$$\big[[V,A],A\big] \text{ is } H\text{-bounded} \tag{5.1}$$

in the sense already described for the virial $[V,A]$. We begin by reviewing the notion of the *absolutely continuous spectral subspace* $\mathcal{H}_{AC}$ of H. By definition,

$$\psi \in \mathcal{H}_{AC} \iff E_N\psi = 0, \tag{5.2}$$

where E is the spectral measure of H and N any Borel null-set. Thus $\mathcal{H}_{AC}$ is a closed subspace of $\mathcal{H}$, invariant under $\exp(-iHt)$ and orthogonal to $\mathcal{H}_B$. As a corollary of the local decay estimate we will have

$$\mathcal{H}_{AC} = \mathcal{H}_C \tag{5.3}$$

(Absence of singular continous spectrum). By the Radon-Nikodym theorem, any continuum state ψ then has an integrable energy density:

$$d(\psi, E_\lambda \psi) = \rho_\psi(\lambda) d\lambda; \quad \rho_\psi \in L^1(R), \tag{5.4}$$

which implies

$$(\psi, e^{-iHt}\psi) = \int \rho_\psi(\lambda) e^{-i\lambda t} \to 0, \quad (t \to \pm\infty).$$

By polarisation this is equivalent to

$$w - \lim_{t\to\pm\infty} e^{-iHt}\psi = 0, \text{ for all } \psi \in \mathcal{H}_C. \tag{5.5}$$

From local compactness it follows that Ruelle's theorem can be sharpened to

$$\lim_{t\to\pm\infty} \|F_R e^{-iHt}\psi\|^2 = 0, \tag{5.6}$$

i.e. the mean limit (1.8) becomes an ordinary limit. In fact, the previous results on the structure of the continuum allow a considerably stronger escape estimate for states with spectral support away from the set

$$\mathcal{S} = \text{ set of all thresholds and eigenvalues of } H.$$

THEOREM 5.1 (THE PERRY-SIGAL-SIMON THEOREM). *Suppose that $V(x)$ and its virial $(x \cdot \nabla)V(x)$ satisfy* A.1 *and* A.2, *and that the second virial $(x\cdot\nabla)^2 V(x)$ is H-bounded. Then, for $\alpha > 1/2$ and any compact $\Delta \subset R \setminus \mathcal{S}$,*

$$\int_{-\infty}^{+\infty} dt \left\|\langle x\rangle^{-\alpha} e^{-iHt}\psi\right\|^2 \le C(\Delta, \alpha)\|\psi\|^2 \tag{5.7}$$

for all $\psi = E_\Delta \psi$, where $\langle x\rangle = (1+x^2)^{1/2}$. (Local Decay)

PROOF OF (5.3). Let $f \in C_0^\infty(R)$ and $\psi = E_\Delta\psi$. Then (5.7) implies

$$\left\|\langle x\rangle^{-\alpha} f(H)\psi\right\| \le (2\pi)^{-1/2} \int dt \left|\hat{f}(t)\right| \left\|\langle x\rangle^{-\alpha} e^{-iHt}\psi\right\| \le \text{const.} \|f\|_2 \|\psi\|.$$

Taking pointwise limits this extends to the characteristic functions f of all bounded Borel sets, showing that $\psi \in \mathcal{H}_{AC}$. Since $R \setminus \mathcal{S}$ is open, the states $\psi = E_\Delta\psi$ with compact $\Delta \subset R\setminus\mathcal{S}$ span $\operatorname{ran}(E_{R\setminus\mathcal{S}})$, which proves $\operatorname{ran}(E_{R\setminus\mathcal{S}} \subset \mathcal{H}_{AC})$. On the other hand, $\operatorname{ran}(E_\mathcal{S}) = \mathcal{H}_B$ since $\mathcal{S}$ is countable and contains all eigenvalues. □

Theorem 5.1 is derived from a resolvent estimate:

THEOREM 5.2. *Let $f \in C_0^\infty(R)$ with $\operatorname{supp}(f) \cap \mathcal{S} = \emptyset$, and $\alpha > 1/2$. Then, under the hypothesis of Theorem* 5.1:

$$\left\|\langle x\rangle^{-\alpha} f(H) R(z) f(H) \langle x\rangle^{-\alpha}\right\| \le \text{const.} \tag{5.8}$$

uniformly in $z \notin R$.

PROOF OF THEOREM 5.1. (5.8) implies that the operator $\langle x\rangle^{-\alpha}f(H)$ is H-*smooth*, i.e that (5.7) holds for all states $\psi = f(H)\phi$ [M. Reed and B. Simon, *Methods of modern mathematical physics* IV, Academic Press 1978, Theorem XIII.25 and Corollary]. Since we may choose $f(x) = 1$ on Δ, (5.7) holds for all $\psi = E_\Delta\psi$. □

The rest of this section is devoted to the proof of Theorem 5.2, which we prepare by estimating the resolvent

$$R_s(z) = (z - H_s)^{-1} \text{ of } H_s = H - isB; \quad B = i[H, A]. \tag{5.9}$$

In these estimates (Lemma 5.3 and 5.4 below) only the first virial appears, so that (5.1) is not yet required. Let $\Delta_1 \subset R \setminus \mathcal{S}$ be compact, and

$$\Omega = \{z \mid 0 < \mathrm{Im}(z) < b; \quad \Re(z) \in \Delta_1\}, \tag{5.10}$$

where $b > 0$ is an arbitrary constant. Since B is H-bounded, Ω intersects the resolvent set $\rho(H_s)$ for small s.

LEMMA 5.3. *There exist constants s_0, c_1 such that $\Omega \subset \rho(H_s)$ for $0 < s < s_0$ and*

$$\|R_s(z)\| \le c_1 s^{-1} \tag{5.11}$$

uniformly in $0 < s < s_0$ and $z \in \Omega$.

PROOF. By hypothesis there is an open set $\Delta \supset \Delta_1$ with a Mourre-estimate

$$E_\Delta B E_\Delta \ge \delta E_\Delta; \quad \delta > 0. \tag{5.12}$$

Let $\overline{E}_\Delta = 1 - E_\Delta$. Then

$$\begin{aligned} \|E_\Delta(z - H_s)u\| &\ge \|E_\Delta(z - H_s)E_\Delta u\| - s\|E_\Delta B\overline{E}_\Delta u\| \\ &\ge s\delta\|E_\Delta u\| - sM_1\|\overline{E}_\Delta u\|. \end{aligned} \tag{5.13}$$

The first term comes from (5.12): $\mathrm{Im}(u, E_\Delta(z - H_s)E_\Delta u) \ge \delta\|E_\Delta u\|^2$. In the second term we have used that BE_Δ is bounded. Next we note that $\|(z - H)\overline{E}_\Delta u\| \ge \varepsilon\|\overline{E}_\Delta u\|$ for some $\varepsilon > 0$, $u \in D(H)$ and $\Re(z) \in \Delta_1$. Therefore:

$$\begin{aligned} \|\overline{E}_\Delta(z - H_s)u\| &= \left\|(z - H_s)\overline{E}_\Delta\big(1 - isR(z)B\big)u\right\| \\ &\ge \varepsilon\left\|\overline{E}_\Delta\big(1 - isR(z)B\big)u\right\| \\ &\ge \varepsilon\|\overline{E}_\Delta u\| - \varepsilon s M_2\|u\|. \end{aligned} \tag{5.14}$$

Here we have used that $BR(z)^*\overline{E}_\Delta$ is bounded since $HR(z)^*\overline{E}_\Delta$ is bounded. Combining (5.13) and (5.14) we arrive at

(5.15)
$$\|E_\Delta(z - H_s)u\| + \|\overline{E}_\Delta(z - H_s)u\| \ge s(\delta - \varepsilon M)\|E_\Delta u\| + (\varepsilon - sM)\|\overline{E}_\Delta u\|,$$

where $M = M_1 + M_2$. We now choose $\varepsilon < \delta/2M$ and $s < 2\varepsilon/3M = s_0$. Then the r.h.s. of (5) is bounded below by $s(\min(\delta, M))\|u\|/2$ for $0 < s < s_0$ and $z \in \Omega$. Therefore

$$\|(z - H_s)u\| \ge c_1 s\|u\|$$

for some constant $c_1 > 0$. (5.11) now follows since Ω intersects $\rho(H_s)$. □

LEMMA 5.4. *For all* $u \in \mathcal{H}$, $\left\|R_s(z)u\right\|^2$ *and* $\left\|R_s^*(z)u\right\|^2$ *are bounded by*

$$c_2\left(s^{-1}\left|\operatorname{Im}\left(u, R_s(z)u\right)\right| + \|u\|^2\right), \tag{5.16}$$

for some constant c_2, *uniformly in* $0 < s < s_0$ *and* $z \in \Omega$.

PROOF. From $(z-H_s)-(z-H_s)^* = 2i\big(\operatorname{Im}(z)+sB\big)$ we find the two estimates:

$$\frac{1}{2is}(R_s^* - R_s) \geq R_s^* B R_s; \quad \frac{1}{2is}(R_s^* - R_s) \geq R_s B R_s^*. \tag{5.17}$$

In the first case the r.h.s. is bounded from below as follows:

$$\begin{aligned}(u, R_s^* B R_s u) = (R_s u, E_\Delta B E_\Delta R_s u) &+ (R_s u, E_\Delta B \overline{E}_\Delta R_s u) \\ &+ (\overline{E}_\Delta R_s u, B R_s u) \\ \geq \delta \|E_\Delta R_s u\|^2 - M\|u\|&\big(\|R_s u\| + \|u\|\big).\end{aligned} \tag{5.18}$$

The first term comes from the Mourre-estimate (5.12). In the remainder we have used that $\overline{E}_\Delta R_s$ is bounded. This follows from $\overline{E}_\Delta R_s = \overline{E}_\Delta R - is\overline{E}_\Delta RBR_s$ and from Lemma 5.3, since $\overline{E}_\Delta RB$ is bounded. From (5.9) and (5.12) we find:

$$\begin{aligned}\|R_s u\|^2 &= \|E_\Delta R_s u\|^2 + \|\overline{E}_\Delta R_s u\|^2 \\ &\geq \text{const.}\left(s^{-1}\left|\operatorname{Im}(u, R_s u)\right| + \|u\|\,\|R_s u\| + \|u\|^2\right),\end{aligned}$$

which implies the bound (5.16) for $\|R_s u\|^2$. The bound for $\|R_s^* u\|^2$ follows from the second inequality (5.17). □

PROOF OF THEOREM 5.2. Let $\Delta_1 \subset R \setminus \mathcal{S}$ be compact with $\operatorname{supp}(f)$ in the interior of Δ_1, and let Ω be defined by (5.10). Defining

$$\rho_s = \langle x\rangle^{-\alpha}\langle sx\rangle^{\alpha-1} f(H)$$

we prove that

$$\left\|\rho_s R_s(z)\rho_s^*\right\| \leq \text{const.} \tag{5.19}$$

uniformly in $z \in \Omega$ and small $s > 0$. For these z (5.8) then follows in the limit $s \to 0$, and the extension to all $z \notin R$ is trivial. From Lemma 5.4 we see that $\|R_s\rho_s^*\|^2$ and $\|\rho_s R_s\|^2 = \|R_s^*\rho_s^*\|^2$ are both bounded by

$$\text{const.}\big(s^{-1}\|\rho_s R_s \rho_s^*\| + 1\big),$$

and by an explicit calculation of the derivative:

$$\left\|\frac{d\rho_s}{ds}\right\| \leq \text{const.}\, s^{\alpha-1}.$$

These estimates give norm-bounds of the form

$$\text{const.}\, s^{\alpha-1}\big(s^{-1}\|\rho_s R_s \rho_s^*\| + 1\big) \tag{5.20}$$

for the first two terms on the r.h.s. of

$$\frac{d}{ds}\rho_s R_s \rho_s^* = \frac{d\rho_s}{ds}R_s\rho_s^* + \rho_s R_s \frac{d\rho_s^*}{ds} + \rho_s \frac{dR_s}{ds}\rho_s^*.$$

The third term is expanded using

$$\frac{dR_s}{ds} = R_s iBR_s = -R_s[H,A]R_s = -R_s[H_s,A]R_s - isR_s[B,A]R_s \\ = [A,R_s] - isR_s[B,A]R_s, \tag{5.21}$$

and requires an estimate of $\rho_s[A,R_s]\rho_s^*$. We will prove below that

$$Af(H)\langle x\rangle^{-1} \in L(\mathcal{H}), \tag{5.22}$$

so that

$$\|\rho_s A\| = \|A\rho_s^*\| \le \|Af(H)\langle x\rangle^{-1}\| \, \|\langle x\rangle^{1-\alpha}\langle sx\rangle^{\alpha-1}\| \le \text{const.}\, s^{\alpha-1}.$$

This shows that $\|\rho_s[A,R_s]\rho_s^*\|$ also has a bound of the form (5.20). In the last term of (5.21) (and only there !) we encounter the double commutator $\big[[H,A],A\big]$. Since it is H-bounded by hypothesis, we have

$$\big\|[B,A]R_s\big\| \le \text{const.}\big(\|R_s\|+1\big) \le \text{const.}\, s^{-1},$$

and therefore by (5.11):

$$s\big\|\rho_s R_s[B,A]R_s\rho_s^*\big\| \le \text{const.}\, \|\rho_s R_s\| \le \text{const.}\big(s^{-1/2}\|\rho_s R_s\rho_s^*\|^{1/2}+1\big).$$

Collecting all these estimates we arrive at the differential inequality:

$$\left\|\frac{d}{ds}\rho_s R_s\rho_s^*\right\| \le c(1+s^{\alpha-1})\big(s^{-1/2}\|\rho_s R_s\rho_s^*\|^{1/2}+1\big). \tag{5.23}$$

Since $\alpha > 1/2$, the function $s^{\alpha-3/2}$ is integrable near $s=0$. Defining

$$M_s = \sup_{s\le t\le s_0} \|\rho_t R_t \rho_t^*\|$$

for $s>0$, we can integrate (5.23) from s to s_0 to obtain

$$M_s \le M_{s_0} + \text{const.}(M_s^{1/2}+1),$$

which shows that M_s remains bounded as $s\to 0$. This proves (5.19). □

PROOF OF (5.22). It suffices to prove $p_k x_k f(H)\big(1+|x|\big)^{-1} \in L(\mathcal{H})$, which follows if $p_k\big[x_k, f(H)\big] \in L(\mathcal{H})$. We first show that $\big[x_k, f(H)\big]$ is H-bounded. Let $R=(i+H)^{-1}$. Then, since $[x_k,H]=ip_k$,

$$\big[x_k, f(H)\big]R = \int_{-\infty}^{+\infty} dt\, \hat f(t) \int_0^t ds\, e^{-iH(t-s)} ip_k R e^{-iHs} \in L(\mathcal{H}). \tag{5.24}$$

Therefore, setting $g(x) = (i+x)f(x)$,

$$\big[x_k, f(H)\big] = \big[x_k, g(H)R\big] = \big[x_k, g(H)\big]R + g(H)[x_k,R] \\ = \big[x_k, g(H)\big]R + ig(H)Rp_kR \in L(\mathcal{H}).$$

Finally, using (5.24) for g in place of f:

$$p_k\big[x_k, f(H)\big] = p_k\big[x_k, Rg(H)\big] = p_k[x_k,R]g(H) + p_kR\big[x_k,g(H)\big] \\ = -ip_kRp_kRg(H) + p_kR\big[x_k,g(H)\big] \\ \in L(\mathcal{H}). \quad □$$

5.1. Using Local Decay. We conclude this section with some practical remarks concerning the use of the local decay estimate (5.7). Δ always stands for a compact $\Delta \subset \mathcal{R} \setminus \mathcal{S}$.

DEFINITION. Let $R = (i+H)^{-1}$. If A is an operator on $\mathcal{H}$, we say that

$$A = \mathcal{O}_m\big(|x|^{-\alpha}\big) \iff AR^m \langle x\rangle^\alpha \in L(\mathcal{H}). \tag{5.25}$$

LEMMA 5.5. *If $A = \mathcal{O}_m\big(|x|^{-\alpha}\big)$ for some $\alpha > 1/2$, then*

$$\int_0^\infty dt\, \big\| Ae^{-iHt} E_\Delta(H)\psi\big\|^2 \le \text{const.}\|\psi\|^2. \tag{5.26}$$

PROOF.

$$\big\|Ae^{-iHt}E_\Delta(H)\psi\big\| \le \big\|AR^m\langle x\rangle^\alpha\big\|\, \big\|\langle x\rangle^{-\alpha} e^{-iHt} E_\Delta(i+H)^m E_\Delta\psi\big\|. \quad \square$$

As an example, let $B = \mathcal{O}_0\big(|x|^{-\beta}\big)$. Then $A = Bp_k = \mathcal{O}_1\big(|x|^{-\alpha}\big)$ for $\alpha = \max(1,\beta)$, since

$$Bp_kR\langle x\rangle^\alpha = B\langle x\rangle^\alpha p_k R + B\big[p_k, \langle x\rangle^\alpha\big]R + Bp_kR\big[\langle x\rangle^\alpha, H\big]R,$$

where all terms are bounded for $\alpha \le 1$. Therefore, (5.26) holds for $\beta > 1/2$.

LEMMA 5.6. *Suppose that A can be factorized into $A = B^*C$, where $B, C = \mathcal{O}_m\big(|x|^{-\alpha}\big)$ for some $\alpha > 1/2$. Then the following limit exists in norm sense for all ψ:*

$$\lim_{T\to\infty} \int_0^T dt\, E_\Delta(H_a) e^{iH_a t} A e^{-iHt} E_\Delta(H)\psi. \tag{5.27}$$

PROOF. Let $\psi(t)$ be the integrand of (5.27). Then

$$\begin{aligned} \left\| \int_T^\tau dt\, \psi(t)\right\|^2 &= \sup_{\|\phi\|=1} \left| \int_T^\tau dt\, \big(\phi, \psi(t)\big)\right|^2 \\ &\le \left(\sup_{\|\phi\|=1} \int_0^\infty dt\, \big\|BE_\Delta(H_a)e^{-iH_a t}\phi\big\|^2 \right) \cdot \int_T^\tau dt\, \big\|Ce^{-iHt}E_\Delta(H)\psi\big\|^2 \end{aligned}$$

by the Schwarz inequality. Since Δ is separated from the thresholds and eigenvalues of H_a, the first factor is bounded, while the second vanishes as $T, \tau \to \infty$. $\square$

6. Asymptotic Completeness

Introduction. Short range potentials are defined by the condition:

$$\big|I_a(x)\big| \le \text{const.}|x|_a^{-\mu}; \quad (\mu > 1) \tag{6.1}$$

for sufficiently large $|x|_a$. Outgoing scattering states ψ are then characterized by the asymptotic behavior:

$$\psi_t = e^{-iHt}\psi \to \sum_{a > \{0\}} e^{-iH_a t}\big(1 \otimes P_B(H^a)\big)\phi_a \tag{6.2}$$

in norm sense as $t \to \infty$, where $P_B(H^a)$ is the projection onto the bound state subspace of H^a. Under the conditions A.1 and (6.1) alone, it is well known that outgoing scattering states ψ exist for any given ϕ_a, and that they form a closed subspace $\mathcal{H}^+ \subset \mathcal{H}_C$.

THEOREM 6.1 (THE SIGAL-SOFFER THEOREM). *Suppose that V satisfies the conditions of Theorem* 5.1 *as well as* (6.1), *and that*

$$|\nabla I_a(x)| \leq \text{const.}\, |x|_a^{-\mu}; \quad (\mu > 1) \tag{6.3}$$

for sufficiently large $|x|_a$. Then $\mathcal{H}^+ = \mathcal{H}_C$, i.e. any orbit ψ_t in $\mathcal{H}_C$ has the asymptotic behavior (6.2) (*Asymptotic Completeness*).

This section is devoted to the proof of Theorem 6.1. Due to the work of Graf and Yafaev, the N-body part of the analysis is reduced to a geometric construction in configuration space. The other ingredients are: the use of positive commutators to derive propagation estimates, Mourre's Theorem, the Perry-Sigal-Simon Theorem, and the notion of Deift-Simon wave operators.

6.1. Positive Commutators and the Basic Propagation Estimate. The following argument describes the use of essentially positive commutators to control the time evolution. Let A, B, R be symmetric operators on $\mathcal{H}$. Suppose that on some H-invariant subspace $\mathcal{H}_\Delta$:

$$\langle i[H, A]\rangle \geq \langle B\rangle - \langle R\rangle \tag{6.4}$$

$$\langle B\rangle \geq 0; \tag{6.5}$$

$$|\langle A\rangle| \leq c_A\|\psi\|^2; \quad \text{and} \tag{6.6}$$

$$\int_0^\infty dt\, |\langle R\rangle_t| \leq c_R\|\psi\|^2 \tag{6.7}$$

under the time evolution $\psi_t = e^{-iHt}\psi$. Then

$$\langle A\rangle_s - \langle A\rangle_0 = \int_0^s dt\, \langle i[H, A]\rangle_t \geq \int_0^s dt\, \langle B\rangle_t - \int_0^s dt\, \langle R\rangle_t,$$

so that

$$\int_0^s dt\, \langle B\rangle_t \leq (2c_A + c_R)\|\psi\|^2.$$

Since $\langle B\rangle_t \geq 0$ we obtain:

$$\int_0^\infty dt\, \langle B\rangle_t \leq \text{const.}\|\psi\|^2 \quad \forall\psi \in \mathcal{H}_\Delta. \tag{6.8}$$

Estimates of this type are called *propagation estimates*: $\langle B\rangle > 0$ describes a region in $\mathcal{H}_\Delta$ which is avoided by the orbit ψ_t for large t.

We now consider a specific case. In what follows, Δ is a fixed compact $\Delta \subset R\backslash\mathcal{S}$ as in Theorem 5.1, and $\mathcal{H}_\Delta = E_\Delta(H)\mathcal{H}$. Let $g(x)$ be a real function on X with the property:

(a) (i) *$g(x)$ is smooth, convex, and homogeneous of degree* 1 *for $|x| > R^+$*

Then:

$$i[H, g] = \frac{1}{2}(\nabla g\cdot p + p\cdot\nabla g) \equiv \gamma; \tag{6.9}$$

$$i[H, \gamma] = pg''p - \frac{1}{2}\Delta^2 g - \nabla g\cdot\nabla V, \tag{6.10}$$

where g'' denotes the Hessian of g:

$$pg''p = p_k g_{kl} p_l \geq 0 \tag{6.11}$$

by convexity. Comparing (6.10) with (6.4), we identify:

$$A = \gamma; \quad B = pg''p; \quad R = -\frac{1}{2}\Delta^2 g - \nabla g \cdot \nabla V. \tag{6.12}$$

We first remark that (6.6) holds since A is H-bounded and since Δ is compact. Also, (6.7) holds for the term $\Delta^2 g$ by local decay because it is of order $|x|^{-3}$ at ∞. To handle the term $\nabla g \cdot \nabla V$, we want $g(x)$ to have the further properties:

(ii) $g(x) =$ *constant for* $|x| < R^-$ (R^- *arbitrary large*)

(iii) $g(x) = g(x_a)$ *near any* $a > \{0\}$, *e.g. on cones*:

$$C_a = \{x \mid |x^a| < \varepsilon_a |x|\}, \quad (0 < \varepsilon_a < 1).$$

The first of these conditions makes $\nabla g = 0$ for $|x| < R^-$, the second $\nabla g \in a$ on C_a, so that

$$\nabla g \cdot \nabla V = \nabla g \cdot \nabla I_a \text{ on } C_a.$$

By an argument given below we can arrange $\varepsilon = \{\varepsilon_a\}$ so that:

$$\overline{C_a} \cap \overline{C_b} \subset C_{a \cap b} \text{ if } a \cap b < a, b. \tag{6.13}$$

Since $C_X = X$, the sets

$$C_a^* \equiv C_a \setminus \bigcup_{b<a} C_b \tag{6.14}$$

then form a disjoint covering of X. Moreover, there exists a constant $\delta(\varepsilon) > 0$ such that

$$|x|_a \geq \delta |x| \text{ on } C_a^*. \tag{6.15}$$

PROOF. It suffices to prove $|x|_a > 0$ on $\overline{C_a^*}$ for $|x| > 0$. Suppose that $|x|_a = 0$, i.e. that $x \in b$ for some b with $b \cap a < a$, which already excludes $b \geq a$. Since $x \in C_b$, $b < a$ is also excluded by (6.14). The same argument excludes $a \cap b < a, b$, since then (6.13) implies $x \in C_{a \cap b}$ □

On $C_a^* \cap \operatorname{supp}(\nabla g)$ we therefore have

$$|x|_a \geq \delta |x| \geq \delta R^-. \tag{6.16}$$

Taking R^- sufficiently large, we see that the condition (6.3) suffices to estimate $\nabla g \cdot \nabla V = \nabla g \cdot \nabla I_a$ on C_a^*. Since ∇g is bounded, we obtain

$$|\nabla g \cdot \nabla V| \leq \text{const.}\, |x|^{-\mu} \tag{6.17}$$

on each C_a^*, and therefore globally on X. Since $\mu > 1$, $R = \nabla g \cdot \nabla V$ satisfies (6.7) as a consequence of local decay. The proof of the basic propagation estimate

$$\int_0^\infty dt\, \langle pg''p \rangle_t \leq \text{const.} \|\psi\|^2 \tag{6.18}$$

is thus reduced to the construction of a function $g(x)$ having the properties (i)–(iii) given above.

6.2. Construction of $g(x)$. The prototype of $g(x)$ is:

$$g(x,R,\sigma) = \max\bigl(R, g_1(x,\sigma)\bigr); \quad g_1(x,\sigma) = \max_a\bigl(\sigma_a|x_a|\bigr), \tag{6.19}$$

where $\sigma = \{\sigma_a\}$, $R, \sigma_a > 0$. This function is convex to begin with. Since $a = \{0\}$ does not contribute, we will tacitly assume $a > \{0\}$. Let

$$C_a(\sigma) = \bigl\{x \mid \sigma_a|x_a| > \sigma_b|x_b| \quad \forall b > a\bigr\}, \tag{6.20}$$

in particular $C_X(\sigma) = X$. To have $C_a(\sigma) \supset a \setminus \{0\}$ for all a, we require

$$1 < \sigma_a < \sigma_b \text{ if } a > b, \tag{6.21}$$

The lower bound 1 implies

$$g(x,R,\sigma) \geq |x|. \tag{6.22}$$

LEMMA 6.2. *There exists σ, given explicitly in the proof, such that for $a \cap b < a, b$:*

$$\overline{C_a}(\sigma) \cap \overline{C_b}(\sigma) \subset C_{a\cap b}(\sigma). \tag{6.23}$$

PROOF. Defining

$$\begin{aligned} U_a &= \bigl\{x \mid \sigma_a|x_a| > \sigma_X|x|\bigr\}; \\ V_a &= \bigl\{x \mid \sigma_a|x_a| > \sigma_b|x|, \forall b > a\bigr\}, \end{aligned}$$

we have $V_a \subset C_a \subset U_a$ and:

$$\begin{aligned} x \in U_a &\iff |x^a|^2 < \left(1 - \frac{\sigma_X^2}{\sigma_a^2}\right)|x|^2; \\ x \in V_a &\iff |x^a|^2 < \left(1 - \frac{\sigma_b^2}{\sigma_a^2}\right)|x|^2 \quad \forall b > a. \end{aligned}$$

Now we construct σ such that

$$\overline{U_a} \cap \overline{U_b} \subset V_{a\cap b} \text{ if } a \cap b < a, b.$$

Let $0 < \varepsilon < 1$; $|a| \equiv \dim(a)$, and

$$1 - \frac{1}{\sigma_a^2} = \varepsilon^{2|a|}. \tag{6.24}$$

Then, for $b > a$:

$$\varepsilon^{2|a|} > \left(1 - \frac{\sigma_b^2}{\sigma_a^2}\right) = \frac{\varepsilon^{2|a|} - \varepsilon^{2|b|}}{1 - \varepsilon^{2|b|}} > \varepsilon^{2|a|}(1 - \varepsilon^2), \tag{6.25}$$

since $|b| \geq |a| + 1$. Now we use that

$$|x^{a\cap b}| \leq M \cdot \max\bigl(|x^a|, |x^b|\bigr) \tag{6.26}$$

for all a, b, x, with a constant M depending only on the lattice L (both sides are seminorms on X, and $x^a = x^b = 0$ implies $x \in a \cap b$ and therefore $|x^{a\cap b}| = 0$). Now let $a \cap b = c < a, b$, and $x \in \overline{U_a} \cap \overline{U_b}$. Then

$$|x^c|^2 \leq M^2 \max(\varepsilon^{2|a|}, \varepsilon^{2|a|})|x|^2 \leq M^2 \varepsilon^{2|c|+2}|x|^2.$$

Therefore, by (6.25), $x \in V_c$ if $M^2\varepsilon^{2|c|+2} \le \varepsilon^{2|c|}(1-\varepsilon^2)$, i.e. if

$$\varepsilon^2 \le (1+M^2)^{-1}. \tag{6.27}$$

□

As a consequence of (6.23), the sets

$$C_a^*(\sigma) = C_a(\sigma) \setminus \bigcup_{b<a} C_b(\sigma) \tag{6.28}$$

form a disjoint covering of X.

LEMMA 6.3. *The function* $g_1(x,\sigma)$ *defined in* (6.19) *is homogeneous of degree* 1 *and satisfies*:

$$g_1(x,\sigma) = \sigma_a|x_a| \text{ on } C_a^*(\sigma); \tag{6.29}$$

$$g_1(x,\sigma) = g_1(x_a,\sigma) \text{ on } C_a(\sigma). \tag{6.30}$$

PROOF. Suppose that $x \in C_a^*$ and $\sigma_b|x_b| > \sigma_c|x_c|$ for all $c \ne b$. To prove (6.29) we must show that $b = a$. Since $x \in C_b$, $b < a$ is excluded by the definition of C_a^*. Similarly, $a \cap b < a, b$ is excluded, since then $x \in C_{a\cap b}$. Finally, $b > a$ is excluded, since then $\sigma_a|x_a| < \sigma_b|x_b|$, which contradicts $x \in C_a$. (6.30) follows from (6.29), since

$$C_a = \bigcup_{b\le a} C_b^*,$$

and from the fact that $x_b = (x_a)_b$ for $b \le a$. □

6.3. Regularization and Decomposition. We now treat the parameters R, σ as variables, restricted to intervals :

$$\begin{gathered} 0 < R^- \le R \le R^+ < \infty; \\ (\sigma_a^-)^2 \equiv (1-\varepsilon^{2|a|})^{-1} \le \sigma_a^2 \le (1-2\varepsilon^{2|a|})^{-1} \equiv (\sigma_a^+)^2. \end{gathered} \tag{6.31}$$

Here $R^\pm$ and ε are constants to be adjusted. We require that $\sigma_a^+ < \sigma_b^-$ for $b < a$, which is true if $\varepsilon < 1/2$. Let

$$C_a^\pm(\sigma) = \{x \mid \sigma_a^\pm|x_a| > \sigma_b^\mp|x_b| \forall b > a\}, \tag{6.32}$$

so that $C_a^- \subset C_a(\sigma) \subset C_a^+$ for all allowed σ. As in the proof of Lemma 6.2 we find that

$$C_a^+ \cap C_b^+ \subset C_{a\cap b}^- \text{ if } a \cap b < a, b, \tag{6.33}$$

provided that

$$\varepsilon^2 \le (2+M^2)^{-1}. \tag{6.34}$$

Now let $m_0 \in C_0^\infty(R^-, R^+)$ and $m_a \in C_0^\infty(\sigma_a^-, \sigma_a^+)$ be nonnegative and normalized to $\int ds\, m(s) = 1$. Then we define:

$$\begin{gathered} m(R,\sigma) = m_0(R)\prod_a m_a(\sigma_a); \\ g(x) = \int dR\, d\sigma\, m(R,\sigma) g(x,R,\sigma). \end{gathered} \tag{6.35}$$

From the decomposition:

$$
\begin{aligned}
g(x,R,\sigma) &= \max\big(R, g_1(x,\sigma)\big) = g_0(x,R,\sigma) + \sum_a g_a(x,R,\sigma);\\
g_0(x,R,\sigma) &= R\prod_a \Theta(R-\sigma_a|x_a|);\\
g_a(x,R,\sigma) &= \sigma_a|x_a|\Theta(\sigma_a|x_a|-R)\prod_{b\neq a}\Theta(\sigma_a|x_a|-\sigma_b|x_b|),
\end{aligned}
\tag{6.36}
$$

we find:

$$
\begin{aligned}
g(x) &= g_0(x) + \sum_a g_a(x);\\
g_0(x) &= \int ds\, m_0(s)\prod_a j_a\left(\frac{s}{|x_a|}\right);\\
g_a(x) &= |x_a|\int ds\, m_a(s) j_0\left(\frac{s}{|x_a|}\right)\prod_{b\neq a} j_b\left(\frac{s|x_a|}{|x_b|}\right),
\end{aligned}
\tag{6.37}
$$

with $j_0(s) = \int_{-\infty}^{s} dt\, m_0(t)$, and similarly for $j_a(s)$. These formulas exhibit the smoothness of the functions $g_0(x), g_a(x)$ and $g(x)$. The other relevant properties follow directly from (6.35) and from the corresponding properties of the unregularized functions:

LEMMA 6.4. *Suppose that ε satisfies* (6.34) *and that* $0 < R^- < R^+ < \infty$. *Then all terms in the decomposition* (6.37) *are in* $C^\infty(X)$ *and have the following properties*: *Properties of* $g(x)$:

(i) $g(x)$ *is convex, and homogeneous of degree* 1 *for* $|x| > R^+$;
(ii) $g(x) = \overline{R} = \int ds\, s m_0(s)$ *for* $|x| < R^-(1-2\varepsilon^2)^{-1/2}$;
(iii) $g(x) = g(x_a)$ *on each* C_a^-;
(iv) $g(x) = \overline{\sigma_a}|x_a|$; $\overline{\sigma_a} = \int ds\, s m_a(s)$; *for* $|x| > R^+$ *and* $x \in C_a^- \setminus \bigcup_{b<a} C_b^+$.

Properties of $g_0(x), g_a(x)$:

(i) $g_0(x) = 0$ *for* $|x| > R^+$;
(ii) $g_a(x) = 0$ *for* $|x| < R^-(1-2\varepsilon^2)^{-1/2}$;
(iii) $g_a(x)$ *is homogeneous of degree* 1 *for* $|x| > R^+$;
(iv) $g_a(x) = g_a(x_b)$ *on each* C_b^-.

PROOF. Properties of $g(x)$: (i) follows directly from (6.35): the regularization preserves convexity and homogeneity. Also, $g(x,R,\sigma) = g_1(x,\sigma)$ if $\sigma_a|x_a| > R$ for some a, in particular if $|x| > R^+$. (ii) follows from $g(x,R,\sigma) = R$ if $\sigma_a|x_a| \le R$ for all a, in particular if $\sigma_a^+|x| \le R^-$ for all a. Note that $\sigma_a^+ \le (1-2\varepsilon^2)^{-1/2}$, since $\dim(a) \ge 1$. (iii) and (ii) follow from Lemma 6.3. The properties of $g_0(x)$ and $g_a(x)$ are proved in the same way. □

Lemma 6.4 shows that $g(x)$ has all the properties required to prove the basic propagation estimate (6.18). In particular, the cones C_a^- play the role of the cones C_a used in that proof: (6.13) is a consequence of (6.33).

Unlike $g(x)$, the functions $g_a(x)$ are not convex. For this reason we will need upper and lower bounds on the Hessian g_a'' in terms of the Hessian of some convex function $\tilde{g}$ which satisfies the propagation estimate (6.18). In the following Lemma, any $g_a(x)$ can take the role of $f(x)$:

LEMMA 6.5. *Let $f(x)$ be a real, smooth function on X, homogeneous of degree 1 for $|x| > R^+$, with*

$$f(x) = f(x_a) \text{ on each } C_a^-. \tag{6.38}$$

Then there exists a convex function $\tilde{g}(x)$, homogeneous of degree 1 for $|x| > R^+$, such that

$$\int_0^\infty dt\, \langle p\tilde{g}''p\rangle_t \leq \text{const.}\|\psi\|^2 \quad \forall \psi \in \mathcal{H}_\Delta; \tag{6.39}$$

and

$$+f''(x) \leq \text{const.}\tilde{g}''(x) \text{ for } |x| > R^+. \tag{6.40}$$

PROOF. By homogeneity it suffices to construct $\tilde{g}$ such that it provides a local bound of the form (6.40) for all x in some neighbourhood of y, where y is any point on a sphere $|y| = R$, $R > R^+$. A global bound is then obtained by summing over finitely many local bounds.

Let y be fixed. Then $y \in (C_a^-)^*$ for some a, so that $f(x) = f(x_a)$ near y. Since f is homogeneous near y we have

$$\pm f''(x) \leq \text{const.}|x_a|^{-1}\pi_a(x), \quad \pi_a \colon \xi \in X \to \xi_a^2 - |x_a|^{-2}(x_a \cdot \xi_a)^2. \tag{6.41}$$

We construct $\tilde{g}$ exactly like g, keeping ε and $R^\pm$ the same, but choosing modified weight functions $\widetilde{m}_a(s)$. This guarantees the propagation estimate (6.39). Let σ be fixed in the allowed intervals, e.g. $\sigma_a = (\sigma_a^- + \sigma_a^+)/2$. By (6.28) $y \in C_b^*(\sigma)$ for some b, which implies $y \in C_b^-$. Since $y \in (C_a^-)^*$, $b < a$ and $a \cap b < a, b$ are excluded. Therefore $b \geq a$ and, by (6.29),

$$\sigma_b|y_b| \geq \sigma_c|y_c| \quad \forall c \neq b.$$

Now we pick $\text{supp}(\widetilde{m}_b) \subset (\sigma_b, \sigma_b^+]$ and $\text{supp}(\widetilde{m}_c) \subset [\sigma_c^-, \sigma_c)$. Then

$$\tilde{\sigma}_b|x_b| \geq \tilde{\sigma}_c|x_c| \quad \forall c \neq b,$$

all x near y and all $\tilde{\sigma} \in \text{supp}(\widetilde{m})$. As a result, $\tilde{g}(x) = \text{const.}|x_b|$ for x near y, so that

$$\tilde{g}''(x) = \text{const.}|x_b|^{-1}\pi_b(x) \geq \text{const.}|x_a|^{-1}\pi_a(x),$$

since $\pi_b(x) \geq \pi_a(x)$ for $b > a$. The local bound now follows from (6.41). □

6.4. Deift-Simon Wave Operators. By (6.37) we have:

$$\begin{aligned} \gamma = i[H, g] = \gamma_0 + \sum_a \gamma_a; \quad \gamma_0 &= \frac{1}{2}(p \cdot \nabla g_0 + \nabla g_0 \cdot p); \\ \gamma_a &= \frac{1}{2}(p \cdot \nabla g_a + \nabla g_a \cdot p). \end{aligned} \tag{6.42}$$

LEMMA 6.6. *The Deift-Simon wave operators*

$$W_a^+ = s - \lim_{t \to \infty} e^{iH_a t}\gamma_a e^{-iHt}E_\Delta(H) \tag{6.43}$$

exist.

PROOF. First we note that the operator

$$f(H_a)\gamma_a - \gamma_a f(H) \tag{6.44}$$

is compact for any $f \in C_0^\infty(R)$, by a proof very similar to the proof of (4.16). Taking f with support in $R \setminus \mathcal{S}$ and with $f = 1$ on Δ, and using the fact that $e^{-iHt}E_\Delta(H) \to 0$ weakly as $t \to \infty$, we find:

(6.45)
$$\begin{aligned} W_a^+ &= s - \lim_{t\to\infty} E_\Delta(H_a)e^{iH_at}\gamma_a e^{-iHt}E_\Delta(H) \\ &= E_\Delta(H_a)\gamma_a E_\Delta(H) + s - \lim_{t\to\infty}\int_0^t ds\, E_\Delta(H_a)e^{iH_as}i(H_a\gamma_a - \gamma_a H) \\ &\qquad \times e^{-iHs}E_\Delta(H), \end{aligned}$$

provided that the last limit exists. From

$$\begin{aligned} i[H_a\gamma_a - \gamma_a H] &= i[H, \gamma_a] - iI_a\gamma_a \\ &= pg_a''p - \frac{1}{2}\Delta^2 g_a - \nabla g_a \cdot \nabla V - iI_a\gamma_a \end{aligned} \tag{6.46}$$

we pick up four contributions to the integral (6.45). They can all be handled by Lemma 5.6. We note that $\Delta^2 g_a = \mathcal{O}_0(|x|^{-3})$ and $\nabla g_a \cdot \nabla V = \mathcal{O}_0(|x|^{-\mu})$ by the argument leading to (6.17). Similarly $|I_a(x)| \le \text{const.}\, |x|^{-\mu}$ on $\text{supp}(\nabla g_a)$ by (6.1), and therefore:

$$I_a\gamma_a = \big(\text{sgn}(I_a)|I_a|^{1/2}\big)\cdot\big(|I_a|^{1/2}\gamma_a\big) = \mathcal{O}_0\big(|x|^{-\mu/2}\big)\cdot\mathcal{O}_1\big(|x|^{-\mu/2}\big). \tag{6.47}$$

This is the only instance where the short range condition (6.1) enters. As a result, by local decay alone, the last three terms in (6.46), give convergent contributions to the integral (6.45). The remaining term $pg_a''p$ cannot be treated by local decay, because $g_a''(x) = O(|x|^{-1})$ at ∞. We first split g_a'' into

$$g_a''(x) = g_a''(x)\Theta(|x| - R^+) + g_a''(x)\Theta(R^+ - |x|).$$

The second term has compact support: its contribution is convergent by local decay. The contribution

$$A = pg_a''(x)\Theta(|x| - R^+)p$$

of the first term to $pg_a''p$ is further decomposed into positive and negative parts, then we use the bound (6.40) for $f = g_a$:

$$A = A_+ - A_-; \quad 0 \le A_\pm \le \text{const.}\, p\tilde{g}''p. \tag{6.48}$$

Now we can use Lemma 5.6 with the factorization

$$A_\pm = (A_\pm)^{1/2}\cdot(A_\pm)^{1/2}.$$

As a result we see $A_\pm$ gives a convergent contribution to the integral (6.45), due to the propagation estimate (6.39) which also holds for the time-evolution generated by H_a. □

6.5. The Asymptotic Observable γ^+. Recalling (6.42) we define:

$$\gamma^+ = s - \lim_{t\to\infty} e^{iHt}\gamma e^{-iHt}E_\Delta(H) \tag{6.49}$$

and, in the same way, the corresponding γ_0^+, γ_a^+. We first note that $\gamma_0^+ = 0$, since γ_0 is H-compact and since $e^{-iHt}E_\Delta \to 0$ weakly as $t \to \infty$. The existence of γ_a^+ is proved like the existence of W_a^+: H_a is replaced by H, so that the term $I_a\gamma_a$ in (6.46) does not arise. Moreover, this proof shows that γ_a^+ maps $\mathcal{H}_\Delta$ into itself. Thus we have

$$\gamma^+ = \sum_a \gamma_a^+ \text{ on } \mathcal{H}_\Delta. \tag{6.50}$$

LEMMA 6.7.

$$(\gamma^+)^2 \geq 2\theta_{\min} > 0,$$

on $\mathcal{H}_\Delta$, where $\theta_{\min}$ is the minimal Mourre constant $\theta(E)$ for $E \in \Delta$.

PROOF. Given any $\varepsilon > 0$ and any $E \in \Delta$ we have a Mourre estimate

$$E_I i[H, A]E_I \geq \big(\theta(E) - \varepsilon\big)E_I \tag{6.51}$$

for some open interval $I \ni E$. Let $x^2(t)$ be the Heisenberg operator

$$x^2(t) = e^{iHt}x^2e^{-iHt}.$$

Then (6.51) implies

$$E_I x^2(t)E_I \geq 2t^2(\theta(E) - \varepsilon)E_I \tag{6.52}$$

in form sense as $t \to \infty$, modulo terms of lower order in t. Since $\gamma = i[H, g]$, the corresponding Heisenberg operators $g(t), \gamma(t)$ satisfy:

$$s - \lim_{t\to\infty} \frac{1}{t}g(t)E_I = s - \lim_{t\to\infty} \left(\frac{1}{t}g(0) + \frac{1}{t}\int_0^t ds\,\gamma(s)\right) E_I = \gamma^+ E_I$$

by (6.49), so that in form sense:

$$E_I(\gamma^+)^2E_I = \lim_{t\to\infty} \frac{1}{t^2}E_I g^2(t)E_I. \tag{6.53}$$

Now, because $g^2(x) > x^2$ on X, we have $g^2(t) \geq x^2(t)$, so that (6.52) and (6.53) imply

$$E_I(\gamma^+)^2E_I \geq 2\big(\theta(E) - \varepsilon\big)E_I \geq 2(\theta_{\min} - \varepsilon)E_I.$$

Since γ^+ commutes with E_I, we can replace E_I by E_Δ and let $\varepsilon \to 0$. □

6.6. Asymptotic Completeness. By Lemma 6.7, γ^+ maps $\mathcal{H}_\Delta$ onto itself. Hence any $\psi \in \mathcal{H}_\Delta$ can be represented as

$$\psi = \gamma^+ \phi = \sum_a \gamma_a^+ \phi.$$

Using the notation: $u(t) \approx v(t)$ for $\|u(t) - v(t)\| \to 0$ as $t \to \infty$, we obtain:

$$\begin{aligned} \psi_t = e^{-iHt}\psi &\approx \sum_a \gamma_a e^{-iHt}\phi \\ &= \sum_a e^{-iH_a t} e^{iH_a t} \gamma_a e^{-iHt} \phi \\ &\approx \sum_a e^{-iH_a t} \psi_a; \quad (\psi_a = W_a^+ \phi). \end{aligned} \tag{6.54}$$

Proceeding by induction, we assume that asymptotic completeness holds for all subsystems $a > \{0\}$, i.e. that

$$e^{-iH_a t}\phi \approx \sum_{b \geq a} e^{-iH_b t}\big(1 \otimes P_B(H^b)\big)\phi_b,$$

for any $\phi \in \mathcal{H}$, which is trivially true for $a = X$. Inserting this into (6.54) we see that ψ_t has the asymptotic behavior (6.2) for any $\psi \in \mathcal{H}_\Delta$ and any compact $\Delta \subset R \backslash \mathcal{S}$. These states span a dense set in $\mathcal{H}_C$, and since the space $\mathcal{H}^+$ of outgoing scattering states is closed, it follows that $\mathcal{H}^+ = \mathcal{H}_C$.

INSTITUT FÜR THEORETISCHE PHYSIK, ETH–HÖNGGERBERG, CH-8093 ZÜRICH, SWITZERLAND
E-mail address: hunziker@itp.ethz.ch

DEPARTMENT OF MATHEMATICS, UNIVERSITY OF TORONTO, TORONTO, ONTARIO, M5S 1A1, CANADA
E-mail address: sigal@math.toronto.edu

Centre de Recherches Mathématiques
CRM Proceedings and Lecture Notes
Volume 8, 1995

Lectures on Large Coulomb Systems

I. M. Sigal

Dedicated to E. H. Lieb and to G. M. Zhislin on the occasion of their 60th *birthdays.*

0. Introduction

The broad goal in these lectures is to understand ground state struture of large molecules. The narrow defined task is to derive asymptotic behaviour of their ground state energies for large values of nuclear charges. I hope these notes give a fair snapshot of the young and fast developing subject. I regret a haphazard state of the references. I have tried to collect as many references as it was possible to do in a short interval of time without attempting to be either complete or systematic.

Consider a molecule consisting of N electrons, each of the charge -1 and M nuclei of charges $Z_1, \dots, Z_M$. The position and spin space of a single electron are $\mathbb{R}^3$ and $\mathbb{Z}_2$, respectively. We assume that the system is neutral, i.e. $N = \sum_{i=1}^M Z_i$, and that the nuclei are infinitely heavy and located at positions $R_1, \dots, R_M$. The Schrödinger operator of such a system in appropriate units is

$$\mathrm{H}(Z,R) = \sum_{i=1}^{N}\Big(-\frac{1}{2}\Delta_i - \mathrm{V}(x_i,Z,R)\Big) + \sum_{i<j}|x_i - x_j|^{-1} \tag{0.1}$$

acting on $\bigwedge_{i=1}^N \mathrm{L}^2(\mathbb{R}^3 \times \mathbb{Z}_2)$. Here $x_i \in \mathbb{R}^3$ is the coordinate of the i-th electron and Δ_i is the Laplacian in x_i, $Z = (Z_1, \dots, Z_M)$, $R = (R_1, \dots, R_M)$ and

$$\mathrm{V}(x,Z,R) = \sum_{i=1}^{M} \frac{Z_i}{|x - R_i|}. \tag{0.2}$$

The potential $\mathrm{V}(x,Z,R)$ describes attraction of an electron to the nuclei situated at positions $R_1, \dots, R_M$. Let

$$\mathrm{E}(Z,R) = \inf \operatorname{spec} \mathrm{H}(Z,R),$$

the ground state energy of $\mathrm{H}(Z,R)$. In these lectures we study the asymptotic behaviour of $\mathrm{E}(Z,R)$ as $Z \to \infty$ in $\mathbb{R}^M$.

Let $\mathrm{E}_{\mathrm{TF}}(Z,R)$ be the ground state (or Thomas-Fermi) energy of the Thomas-Fermi gas (see a review in the next section) and let $|Z| = \sum_{j=1}^M Z_j$. It is shown there that it has the following scaling property

$$\mathrm{E}_{\mathrm{TF}}(Z,R) = |Z|^{7/3}\, \mathrm{E}_{\mathrm{TF}}(\lambda, y), \tag{0.3}$$

1991 *Mathematics Subject Classification.* Primary: 35J10, 81V70; Secondary: 35Q40, 81V55, 81Q15.

Revised, in parts abridbed and in parts expanded version, when ready, will appear elsewhere.

TABLE 1

System\Term	$\lvert Z\rvert^{7/3}$ **[TF '27]**	$\lvert Z\rvert^{2}$ **[Scott '53]**	$\lvert Z\rvert^{5/3}$ **[Dirac '31]** **[Schwinger '81]**
Atoms	**[Lieb-Simon '77]**	**[H-S-W '86–89]**	**[Fefferman-Seco '92]**
Molecules	"	**[Ivrii-Sigal '92]**	Open

where $\lambda = |Z|^{-1}Z$ and $y = |Z|^{1/3}R$. Denote $a_{\mathrm{TF}} = -\mathrm{E}_{\mathrm{TF}}(\lambda, y) \geq 0$. The main result discussed in these lectures is

THEOREM 0.1. *Let Z and R be s.t.* (a) $|Z| \geq 1$ *and* (b) $\min Z_j \geq \delta_1 |Z|$ *for some $\delta_1 > 0$ independent of Z and R and* (c) *the mutual distances between R_j (which are allowed to depend on Z) are bounded below by $|Z|^{-5/9+\varepsilon}$ with $\varepsilon > 0$. Then for each $\delta > 0$*

$$\mathrm{E}(Z, R) = -a_{\mathrm{TF}}|Z|^{7/3} + \frac{1}{2}\sum Z_j^2 + \mathrm{O}\big(a^{-1}|Z|^{16/9+\delta}\big), \tag{0.4}$$

where $a = \min\big(|Z|^{1/3}|R_i - R_j|,\ i \neq j; 1\big)$, with the remainder estimate uniform in Z and R.

It will follow from the analysis below that the leading term on the r.h.s. represents the quasiclassical energy of the bulk of electrons and the second term, the quantum spectrum of Coulomb singularities. Conjectures and results are summed up in the Table 1. (The Scott conjecture for molecules was formulated in [**50**].)

The approach in [**23–24, 39, 75–77**] uses the fact that the electron interaction with the nucleus $\mathrm{V}(x, Z, R)$, is spherically symmetric in atoms. (See a review in [**79**] and [**22**].) The proof in [**45**] is rather general.

Energy asymptotics yield also information about various quantities pertaining to the ground state. For instance, an (easy) part of the proof of Theorem 0.1 implies as a by-product an estimate on the one-electron density for the ground state (Corollary 5.1), while a result from the remaining part of the proof and the Feynman-Hellmann technique yield estimates on the one-electron density matrices (Theorem 5.2 and eq. (5.8)). (See [**24, 34, 56, 73**] for some of the first results in this direction.) Of course, atoms and molecules with $Z_j \geq 120$ is a mathematical abstraction. However, a brief look at the table above shows that the first two terms on the r.h.s. of (0.4) agree with the true (or Hartree-Fock) ground state energies of atoms with amazing accuracy already for $Z > 5$. (The three term asymptotics yields an agreement with the Hartree-Fock calculations ranging between 0.6% and 0.003% as Z changes from 6 to 120.) It is not a rare occurance in Physics that asymptotic and perturbation expressions agree with the experimental values of the observables they represent with astonishing degree. I do not know a single case when the reason for such an agreement or estimates reflecting it were rigorously established. Resolving this puzzle for the ground state energies of atoms, i.e. understanding why such a simple picture as outlined in Section 4 relects so complex objects as molecules, is a question going to the heart of the theory of Coulomb systems. (For more discussion see Section 8, Problem VI.)

In these notes (Sections 1–4) we outline the proof of Theorem 0.1 following [**45**]. This proof uses and extends a fair number of essential techniques from the theory of Coulomb systems and from PDE's and hopefully will be useful beyond the scope of the present theorem. In Sections 5 and 6 we review a recent work of Bach

TABLE 2. Comparison of the Hartree-Fock energy, $E_{\rm HF}$, with $E_{\rm TFS} = -a_{\rm TF}Z^{7/3} + Z^2/2$ ($a_{\rm TF} = 0.768$) and with the experimentally measured energy for $E_{\rm exp}$, for various values of Z (cf. [**20**]). The experimental values are computed from [**62**]. They are measured as the differences of masses of atoms and of their nuclei (i.e. as small differences of two very large numbers).

Z	$E_{\rm HF}/(Z^2/2)$	$-E_{\rm TFS}/(Z^2/2)$	$-E_{\rm exp}/(Z^2/2)$	$(1 - E_{\rm HF}/E_{\rm TFS})\%$
2	1.4308	0.9352	1.4449	34.5059
4	1.8216	1.4382	1.8245	20.9162
6	2.0922	1.7911	2.0925	14.3284
8	2.3365	2.0720	2.3348	11.2420
10	2.5709	2.3092		10.0555
12	2.7724	2.5166		9.1059
16	3.1053	2.8705		7.4473
20	3.3838	3.1693		6.2180
24	3.6220	3.4306		5.1818
28	3.8439	3.6642		4.5612
32	4.0534	3.8765		4.2486
36	4.2470	4.0718		4.0104
42	4.5073	4.3392		3.6153
48	4.7440	4.5822		3.2975
54	5.3450	4.8057		3.0032
60	5.1577	5.0132		2.6881
66	5.3450	5.2073		2.4632
72	5.5252	5.3900		2.3341
80	5.7528	5.6184		2.2245
88	5.9644	5.8321		
96	6.1617	6.0331		
104	6.3492	6.2233		
112	6.5271	6.4040		
120	6.6948	6.5762		

and Graf and Solovej on one-particle density matrix estimates and on correlation estimates, in Section 7 we sketch briefly results about binding and their connection with the ground state estimate and in Section 8 we formulate some of the open problems. A different approach to the one outlined in Section 3 is proposed in [**1**]. For asymptotics of the ground state energies in the presence of magnetic fields see [**57, 83, 94**].

Finally note that it is immaterial for Theorem 0.1 whether the ground state energy is an (discrete) eigenvalue or not. The latter question which is of a great interest on its own is left out of these notes. The interested reader is referred to works of Zhislin on atoms [**95**] and Lieb and Thirring [**59**] and [**33**] on molecules (see also [**41**]) and to Section 7 of these notes. See also [**40, 92, 93**] for asymptotic behaviour of atomic eigenvalues as they approach the bottom of the continuum.

NOTATION. We use the following standard conventions for the derivatives: $\partial_x = \text{grad}_x$ (the gradient in x), $\partial_t = (\partial)/(\partial t)$, etc., and $\partial_x^\alpha = \prod_{i=1}^{\dim} \partial_{x_i}^{\alpha_i}$ for a multi-index $\alpha = (\alpha_1, \alpha_2, \dots)$. Moreover, ∇ will stand for the gradient in all the variables involved. Let $\langle x \rangle = (1+|x|^2)^{1/2}$, $\mathrm{e}_i(\mathrm{A})$ and $E_A(\mu)$ will denote for the i-th eigenvalue (counting the multiplicities) and the spectral projection on the interval $(-\infty, \mu]$ for a self-adjoint operator A.

1. Thomas-Fermi Theory

We review basic facts about the Thomas-Fermi theory of [**56**] and prove some additional results needed later. We refer the reader to [**51**] for an extensive review of the subject.

L. H. Thomas and E. Fermi have suggested in 1927 that a large Coulomb system (atom or molecule) in the ground state (= eigenfunction corresponding to the lowest eigenvalue) looks like a classical gas but with the Pauli principle, namely, there could be at most 2 electrons per volume $(2\pi)^3$ in the phase space. Such an object is called now the *Thomas-Fermi gas*. Its states are described by the *electron density* $\rho \geq 0$ on $\mathbb{R}^3$ normalized as

$$\int \rho = N = \#\text{ of electrons}. \tag{1.1}$$

The energy of the Thomas-Fermi gas is given by a non-linear functional

$$\mathcal{E}^{\mathrm{TF}}(\rho) = \gamma \int \rho^{5/3} - \int \mathrm{V}\rho + \frac{1}{2}\int \rho(|x|^{-1} * \rho), \tag{1.2}$$

where $\gamma = 3(3\pi^2)^{2/3}/10$ and $\mathrm{V}(x, Z, R)$ is given by (0.2). This functional is finite for $\rho \in \mathrm{L}^{5/3} \cap \mathrm{L}^1$ and is bounded from below on the set $\mathrm{S}_N = \{\rho \in \mathrm{L}^{5/3} \cap \mathrm{L}^1 \mid \rho \geq 0 \text{ and } \int \rho = N\}$.

Existence and uniqueness. In order to define the first term in $\mathcal{E}^{\mathrm{TF}}(\rho)$ one has to take $\rho \in \mathrm{L}^{5/3}$. The physical origin of ρ dictates $\rho \in \mathrm{L}^1$. This already suffices for $\mathcal{E}^{\mathrm{TF}}(\rho)$ to be well defined and bounded from below. To see this one observes that $|x|^{-1} \in \mathrm{L}^{5/2} + \mathrm{L}^\infty$ which implies that

$$\int |x|^{-1}\rho \leq \mathrm{C}_1\|\rho\|_{5/3} + \mathrm{C}_2\|\rho\|_1. \tag{1.3}$$

The latter inequality yields that the second and the third terms in $\mathcal{E}^{\mathrm{TF}}(\rho)$ are finite in absolute value as well as that $\mathcal{E}^{\mathrm{TF}}(\rho)$ is bounded from below as

$$\mathcal{E}^{\mathrm{TF}}(\rho) \geq \gamma\|\rho\|_{5/3}^{5/3} - \mathrm{C}_1|Z|\,\|\rho\|_{5/3} - \mathrm{C}_2|Z|\,\|\rho\|_1$$

with C_1 and C_2 independent of Z, R and M. Furthermore, it is readily shown that $\mathcal{E}^{\mathrm{TF}}(\rho)$ defined on the set $\mathrm{L}^1 \cap \mathrm{L}^{5/3}$, $\rho \geq 0$ and $\int \rho \leq \lambda$, where λ is an arbitrary positive number, is lower semicontinuous in the weak topology of $\mathrm{L}^{5/3}$. This and Banach-Alaoglu theorem ($\equiv$ every bounded subset in $\mathrm{L}^p, p \neq 1$, is weakly compact) imply that the infimum of $\mathcal{E}^{\mathrm{TF}}(\rho)$ with the side conditions $\int \rho \leq \lambda$ and $\rho \geq 0$ is attained in $\mathrm{L}^1 \cap \mathrm{L}^{5/3}$. The inequality (1.3) can be sharpened to $\int |x|^{-1} \leq C\|\rho\|_1^{1/6}\|\rho\|_{5/3}^{5/6}$ (see the proof of Lemma 4.4). This implies $\mathcal{E}^{\mathrm{TF}}(\rho) \geq \gamma\|\rho\|_{5/3}^{5/3} - C|Z|\,\|\rho\|_1^{1/6}\|\rho\|_{5/3}^{5/6}$, which shows that the $\mathrm{L}^{5/3}$-norm of the ρ minimizing $\mathcal{E}^{\mathrm{TF}}(\rho)$ is bounded by $\mathrm{C}|Z|^{6/5}N^{1/5}$ with C independent of R and M, provided $\lambda = N$.

Next, analyzing directly expression (1.2) for $\mathcal{E}^{\mathrm{TF}}(\rho)$, one sees that this functional is strictly convex

$$\mathcal{E}^{\mathrm{TF}}\big(t\rho_1 + (1-t)\rho_2\big) < t\mathcal{E}^{\mathrm{TF}}(\rho_1) + (1-t)\mathcal{E}^{\mathrm{TF}}(\rho_2),$$

provided $0 < t < 1$ and $\rho_1 \neq \rho_2$. Since the set on which we minimize $\mathcal{E}^{\mathrm{TF}}(\rho)$ is also convex, we have that the minimizer of $\mathcal{E}^{\mathrm{TF}}(\rho)$ is unique.

We observe that $\mathcal{E}(\nu) \equiv \inf_{\int \rho \le \nu} \mathcal{E}(\rho)$ is convex in ν. This can be shown by using the test density $\alpha_1\rho^{(\nu_1)} + \alpha_2\rho^{(\nu_2)}$ with $\alpha_1 + \alpha_2 = 1$ and $\alpha_1, \alpha_2 \ge 0$, where $\rho^{(\nu)}$ is the minimizer for $\mathcal{E}(\rho)$ with the side condition $\int \rho \le \nu$.

Now we show that $\int \rho^{(\nu)} = \nu$, provided $\nu \le |Z|$. Indeed, let on the contrary there is $\nu_0 \le |Z|$ s.t. $\nu_1 \equiv \int \rho^{(\nu_0)} < \nu_0$. Then $\mathcal{E}(\nu_1) = \mathcal{E}(\nu_0)$ and therefore, by the convexity, $\mathcal{E}(\nu)$ is independent of ν as long as $\nu \ge \nu_1$ and is, consequently, equal to the absolute infimum of $\mathcal{E}(\rho)$. A standard variational calculus yields that for $\nu \ge \nu_1$, the minimizer $\rho^{(\nu)}$ satisfies the Euler-Lagrange equation, which can be written as $\rho^{(\nu)} = 2^{3/2}\big(\phi^{(\nu)}\big)_+^{3/2}/(3\pi^2)$, where $\phi^{(\nu)}$ is the Thomas-Fermi potential for $\rho^{(\nu)}$: $\phi^{(\nu)} = \mathrm{V} - |x|^{-1} * \rho^{(\nu)}$. We claim that $\phi^{(\nu)} \ge 0$. Indeed, let $B = \{x \mid \phi^{(\nu)} < 0\}$, an open subset of $\mathbb{R}^3$. This set is clearly disjoint from $R_1, \dots, R_M$ and $\rho^{(\nu)} = 0$ on B. Hence $\Delta\phi^{(\nu)} = 0$ on B. Therefore $\phi^{(\nu)}$ attains its minimum either on ∂B or at ∞. Consequently, $\phi^{(\nu)} \ge 0$ on B, which implies that $B = \emptyset$, i.e. $\phi^{(\nu)} \ge 0$ on $\mathbb{R}^d$.

Now compute the spherical average

$$\phi_{\mathrm{av}}^{(\nu)}(r) = \frac{1}{4\pi}\int_{|e|=1} \phi^{(\nu)}(re)\,\mathrm{d}^2 e,$$

of $\phi^{(\nu)}$. We have for $r \ge \max|R_k|$

$$\phi_{\mathrm{av}}^{(\nu)}(r) = \frac{|Z|}{r} - \int \min\left(\frac{1}{r}, \frac{1}{|y|}\right)\rho_{\mathrm{av}}^{(\nu)}(y)\,\mathrm{d}y.$$

This yields

$$\phi_{\mathrm{av}}^{(\nu)}(r) \ge \frac{Z - \nu_1}{r} \quad \text{for } r \ge \max|R_k|.$$

On the other hand by the convexity of $\phi \to \phi^{3/2}$

$$\rho_{\mathrm{av}}^{(\nu)} = \frac{2^{3/2}}{3\pi^2}\big(\phi^{(\nu)3/2}\big)_{\mathrm{av}} \ge \frac{2^{3/2}}{3\pi^2}\phi_{\mathrm{av}}^{(\nu)3/2},$$

which yields

$$\rho_{\mathrm{av}}^{(\nu)} \ge \frac{2^{3/2}}{3\pi^2}\left(\frac{|Z| - \nu_1}{r}\right)^{3/2} \quad \text{for } r \ge \max|R_k|.$$

Since $\int \rho_{\mathrm{av}}^{(\nu)} = \int \rho^{(\nu)} \le \nu$, the latter inequality forces $\nu_1 \ge |Z|$. Since $|Z| \ge \nu_0$, we arrive at a contradiction. Thus we have shown that $\rho^{(\nu)} = \nu$, provided $\nu \le |Z|$, and, in particular $\int \rho_Z = |Z|$.

Thomas-Fermi equation. According to Thomas and Fermi, for large N the electron density in the ground state is well approximated by some mean electron density, namely the one, $\rho_{\mathrm{TF}}(x, Z, R)$, which minimizes the Thomas-Fermi functional, $\mathcal{E}^{\mathrm{TF}}(\rho)$ with the side condition $\int \rho = N$. (Remember we assume for simplicity $N = |Z|$.) Hence the potential experienced by any one electron is approximately

$$-\phi_{\mathrm{TF}}(x, Z, R) = -\mathrm{V}(x, Z, R) + |x|^{-1} * \rho_{\mathrm{TF}}, \tag{1.4}$$

i.e. one produced by the nuclei screened by this mean electron density. ϕ_Z is called the Thomas-Fermi potential; it will play an important role in our analysis.

An important fact used below is that the minimizer ρ_Z satisfies the Thomas-Fermi equation

$$\rho_{\mathrm{TF}} = \frac{2^{3/2}}{3\pi^2}\phi_{\mathrm{TF}}^{3/2} > 0. \tag{1.5}$$

Note first of all Variational Calculus implies that the Euler-Lagrange equation for ρ_{TF} is $\rho_{\mathrm{TF}} = 2^{3/2}(\phi_{\mathrm{TF}} + \mu)_+^{3/2}/(3\pi^2)$ for some $\mu \leq 0$. So the point of our exercise is to show that $\mu = 0$. (Note aside that μ is the chemical potential, i.e. the energy per particle, see [**50**]).

If $\mu > 0$, then ρ_{TF}, due to the Euler-Lagrange equation, has a compact support. Consequently,

$$\phi_{\mathrm{TF,av}}(r, Z, R) = 0 \quad \text{for large } r.$$

This implies $\phi_{\mathrm{TF}}(x) = 0$ for large $|x|$ and therefore for all x outside $\operatorname{supp}\rho_{\mathrm{TF}}$. The latter contradicts to $\phi_{\mathrm{TF}} = -\mu > 0$ on the boundary of $\operatorname{supp}\rho$ and the continuity of ϕ_{TF}. Thus $\mu = 0$, which implies (1.5).

Note aside that the fact that ρ_{TF} satisfies (1.5) shows that ρ_{TF} minimizes $\mathcal{E}(\rho)$ with no side conditions. Hence $\inf_{\int\rho\leq\lambda}\mathcal{E}(\rho)$ is independent of λ for $\lambda > |Z|$ (see [**56**, Thms. II.20 and IV.3]).

Scaling. The Thomas-Fermi theory has a simple scaling property which we will use later. Using that

$$\mathrm{V}(x, Z, R) = \beta^{-4}\,\mathrm{V}(\beta^{-1}x, \beta^3 Z, \beta^{-1}R) \tag{1.6}$$

one concludes readily that the Thomas-Fermi functional for the potential $\mathrm{V}(x, Z, R)$ evaluated at a density $\rho(x)$ equals β^{-7} times the Thomas-Fermi functional for the potential $\mathrm{V}(x, \beta^3 Z, \beta^{-1}R)$ evaluated at the density $\beta^6\rho(\beta x)$. Using the uniqueness of the minimizer, we conclude that

$$\rho_{\mathrm{TF}}(x, Z, R) = \beta^{-6}\rho_{\mathrm{TF}}(\beta^{-1}x, \beta^3 Z, \beta^{-1}R) \tag{1.7}$$

and that (0.3) holds. Using relation (1.7) together with either (1.4) and (1.6) or with (1.5), we conclude that

$$\phi_{\mathrm{TF}}(x, Z, R) = \beta^{-4}\phi_{\mathrm{TF}}(\beta^{-1}x, \beta^3 Z, \beta^{-1}R). \tag{1.8}$$

Ground state energy. The ground state (or Thomas-Fermi) energy namely, $\mathrm{E}_{\mathrm{TF}}(Z, R)$, is the infimum (in fact, minimum if $N \leq \Sigma Z_j$) of this functional on S_N. It was shown to have the scaling property (0.3). This formula determines the dependence of $\mathrm{E}_{\mathrm{TF}}(Z, R)$ on $|Z|$ in the case when λ and y are fixed. In the general case eq. (0.3) and [**56**, Thms. V.3 and V.4] yield that

$$(\Sigma Z_j)^{7/3}\,\mathrm{E}_{\mathrm{TF}}^{\mathrm{atom}}(1) \leq \mathrm{E}_{\mathrm{TF}}(Z, R) \leq (\Sigma Z_j^{7/3})\,\mathrm{E}_{\mathrm{TF}}^{\mathrm{atom}}(1), \tag{1.9}$$

where $\mathrm{E}_{\mathrm{TF}}^{\mathrm{atom}}(\lambda)$ is the infimum of $\mathcal{E}^{\mathrm{TF}}(\rho)$ for $\mathrm{V}(x) = \lambda|x|^{-1}$ on the set S_λ, i.e. the Thomas-Fermi energy of the "λ-atom". Note that $\mathrm{E}_{\mathrm{TF}}^{\mathrm{atom}}(1)$ is independent of Z and R. If $Z \to \infty$ in a fixed direction and R obeys $\min_{i\neq j}|R_i - R_j| \geq |Z|^{-\nu}$ with $\nu < 1/3$, then

$$\mathrm{E}_{\mathrm{TF}}(Z, R) - \sum_{j=1}^{M}\mathrm{E}_{\mathrm{TF}}^{\mathrm{atom}}(Z_j) = \mathrm{O}\big(|Z|^{2+\nu}\big),$$

This follows from (0.3) and [**50**, Thm. 4.13].

Connection to quasiclassics. The Thomas-Fermi energy $\mathrm{E_{TF}}(Z,R)$ is, in fact, the quasiclassical energy of a gas of $|Z|$ non-interacting Fermions in the external potential $-\phi_Z(x,R)$. Indeed, the quasiclassical Fermi energy of such a gas in the ground state is 0 as follows from:

$$2\int_{p\leq 0} \mathrm{d}x\,\mathrm{d}\xi = 2^{3/2}(3\pi^2)^{-1}\int \phi_{\mathrm{TF}}^{3/2}\,\mathrm{d}x = \int \rho_{\mathrm{TF}}\,\mathrm{d}x = |Z|, \tag{1.10}$$

where $p(x,\xi) = |\xi|^2/2 - \phi_{\mathrm{TF}}(x,Z,R)$, the classical Hamiltonian of a single Fermion, ξ is the variable dual to x and we have used the convention that $\mathrm{d}\xi$ denotes the Lebesgue measure times $(2\pi)^{-d}$ (we keep this convention in the rest of the notes). Here we have used Thomas-Fermi equation (1.5) and the normalization of ρ_{TF}. Hence the quasiclassical energy of this gas is

$$2\int_{p\leq 0} p\,\mathrm{d}x\,\mathrm{d}\xi = -\frac{2}{15\pi^2}\int \phi_{\mathrm{TF}}^{5/2}.$$

Now the Thomas-Fermi equation (1.4) implies

$$\int \phi_{\mathrm{TF}}^{5/2} = (3\pi^2)^{5/3}\int \rho_{\mathrm{TF}}^{5/3} \quad \text{and} \quad \int \phi_{\mathrm{TF}}\rho_{\mathrm{TF}} = (3\pi^2)^{2/3}\int \rho_{\mathrm{TF}}^{5/3}.$$

The last three equations together with the definition of $\mathrm{E_{TF}}(Z,R)$ yield

$$2\int_{p\leq 0} p\,\mathrm{d}x\,\mathrm{d}\xi = \mathrm{E_{TF}}(Z,R) + \mathrm{D_{TF}}. \tag{1.11}$$

Here $\mathrm{D_{TF}}$ is the number compensating for overcounting the electron-electron interaction in ϕ_Z.

$$\mathrm{D_{TF}} = \frac{1}{2}\iint \frac{\rho_{\mathrm{TF}}(x,Z,R)\rho_{\mathrm{TF}}(y,Z,R)}{|x-y|}\,\mathrm{d}x\,\mathrm{d}y.$$

This relation is, in fact, the Virial theorem of the Thomas-Fermi theory (see [**56**, Thm. II.23, 21, Thm. 2.14]) and it is a consequence of a scaling property of the Thomas-Fermi potential under the map $\rho(x) \to t^3\rho(tx)$.

Estimates on ϕ_{TF}. Our next task is to derive estimates on $\phi_{\mathrm{TF}}(x,Z,R)$ needed later. These estimates extend somewhat earlier results of [**56**, Thms. IV.5, IV.10] and their proof proceeds along similar lines. To make future references more convenient we change the symbols for parameters and in the next theorem deal with $\phi_{\mathrm{TF}}(x,\lambda,y)$ instead of $\phi_{\mathrm{TF}}(x,Z,R)$. We introduce the scale functions

$$l(x) \equiv l(x,y) \equiv \min_j |x-y_j|, \tag{1.12}$$

$$f(x) = l(x)^{-1/2}\langle l(x)\rangle^{-3/2}. \tag{1.13}$$

For simplicity we fix M and λ.

THEOREM 1.1. *The Thomas-Fermi potential $\phi_{\mathrm{TF}}(x,\lambda,y)$ is smooth (in x) outside the points $y_1, \ldots, y_M$ and, if $\sum \lambda_j = 1$, obeys the estimates*

$$\left|\partial^\nu \phi_{\mathrm{TF}}(x,\lambda,y)\right| \le \mathrm{C}_\nu f(x)^2 l(x)^{-|\nu|} \tag{1.14}$$

for any multi-index ν and uniformly in y, M and in λ. Here $l(x)$ and $f(x)$ are defined in (1.12) *and* (1.13), *respectively.*

PROOF. We prove here only the case $\nu = 0$ (cf. [**56**, Thm. V.12]). For the general case see [**45**]. We want to prove that

$$\phi_{\mathrm{TF}}(x,Z,R) \le \left(\sum \lambda_j \phi^{\mathrm{at}}_{\lambda=1}(x-y_j)^{3/2}\right)^{2/3}, \tag{1.15}$$

where $\phi^{\mathrm{at}}_{\lambda=1}(x)$ is the Thomas-Fermi potential for a $\lambda = 1$ atom.

We omit the subindex TF at ρ and ϕ. Note first that by the definition of $\phi(x,y)$

$$\Delta\phi = 4\pi\left(-\sum_{j=1}^{M} \lambda_j \delta(x-y_j) + \rho(x)\right).$$

Let $\mathrm{B} = \{x \in \mathbb{R}^d \mid \rho(x) \ge \sum \lambda_j \rho_j(x)\}$, where $\rho_j(x) = \rho^{\mathrm{at}}(x-y_j)$ with $\rho^{\mathrm{at}}(x)$, the Thomas-Fermi density of the $(z=1)$-atom. Let $\psi = \phi - \sum \lambda_j \phi_j$, where $\phi_j = \phi^{at}_{z=1}(x-y_j)$. Then $-\Delta\psi = 4\pi(-\rho + \sum \lambda_j \rho_j)$ and therefore $-\Delta\psi < 0$ on B, i.e. ψ is subharmonic on B. Hence it attains its maximum either on $\partial\mathrm{B}$ or at infinity. Since $\psi = 0$ on $\partial\mathrm{B}$ and at ∞, we conclude that $\psi \le 0$ on B. On the other hand, $\mathrm{B} \subseteq \{x \in \mathbb{R}^d \mid \psi(x) > 0\}$ and therefore $\mathrm{B} = \phi$. The latter fact is equivalent to the inequality $\rho \le \sum \lambda_j \rho_j$, which, due to the Thomas-Fermi equation (1.5), can be written as the upper bound in (1.15).

Now we estimate $\phi^{\mathrm{at}}_{\lambda=1}(x)$. First of all, by the definition, $|x|^{-1} \ge \phi^{\mathrm{at}}_{\lambda=1}(x)$. Estimation of $\phi^{at}_{\lambda=1}(x)$ for $|x| \ge R$ follows [**56**]. First, note that $\phi^{\mathrm{at}}_{\lambda=1}(x) \to 0$ as $|x| \to \infty$ and $\inf_{|x|=R} \phi^{\mathrm{at}}_{\lambda=1}(x) > 0$. The Thomas-Fermi equation and the potential equation written out above imply that $\phi = 2^3/(9\pi^4)\phi^{\mathrm{at}}_{\lambda=1}$ obeys $\Delta\phi = 4\pi\phi^{3/2}$ for $x \ne 0$. Let $\psi_c(x) = c|x|^{-4}$. Then for $c = \max\left(\sup_{|x|=R}\phi(x), 9\pi^{-2}\right)$ we have $\psi_c \ge \phi$ on $|x| = R$ and $\Delta\psi_c \le 4\pi\psi_c^{3/2}$. Hence, by the subharmonic argument as in the proof above, $\psi_c \ge \phi$ for $|x| \ge R$. We summarize the above estimates as

$$\phi^{at}_{\lambda=1}(x) \le \mathrm{C}\min\left(|x|^{-1}, |x|^{-4}\right) \tag{1.16}$$

for some (absolute) constant $0 < \mathrm{C} < \infty$. The latter inequality combined with (1.15) yields (1.14) for $\nu = 0$. □

2. Mean Field Theory

The mean field theory is the next step toward describing the full molecular Hamiltonian. It treats the electrons as independent but moving in an effective (mean) potential. This potential is the nuclear potential screened by a "mean" electron density. The latter is then found self-consistently by optimizing the obtained result. In our case the mean electron density is the Thomas-Fermi density and consequently the resulting potential is the Thomas-Fermi potential, $-\phi_{\mathrm{TF}}(x,Z,R)$.

The Schrödinger operator of N independent Fermions moving in the effective external potential $-\phi_{\mathrm{TF}}(x, Z, R)$ is

$$\mathrm{H}^{\mathrm{ind}}(Z,R) = \sum_{i=1}^{N}\Big(-\frac{1}{2}\Delta_{x_i} - \phi_{\mathrm{TF}}(x_i, Z, R)\Big) - \mathrm{D}_{\mathrm{TF}} \tag{2.1}$$

acting on the Fermi space $\bigwedge_{i=1}^{N} L^2(\mathbb{R}^3 \times \mathbb{Z}_2)$. This is the mean field Hamiltonian. We are interested in the asymptotic behaviour of the ground state energy,

$$\mathrm{E}_{\mathrm{ind}}(Z,R) = \inf\operatorname{spec}\mathrm{H}^{\mathrm{ind}}(Z,R), \tag{2.2}$$

of $\mathrm{H}^{\mathrm{ind}}(Z,R)$ as $Z \to \infty$.

The key to $\mathrm{E}_{\mathrm{ind}}(Z,R)$ is understanding the one-particle Schrödinger operator

$$\mathrm{P} = -\frac{1}{2}\Delta_x - \phi_{\mathrm{TF}}(x, Z, R) \text{ on } \mathrm{L}^2(\mathbb{R}^3 \times \mathbb{Z}_2). \tag{2.3}$$

Indeed, since $\mathrm{H}^{\mathrm{ind}}(Z,R)$ acts on $\bigwedge_{i=1}^{N} \mathrm{L}^2(\mathbb{R}^3 \times \mathbb{Z}_2)$ and since the variables $x_1, \ldots,$ x_N in it can be separated, we have

$$\mathrm{E}_{\mathrm{ind}}(Z,R) = \sum_{i=1}^{N} E_i - \mathrm{D}_{\mathrm{TF}}, \tag{2.4}$$

where E_i are the eigenvalues of P labelled in order of their increase and counting their multiplicity, if $i \leq$ the total number of negative eigenvalues of P, and are zeros otherwise. This is a well known relation in Quantum Physics and is a consequence of the Pauli principle: at most two electrons (the double degeneracy corresponding to $\mathbb{Z}_2$) for an energy level. The asymptotics expansion of the sum of the first N eigenvalues of P is given in

THEOREM 2.1. *Under assumption and with the notation of Theorem* 0.1

$$\sum_{i=1}^{N} E_i = \mathrm{Weyl} + \mathrm{Scott} + \mathrm{O}\big(a^{-1}|Z|^{16/9+\delta}\big), \tag{2.5}$$

where the remainder estimate is uniform in the $|Z|^{1/3}R_j$*'s and in* $Z_j|Z|^{-1}$*'s,*

$$\mathrm{Weyl} = \iint_{p\leq 0} p \,\mathrm{d}x\,\mathrm{d}\xi \tag{2.6}$$

and

$$\mathrm{Scott} = \frac{1}{4}\sum Z_j^2. \tag{2.7}$$

This result and eq. (2.4) implies the desired asymptotic relation for $\mathrm{E}^{\mathrm{ind}}(Z,R)$.

Besides of $\mathrm{E}^{\mathrm{ind}}(Z,R)$ we have to understand also the asymptotics of the Fermi energy and of the particle density as $|Z| \to \infty$. We begin with the study of the asymptotics in Z of the number of particles as a function of the chemical potential.

The number of particles in the ground state with the chemical potential $\mu \leq 0$ is $\#\{E_i \leq \mu\}$ (counting the multiplicities). The fact that μ is the chemical potential can be infered from the following formal relation

$$\frac{\mathrm{d}}{\mathrm{d}\lambda}\varepsilon^{\mathrm{MF}}(\lambda) = \frac{\mathrm{d}\mu}{\mathrm{d}\lambda}\mu\frac{\mathrm{d}}{\mathrm{d}\mu}\#\{E_i \leq \mu\} = \mu,$$

where $\varepsilon^{\mathrm{MF}}(\lambda) = \int_0^\mu E \frac{\mathrm{d}}{\mathrm{d}E} \#\{E_i \leq E\} \mathrm{d}E$, the mean field energy for the "number of particles λ" and μ as a function of λ is determined by the relation $\#\{E_i \leq \mu\} = \lambda$.

The asymptotics of $\#\{E_i \leq \mu\}$ in $|Z|$ is given in

THEOREM 2.2. *Under the assumptions of Theorem* 0.1 *the number of eigenvalues of* P *not greater than* $\mu \leq -\beta|Z|^{-A}$ *for some* $A \geq 0$ *verifies the following asymptotic expansion*

$$\#\{E_i \leq \mu\} = \iint_{p \leq \mu} \mathrm{d}x \, \mathrm{d}\xi + \mathrm{O}(|Z|^{2/3}). \tag{2.8}$$

Here the remainder estimate is uniform in μ, $|Z|^{1/3}R$ *and in* $Z|Z|^{-1}$.

This result must be compared with analogous results in [**13, 35, 36, 43, 67, 89**] for assumptions on the potentials which are somewhat stronger than ours.

Using this result, eq. (1.12) and Thomas-Fermi equation (1.5), we estimate the Fermi energy as $E_N = O(|Z|^{8/9})$, where, recall, $N = |Z|$, the number of electrons and the r.h.s. is uniform in $|Z|^{1/3}R$ and $Z|Z|^{-1}$.

Denote by $\mathrm{e}(x, y, \mu, \mathrm{P})$ the Schwartz kernel of the spectral projection $\mathrm{E_P}(\mu)$. Its diagonal element $\mathrm{e}(x, x, \mu, \mathrm{P})$ is the particle density in the ground state in the mean field theory for the chemical potential $\mu \leq 0$. Its Weyl component is given by

$$\begin{aligned} \mathrm{e}_0(x, \mu, \mathrm{P}) &= \int_{p \leq \mu} \mathrm{d}\xi \\ &= \frac{2^{3/2}}{3\pi^2} (\phi_{\mathrm{TF}} + \mu)_+^{3/2}. \end{aligned} \tag{2.9}$$

Using (2.9) and the Thomas-Fermi equation (1.4), we find

$$\mathrm{e}_0(x, 0, \mathrm{P}) = \rho_{\mathrm{TF}}(x, Z, R). \tag{2.10}$$

This shows the relation between the Weyl asymptotics of the mean field particle density (the l.h.s.) and the Thomas-Fermi density. The asymptotics of $\mathrm{e}(x, x, \mu, \mathrm{P})$ is described in

THEOREM 2.3. *Assume* $\min Z_j \geq \delta|Z|$ *for some* δ *independent of* Z. *Then for any* $\mu \leq 0$

$$\left(\int \left| \mathrm{e}(x, x, \mu, \mathrm{P}) - \mathrm{e}_0(x, \mu, \mathrm{P}) \right|^p dx \right)^{1/p} \leq \mathrm{C}|Z|^{2/3} \tag{2.11}$$

with C *independent of* Z *and* R *and* μ *and, recall,* $1 < p < 3/2$.

As the leading-Weyl-term in the expansion above suggest, this result is quasi-classical. In order to see it explicitely, we rescale the operator P using the scaling property of (1.7) of the potential ϕ_{TF}. Namely let

$$\big(\mathrm{U}(\beta)\psi\big)(x) = \beta^{3/2}\psi(\beta x),$$

the unitary group scaling $x \to \beta x$. Then for $Z = \beta^{-3}\lambda$ and $R = \beta y$

$$\mathrm{U}(\beta)\mathrm{P}\,\mathrm{U}(\beta)^{-1} = \beta^{-4}\mathrm{K}_\beta, \tag{2.12}$$

where K_β is the Schrödinger operator acting on $\mathrm{L}^2(\mathbb{R}^3 \times \mathbb{Z}_2)$ and given by

$$\mathrm{K}_\beta = -\frac{1}{2}\beta^2\Delta_x - \phi_{\mathrm{TF}}(x, \lambda, y). \tag{2.13}$$

TABLE 3

P.D.E.	Tauberian Technique
Heat eq.	Laplace Transform
Wave eq.	Sin Transform
Schrödinger eq.	Fourier Transform
Stationary Schrödinger eq.	Hilbert Transform
Powers H^z	Mellin Transform

Of course, K_β is self-adjoint as well. Thus the large Z problem for P is the quasiclassical problem (for K_β) with the quasiclassical parameter $\beta = |Z|^{-1/3}$.

3. Weyl Asymptotics

Now we study quasiclassical asymptotics for the sum of the first N eigenvalues of P, or, in general, a class of one-particle Schrödinger operators whose potentials have Coulomb singularities. The Weyl term of the asymptotics will be identified with the Thomas-Fermi contribution to the original ground state energy while the second term, with the Scott correction. Physics of the problem suggests that the second term should come from the singularities of the potential. We describe basic ingredients in our approach.

Microlocal spectral analysis. We consider a self-adjoint Schrödinger operator $H_\alpha = -\alpha^2\Delta_x/2 - W(x)$ on $L^2(\mathbb{R}^d)$. The potential $W(x)$ is allowed to have singularities. Consequently standard pseudo-differential calculus is not applicable to the investigation of various spectral characteristics of H_α. We go around this problem by estimating global quantities through local ones. For instance, we study the local sum of negative eigenvalues of P, namely,

$$\operatorname{tr}\big(\psi(x, D_x)g(H_\alpha)\big) \tag{3.1}$$

where $D_x = -i\alpha\partial_x$ and $g(\lambda) = \lambda$ for $\lambda \leq 0$ and 0 for $\lambda \geq 0$. If $\psi \equiv 1$, then the trace above is just the sum of negative eigenvalues of H_α. We take for ψ smooth functions localized outside of the singularities of the potentials and of the flow (i.e. critical points of the symbol $h(x,\xi) = |\xi|^2/2 - W(x)$). Then it is not difficult to obtain asymptotic expansion in the quasiclassical parameter α of the trace (3.1).

The traditional approach to obtaining spectral information is, first, to study a particular P.D.E. $f(H_\alpha, t)$, and then use a tauberian theorem to pass information to the eigenvalues. Different P.D.E.'s and the corresponding tauberian techniques that one can use are summarized in Table 3:

The tauberian analysis is generally easy and sharp, while the P.D.E. analysis is generally hard and subject to improvement. This justifies the choice in our approach of the best tauberian technique (Fourier Transform) and the corresponding-hardest-P.D.E., the Schrödinger Equation. In PDE, this equation for second order operators, which is our case, is avoided due to the infinite speed of the propagation of singularities. However, microlocally, we have

$$|\xi| \sim \sqrt{E + W(x)}$$

and since we are interested only in the negative eigenvalues, we obtain the bound—always microlocally—

$$|\xi| \le \sup\big|W_+(x)\big|^{1/2},$$

where the sup ranges over the support of ψ and $W_+(x) = \max\big(W(x), 0\big)$. This gives us a bound for speed of propagation of singularities. And as a result we are left with a domain in phase space given by

$$\big\{(x,\xi) \mid |\xi| \le \big|W_+(x)\big|^{1/2}, x \in \operatorname{supp}\psi\big\}$$

outside of which we get a negligible contribution to the negative part of the spectrum. We denote by $\varphi(x,\xi)$ a smooth cut-off function adapted to a small neighborhood of such domain.

F.I.O. for the Schrödinger equation. In order to carry out our program, we adapt a standard approach to the spectral and quasiclassical asymptotics (see [**13, 37, 46, 48, 63**] for original works and [**38**] for more recent developments). One represents $g(H_\alpha)$ as

$$g(H_\alpha) = \frac{1}{2\pi\alpha}\int \hat{g}(t)e^{-iH_\alpha t/\alpha}\,dt$$

where $\hat{g}(t)$ is the α-Fourier transform of g,

$$\hat{g}(t) = \int e^{it\lambda/\alpha} g(\lambda)\,d\lambda.$$

The evolution operator $e^{-iH_\alpha t/\alpha}$ is then approximated locally to any power in α by Fourier integral operators in the spirit of the geometrical optics

$$\psi e^{-iH_\alpha t/\alpha} \approx \psi F_t$$

provided $\varphi \equiv 1$ on the support of ψ. The Fourier integral operators F_t are given by

$$F_t u = \alpha^{-d}\iint e^{i(S(t,x,\xi)-z\cdot\xi)/\alpha} a(t,x,\xi,\alpha)u(z)\,dz\,d\xi, \tag{3.2}$$

where the phase S satisfies the Hamilton-Jacobi equation

$$\begin{cases} \partial_t S + h(x, \nabla_x S) = 0 \\ S(0,x,\xi) = x\cdot\xi \end{cases}, \tag{3.3}$$

in which $h(x,\xi) = |\xi|^2/2 - W(x)$ is a symbol of H_α, and the symbol a is represented as

$$a = \sum_{j=0}^{N} \alpha^j a_j(t,x,\xi),$$

where the a_j satisfy the transport equations

$$\begin{cases} \partial_t a_j + \nabla_x S\cdot\nabla_x a_j + \frac{1}{2}\Delta_x S\, a_j = -\frac{i}{2}\Delta_x a_{j-1} \\ a_{-1} \equiv 0 \\ a_j|_{t=0} = \delta_{j,0}\varphi \end{cases}. \tag{3.4}$$

To be more precise, let $W(x)$ be s.t. H_α is self-adjoint and let it satisfy

$$\big|\partial^\alpha W(x)\big| \le C_\alpha \text{ on } \mathrm{B}(0,2) \tag{3.5}$$

Let furthermore φ from (3.4) to be in $C_0^\infty(\mathrm{B}(0,5/4)\times\mathrm{B}(0,K))$ with an constant K defined by $K=2(\sup_{\mathrm{B}(0,2)}W)_+^{1/2}+1$. The Hamilton-Jacobi equation (3.3) and the transport equations (3.4) can be solved by the method of characteristics (see e.g. [**2, 13, 47**]). For instance, the unique solution of the Hamilton-Jacobi equations is given for $|t|$ sufficiently small by the action function

$$\mathrm{S}(t,x,\eta)=z\cdot\eta+\int(\xi\cdot dx-h\,dt)$$

along the classical trajectory for the Hamiltonian $h(x,\xi)$ with the momentum η at time $s=0$ and the position x at time $s=t$. Here z is the position at time $s=0$ as a function of t, x and η. Fix the degree of approximation N. Let $T_1>0$ (depending of K) be such that the Hamiltonian flow for $h(x,\xi)$ starting in $\mathrm{B}(0,3/2)\times\mathrm{B}(0,K+2)$ (resp. $\mathrm{B}(0,5/4)\times\mathrm{B}(0,K)$) exists for $|t|\le T_1$ and stays during this period inside of $\mathrm{B}(0,2)\times\mathbb{R}^d$ (resp. $\mathrm{B}(0,4/3)\times\mathrm{B}(0,K+1)$). Then decreasing T_1 depending on N, if necessary, we conclude that (3.4) and (3.5) with $j\le N$ have unique solutions in $\mathrm{B}(0,3/2)$ for $\xi\in\mathrm{B}(0,K+2)$ and $\mathrm{a}_j\in\mathrm{C}_0^\infty(\mathrm{B}(0,4/3)\times\mathrm{B}(0,K+1))$ for $|t|\le T_1$. Hence $\mathrm{a}\in\mathrm{C}_0^\infty(\mathrm{B}(0,4/3)\times\mathrm{B}(0,K+1))$ for $|t|\le T_1$. Moreover, S is smooth in t, x, η on $\mathrm{B}(0,4/3)\times\mathrm{B}(0,K+1)$. Thus F_t is well defined for $|t|\le T_1$ and obeys $F_0=\varphi(x,D_x)$.

To sum up, for sufficiently small time, the classical trajectories (bicharacteristics) starting in $\operatorname{supp}\varphi$ do not reach the singularities of $W(x)$. Thus, the singularities of $W(x)$ are decoupled, and all the objects we are dealing with are smooth. Such an approximation is possible due to finite speed of propagation of singularities (properly defined) for the Schrödinger equation, provided the energy is bounded from above (i.e. $\sup(\operatorname{supp}g)<\infty$) and the coordinate is localized in a domain in which the potential is bounded from below (cf. [**80**]).

We emphasize that F_t depends only on the restriction of $\mathrm{W}(x)$ to the ball $\mathrm{B}(0,2)$. Furthermore, any finite number of the constants in the estimates of the S and a and their derivatives depends only on K and on the corresponding finite number of constants in (3.5) and in similar inequalities for φ. Moreover, F_t has a C_0^∞ symbol and its analysis is fairly simple.

Expressions like the one in (3.2) allow us to analyze the trace in (3.1) by the method of stationary phase. (cf. [**3**]). Note that after the localization introduced with φ, all integrals are absolutely convergent, so that the analysis involved is rather easy. Since in this part of the analysis we stay away from the singularities, this analysis is quasiclassical in nature and in our asymptotics we will obtain only the contribution to the Weyl term, with a good error bound, provided $W(x)$ obeys (3.5). Introduce the function

$$g_s(\lambda)=\begin{cases}(-\lambda)^s & \text{if } \lambda<0\\ 0 & \text{otherwise}\end{cases}. \tag{3.6}$$

We obtain

THEOREM 3.1. *Let g_s be as specified above with $0\le s\le 1$. Let a cut-off function $\psi\in\mathrm{C}_0^\infty(\mathrm{B}(0,1))$ and let 0 be a regular value of the function h restricted to* $\operatorname{supp}\psi\times\mathbb{R}^d$. *Then for $\alpha\le 1$*

$$\operatorname{tr}(\psi g_s(\mathrm{H}_\alpha))=\alpha^{-d}\iint\psi(x)g_s(h(x,\xi))\,\mathrm{d}x\,\mathrm{d}\xi+\mathrm{O}(\alpha^{s+1-d}). \tag{3.7}$$

Here $\mathrm{O}(\cdot)$ *depends only on a finite number of constants in* (3.5) *and similar constants for* ψ *and on* $|\nabla h|^{-1}\big|_{h=0}$.

In the problem at hand we need an estimate of the type of (3.7) but with ψ supported on various balls (covering $\mathbb{R}^d$), without the restriction on critical points of h and with a good control of dependence of the remainder on the size of potential and its derivatives on $\operatorname{supp}\psi$. In order to convert estimate (3.7) into the desired estimates described above we use the second ingredient in our analysis, the theory described in the next section.

Multiscale analysis. Go back to a singular operator $\mathrm{K}_\beta = \beta^2\Delta_x/2 - \phi(x)$, where $\phi(x)$ is a real, but possibly singular potential, not necessary $\phi_{\mathrm{TF}}(x, \lambda, y)$, s.t. K_β is self-adjoint. There are three scales in the problem: action scale determined by the quasiclassical parameter $\beta = \mathrm{O}(|Z|^{-1/3})$, length scale, $\ell(x)$, determined by the behavior of the potential under differentiation, and the momentum scale, $f(x)$, determined by the size of the potential. The first scale is constant while the other two depend on x. In general we assume that functions $f(x)$ and $\ell(x)$ are positive and such that (cf. [**9, 10**])

(i) $\big|\nabla\ell(x)\big| \le M$.

(ii) $cf(y) \le f(x) \le Cf(y)$ for $x \in B\big(y, \ell(y)\big)$.

(iii) $\big|\partial_x^\alpha\phi(x)\big| \le C_\alpha f(x)^2\ell(x)^{-|\alpha|}$.

The reason for introducing the continuum of scales in our case is to be able to deal with singularities. There are two types of singularities in our problem. First of all it is the singularities of the potential ϕ. Secondly, it is singularities of the flow, namely the fixed points of the flow or critical points of the Hamiltonian function $k(x,\xi) = |\xi|^2/2 - \phi(x)$ at certain energies. Indeed, as Theorem 3.1 shows, the basic result, we depart from, assumes that the flow is not singular in the sense described above. The choice of $\ell(x)$ and $f(x)$ is different for these two types of singularities. In dealing with the singularities of $\phi(x)$ we set

$$\ell(x) = \text{dist of } x \text{ to the singularities} \tag{3.8}$$

and $f(x) = \ell(x)^{-1/2}$ (assuming that ϕ behaves similarly to ϕ_{TF}). Note that due to Theorem 1.1, the conditions (i)–(iii) above are satisfied for such a choice of $\ell(x)$ and $f(x)$ and for the Thomas-Fermi potential ϕ_{TF}. In dealing with the singularities of the flow, say at the energy 0, we set

$$\ell(x) = \delta\big|\nabla k(x,\xi)\big|\Big|_{k(x,\xi)=0}, \tag{3.9}$$

for some appropriate constant δ, and $f(x) = \ell(x)$.

At each point outside of the singularities we rescale the problem using the scales at this point in such a way that the problem is mapped into a model one, i.e. the one with a potential W obeying $\big|\partial^\alpha W(x)\big| \le C_\alpha$ on a unit ball, whose flow on this ball is not singular, which has the effective quasiclassical parameter

$$\alpha_{\mathrm{eff}}(x) = \frac{\beta}{\ell(x)f(x)} \tag{3.10}$$

which depends on all the scales. Namely, consider a point z outside of the singularities and the ball $\mathrm{B}(z, \ell)$ of the radius $\ell = \ell(z)$ centered at this point. Assume that $\psi_z \in C_0^\infty(\mathrm{B}(z,\ell))$ and satisfies the estimate $|\partial^\alpha \psi_z(x)| \le C_\alpha \ell^{-|\alpha|}$. Let $U(\ell)$ be the unitary operator realizing the transformtion $x \to \ell^{-1}(x-z)$, i.e.

$$\big(U(\ell)\psi\big)(x) = \ell^{-d/2}\psi\big(\ell^{-1}(x-z)\big).$$

Thus $U(\ell)$ maps $\mathrm{B}(z,\ell)$ into $\mathrm{B}(0,1)$. Now we do the scale transformation on K_β as follows

$$f^{-2}U(\ell)K_\beta U(\ell)^* = H_\alpha, \tag{3.11}$$

where $f = f(z)$ and $\alpha = \beta/f\cdot\ell$. The operator H_α is of the form described in the previous paragraph: it is self-adjoint and its potential obeys (3.5). Using the relation above, we obtain

$$\mathrm{tr}\big(\psi_z g_s(\mathrm{K}_\beta)\big) = f^{2s}\,\mathrm{tr}\big(\psi g_s(H_\alpha)\big) \tag{3.12}$$

where $\psi \equiv U(\ell)\psi_z U(\ell)^* = \psi_z\big(\ell^{-1}(x-z)\big)$. Note that also ψ satisfies the conditions of Theorem 3.1, i.e. $\psi \in C_0^\infty\big(\mathrm{B}(0,1)\big)$ and $|\partial^\alpha\psi| \le C_\alpha$. Now we apply Theorem 3.1 to the r.h.s. of (3.12) and then rescale the result, eq. (3.7), back to the original set up. Finally, using an appropriate partition of unity in order to patch up estimates on various balls, $B\big(z,\ell(z)\big)$, we obtain an expansion for the original problem outside of the singularities.

The quasiclassical domain in which we conduct the above analysis is determined by the inequality $\alpha_{\mathrm{eff}}(x) \le 1$. The region in which $\alpha_{\mathrm{eff}}(x) > 1$ is quantum and is dealt with differently. For a set $\Omega \subset \mathbb{R}^d$, define Ω_ℓ to be the $\ell(x)$-neighbourhood of Ω, i.e. $\Omega = \bigcup_{x\in\Omega} B\big(x,\ell(x)\big)$. The argument explained above yields

THEOREM 3.2. *Consider a self-adjoint Schrödinger operator* $\mathrm{K}_\beta = -\beta^2\Delta_x/2 - \phi(x)$ *in* $\mathbb{R}^d$, $d \ge 2$. *Assume conditions* (i)–(iii) *above are obeyed. Then with* g_s *defined in* (3.6) *for* $0 \le s \le 1$

$$\left|\mathrm{tr}\left(\psi g_s(\mathrm{K}_\beta)\right) - \beta^{-d}\iint \psi(x)\, g_s(k(x,\xi))\,dx\,d\xi\right| \le C\beta^{-d+1+s}\int_{(\mathrm{supp}\,\psi)_\ell} f(x)^{d+s-1}\ell(x)^{-s-1}\,dx \tag{3.13}$$

for any ψ *smooth, supported in* $\{x \mid \ell(x)f(x) \ge \beta\}$ *and such that* $\partial_x^\alpha\psi(x) = \mathrm{O}\big(\ell(x)^{-|\alpha|}\big)$.

A few comments here are in order. The absence of the β^{1-d}-term in expansion (3.13) for $s > 0$ is due to the fact that the subprincipal symbol of K_β is 0.

Contrary to our discussion preceeding Theorem 3.2, this theorem has no restriction on critical points of the Hamiltonian function $k(x,\xi)$ (singularities of the flow). The reason is that this condition is "scaled away" for $d \ge 2$. More precisely, the scaling argument above yields Theorem 3.2 with ψ supported outside the set of the critical points of the potential $\phi(x)$ (which contains the set of critical points of $k(x,\xi)$ subject to the restriction $k(x,\xi) = 0$). We use this "imperfect" theorem with K_β replaced by H_α, with ψ supported in $\mathrm{B}(0,2)$ and with $f(x) = \ell(x)$ given by (3.9) in order to obtain estimate (3.8) for ψ supported in the (quasiclassical) set

$$\{x \mid \ell(x) \ge \sqrt{\alpha}\} \cap \mathrm{B}(0,1)$$

(with no other restriction on the critical points of k for $d \geq 2$). In the complementary (quantum) set

$$\{x \mid \ell(x) \leq \sqrt{2\alpha}\} \cap \mathrm{B}(0,1) \tag{3.14}$$

we rescale the problem, using the scales $\ell(x) = f(x) = \sqrt{\alpha}$, to the one which is essentially independent of α and then use very rough operator estimates to derive (3.7) for ψ supported in (3.14). This completes the proof of Theorem 3.1 with no restriction on critical points of h whatsoever if $d \geq 2$. Using this improved theorem, we prove Theorem 3.2 in the way discussed above (see the paragraph containing eq. (3.12)) as it stands, with no condition on critical points of k. Note that for $d = 1$ our result is somewhat weaker but it was improved recently to a practically sharp one in [**44**].

In the bootstrapping argument described above, the important role is played by the fact that the quantities of interest (eigenvalues, traces, densities) are transformed under rescaling in a self-similar way.

Applying Theorem 3.2 with $\ell(x) = f(x)^{-2}$ given in (3.8) to the operator K_β given in (2.13) (remember that due to Theorem 1.1 conditions (i)–(iii) hold), we obtain the desired quasiclassical expansion outside of balls of the radius $O(\beta^2)$ around the singularities of $\phi_{\mathrm{TF}}(x, \lambda, y)$ (at $y_1, \ldots, y_m$).

Inside those balls around the Coulomb singularities, we analyze the problem differently. Namely, we replace the potential by its leading term near the singularity, which is pure Coulomb (see (1.3)), and solve the quasiclassical problem for this truncated operator more accurately. We use the spherical symmetry of the new potential to decompose our local trace into the angular momentum channels, use the explicit form of the eigenvalues to sum up the low angular momentum channels and our original quasiclassical expansion (in dimension one), in order to treat the high ones. Leading terms near and outside the singularities combine into a single Weyl term over $\mathbb{R}^3$ which gives the Thomas-Fermi energy. Precise expansion of low-lying eigenvalues (corresponding to low angular momentum channels) of the truncated problem near the singularities yields the Scott correction.

4. Mean Field Approximation and Proof of Theorem 0.1

In this section, we estimate the difference between the ground state energies of $\mathrm{H}(Z, R)$ and $\mathrm{H}^{\mathrm{ind}}(Z, R)$. Recall that $\mathrm{H}^{\mathrm{ind}}(Z, R)$ describes N-independent particles moving in the Thomas-Fermi potential and asymptotics of its ground state was derived in the Section 2. Thus, the result of the present section combined with the result of Section 2, will produce the proof of main Theorem 0.1. In this section we use the shorthand

$$\iint \frac{\rho(x)\sigma(y)}{|x-y|}\,\mathrm{d}x\,\mathrm{d}y \equiv \iint \frac{\rho\sigma}{|x-y|} \quad \text{and} \quad \|f\|_{\mathrm{Coul}} = \left(\frac{1}{2}\iint \frac{ff}{|x-y|}\right)^{1/2}$$

The following theorem, closely related to results of [**25–31, 37, 65, 75, 76**] obtained for atoms, is the main result of this section.

THEOREM 4.1. *As* $|Z| \to \infty$ *so that* $\min Z_j \geq \delta|Z|$ *with* δ *independent of* Z, *the following relation holds uniformly in* $R|Z|^{1/3}$ *and in* $Z|Z|^{-1}$:

$$\mathrm{E}(Z, R) = \mathrm{E}^{\mathrm{ind}}(Z, R) + \mathrm{O}\big(|Z|^{5/3}\big). \tag{4.1}$$

PROOF. Recall that given a self-adjoint operator A, $E_A(\lambda)$ denotes its spectral projection and $e(x, y, \lambda, A)$ stands for the integral kernel of the latter. We recall the one-particle self-adjoint Schrödinger operator

$$\mathrm{P} = -\frac{1}{2}\Delta_x - \phi_{\mathrm{TF}}(x, Z, R) \text{ on } \mathrm{L}^2(\mathbb{R}^3 \times \mathbb{Z}_2). \tag{4.2}$$

Recall that E_1, E_2, ... are the negative eigenvalues of P labelled counting their multiplicities in order of their increase and that we set $E_i = 0$ for $i >$ the total number of negative eigenvalues of P. Denote

$$\rho_{\mathrm{MF}}(x) = \mathrm{e}(x, x, E_N, \mathrm{P}) \tag{4.3}$$

and

$$\gamma_{\mathrm{MF}}(x, y) = \mathrm{e}(x, y, E_N, \mathrm{P}). \tag{4.4}$$

Note that ρ_{MF} and γ_{MF} are the one-particle density and one-particle density matrix, respectively for the ground state of $\mathrm{H}^{\mathrm{ind}}(Z, R)$ i.e. in the mean field approximation. We begin with a result known to experts (see [**49, 65, 75–76**]). Let ψ be any function in $\bigwedge_{i=1}^N \mathrm{L}^2(\mathbb{R}^3 \times \mathbb{Z}_2)$, normalized as $\sum_{\text{spins}} \int |\psi|^2 = 1$. (Remember $N = |Z|$.) We associate with it the one-electron density

$$\rho_\psi(x_1) = N \sum_{\sigma_1, \ldots, \sigma_N} \int |\psi(x_1, \sigma_1, \ldots, x_N, \sigma_N)|^2 \, dx_2 \ldots dx_N. \tag{4.5}$$

Note that $\int \rho_\psi = N$.

THEOREM 4.2. *Let $\mu = \rho_{\mathrm{MF}} - \rho_{\mathrm{TF}}$. Then there is a constant* C *independent of Z and R s.t.*

$$\begin{aligned} \frac{1}{2}\iint \frac{\mu\mu}{|x-y|} - \frac{1}{2}\iint \frac{|\gamma_{\mathrm{MF}}|^2}{|x-y|} &\geq \mathrm{E}(Z, R) - \mathrm{E}^{\mathrm{ind}}(Z, R) \\ &\geq \|\rho_\psi - \rho_{\mathrm{TF}}\|_{\mathrm{Coul}}^2 - \mathrm{C}|Z|^{5/3}, \end{aligned} \tag{4.6}$$

where ψ is the ground state of $\mathrm{H}(Z, R)$.

Upper bound (Slater determinants). Let for simplicity $N \leq$ and the number of negative eigenvalues of P. Then by the variational principle and our convention about the E_i's and ψ_i's.

$$\mathrm{E}(Z, R) \leq \varepsilon^{\mathrm{HF}}(\pi), \tag{4.7}$$

where π is any rank K projection and $\varepsilon^{\mathrm{HF}}$ is the Hartree-Fock functional:

$$\begin{aligned} \varepsilon^{\mathrm{HF}}(\pi) = \operatorname{tr}[(-\frac{1}{2}\Delta - \mathrm{V}(\cdot, Z, R))\pi] \\ + \frac{1}{2}\iint \frac{\pi(x,x)\pi(y,y)}{|x-y|} - \frac{1}{2}\iint \frac{|\pi(x,y)|^2}{|x-y|} \end{aligned} \tag{4.8}$$

with $\pi(x, y)$ the integral kernal of π. Inequality (4.7) follows from the variational inequality

$$\mathrm{E}(Z, R) \leq \langle \psi, \mathrm{H}(Z, R)\psi \rangle,$$

where $\psi = \frac{1}{\sqrt{N!}}\psi_1 \wedge \cdots \wedge \psi_N$ (the Slater determinant of the ψ_i's) with $\{\psi_i\}$ a basis in $\mathrm{Ran}\,\pi$. We will use (4.7) with $\pi = \mathrm{E_P}(E_N) \equiv \pi^{\mathrm{MF}}$. Then by Definition (4.2) of P

$$\varepsilon^{\mathrm{HF}}(\pi^{\mathrm{MF}}) = \mathrm{tr}(\mathrm{P}\pi^{\mathrm{MF}}) - \iint \frac{\rho_{\mathrm{TF}}\rho_{\mathrm{MF}}}{|x-y|} + \frac{1}{2}\iint \frac{\rho_{\mathrm{MF}}\rho_{\mathrm{MF}}}{|x-y|} - \frac{1}{2}\iint \frac{|\gamma_{\mathrm{MF}}|^2}{|x-y|}, \tag{4.9}$$

where ρ_{MF} and γ_{MF} are defined in (4.3) and (4.4). We transform

$$\frac{1}{2}\iint \frac{\rho_{\mathrm{MF}}\rho_{\mathrm{MF}}}{|x-y|} = \frac{1}{2}\iint \frac{\mu\mu}{|x-y|} - \frac{1}{2}\iint \frac{\rho_{\mathrm{TF}}\rho_{\mathrm{TF}}}{|x-y|} + \iint \frac{\rho_{\mathrm{TF}}\rho_{\mathrm{MF}}}{|x-y|}, \tag{4.10}$$

where, recall, $\mu = \rho_{\mathrm{MF}} - \rho_{\mathrm{TF}}$. Substituting this expression into (4.9), recalling the definition (see (1.11)) of $\mathrm{D_{TF}}$ and remembering that π^{MF} is the spectral projection of P corresponding to the interval $(-\infty, E_N]$, we obtain

$$\varepsilon^{\mathrm{HF}}(\pi^{\mathrm{MF}}) = \sum_{i=1}^{N} E_i - \mathrm{D_{TF}} + \frac{1}{2}\iint \frac{\mu\mu}{|x-y|} - \frac{1}{2}\iint \frac{|\gamma_{\mathrm{MF}}|^2}{|x-y|}. \tag{4.11}$$

This together with (4.7) and (4.8) yields the upper bound from (4.5). □

Lower bound discussion. In this section we discuss the main ingredient in proving the lower bound in (4.6). The point is that we have to estimate $\mathrm{H}(Z,R)$ by an operator which is a sum of one particle operators (namely, $\mathrm{H}^{\mathrm{ind}}(Z,R) = \sum_{i=1}^{N} P_{x_i}$). All the terms in $\mathrm{H}(Z,R)$ are one-particle, except of the elctron- electron interaction, $1/2\sum_{i\neq j}|x_i - x_j|^{-1}$, which is two-particle. Thus we desire a bound of the form

$$\frac{1}{2}\sum_{i\neq j}|x_i - x_j|^{-1} \geq \sum_{i=1}^{N} V(x_i) - C \tag{4.12}$$

with a constant C not too large and $V(x)$ essentially $\phi(x,Z,R)$. We show in the next section how such a bound can be obtained.

Correlation estimate. The key to the lower bound in (4.6) is the following electrostatic estimate due to Lieb [**49**] (with the constant considerably improved in [**54**]):

LEMMA 4.3 [Lieb-Oxford inequality]. *Let $\psi \in \mathrm{L}^2(\mathbb{R}^{3N})$ and let ρ_ψ be the one-electron density, $\rho_\psi(x) = \sum_{i=1}^{N}\int |\psi(x_1,\dots,x_N)|^2\delta(x_i - x)\,\mathrm{d}x_1\dots\mathrm{d}x_N$. Then*

$$\left\langle \psi, \sum_{i<j}|x_i - x_j|^{-1}\psi \right\rangle \geq \frac{1}{2}\iint \frac{\rho_\psi\rho_\psi}{|x-y|} - \mathrm{C}\int \rho_\psi^{4/3} \tag{4.13}$$

for some explicit numerical constant C.

PROOF. The proof below was communicated to me by G.-M. Graf. Following [**5**] it combines Fefferman-de la Llave formula with a maximal function estimate of [**49**]. Let χ_{B} be the characteristic function of a set B and let $\mathrm{B}(z,r)$ be the ball centered at z of the radius r. Start with the elementary identity, due to [**21**],

$$\frac{1}{|x-y|} = \frac{1}{\pi}\iint \chi_{\mathrm{B}(z,r)}(x)\chi_{\mathrm{B}(z,r)}(y)\frac{dz\,dr}{r^5}, \tag{4.14}$$

which, except for the constant $1/\pi$, is forced on us by dilation, rotation and translation invariance. Summing over pairs of electrons, we arrive at

$$\frac{1}{2}\sum_{i\neq j}\frac{1}{|x_i-x_j|}=\frac{1}{2\pi}\iint \mathcal{N}\,(\mathcal{N}-1)\,\frac{dz\,dr}{r^5} \tag{4.15}$$

where $\mathcal{N}$ is the random variable (in the probability measure given by the expression $|\psi(x_1,\ldots,x_N)|^2\,dx_1\ldots\,dx_N$) of the number of electrons in the ball $\mathrm{B}(z,r)$

$$\mathcal{N}=\mathcal{N}(x_1,\ldots,x_N;z,r)=\sum_{i=1}^{N}\chi_{\mathrm{B}(z,r)}(x_i). \tag{4.16}$$

Note that the integrand on the r.h.s. of (4.15) is the number of pairs of electrons in $\mathrm{B}(z,r)$.

We expand the r.h.s. of (4.15) around an approximate mean value $\bar{\mathcal{N}}$:

$$\frac{1}{2}\sum_{i\neq j}\frac{1}{|x_i-x_j|}=\frac{1}{2\pi}\iint\left((\mathcal{N}-\bar{\mathcal{N}})^2+\mathcal{N}\,(2\bar{\mathcal{N}}-1)-\bar{\mathcal{N}}^2\right)\frac{dz\,dr}{r^5} \tag{4.17}$$

Since $\bar{\mathcal{N}}$ is independent of the x_i, of the three terms in the integrand above, the first is a two-body potential, the second is one-body and the last is a constant. The two-body part is hard to understand, but its contribution is non-negative. Therefore, dropping the two-body part, we arrive at (4.12), where V is essentially ϕ_{TF}, provided we pick $\bar{\mathcal{N}}(z,r)=\int_{\mathrm{B}(z,r)}\rho_{\mathrm{TF}}\,dx$. However a better choice of $\bar{\mathcal{N}}$ which also leads to Lieb's inequality is

$$\bar{\mathcal{N}}(z,r)=\langle\mathcal{N}\rangle_\psi=\int_{\mathrm{B}(z,r)}\rho_\psi\,dx.$$

Below we use the notation $\langle A\rangle_\psi=\langle\psi,A\psi\rangle$. Denote

$$I_\psi=\left\langle\frac{1}{2}\sum_{i\neq j}|x_i-x_j|^{-1}\right\rangle_\psi \tag{4.18}$$

Restricting the integration in (4.17) to $\{r\geq \mathrm{R}(z)\}$, with $\mathrm{R}(z)$ to be chosen later, and dropping the first term of the sum on the r.h.s., we obtain

$$I_\psi\geq\frac{1}{2\pi}\iint_{r\geq \mathrm{R}(z)}\left(\langle\mathcal{N}\rangle_\psi^2-\langle\mathcal{N}\rangle_\psi\right)\frac{dz\,dr}{r^5} \tag{4.19}$$

Using that $\langle\mathcal{N}\rangle_\psi=\int_{\mathrm{B}(z,r)}\rho_\psi$, we find

$$\frac{1}{2\pi}\iint\langle\mathcal{N}\rangle_\psi^2=\frac{1}{2}\iint\frac{\rho_\psi(x)\rho_\psi(y)}{|x-y|}. \tag{4.20}$$

Hence

$$I_\psi\geq\frac{1}{2}\iint\frac{\rho_\psi(x)\rho_\psi(y)}{|x-y|}-A, \tag{4.21}$$

where

$$A=\frac{1}{2\pi}\iint_{r\leq \mathrm{R}(z)}\langle\mathcal{N}\rangle_\psi^2+\frac{1}{2\pi}\iint_{r\geq \mathrm{R}(z)}\langle\mathcal{N}\rangle_\psi \tag{4.22}$$

Using the definition of maximal function (see [**88**]) and notation f^* for the maximal function for a function f, we estimate

$$\langle \mathcal{N} \rangle_\psi \le \frac{4\pi}{3} r^3 \rho_\psi^*(z)$$

This yields

$$A \le \frac{1}{2\pi} \int \mathrm{d}z \Big[\int_0^{\mathrm{R}(z)} \frac{\mathrm{d}r}{r^5} \Big(\frac{4\pi}{3} r^3 \rho_\psi^*(z) \Big)^2 + \int_{\mathrm{R}(z)}^\infty \frac{\mathrm{d}r}{r^5} \frac{4\pi}{3} r^3 \rho_\psi^*(z) \Big].$$

Integrating out the r.h.s. and choosing $\mathrm{R}(z) = \big(4\pi \rho_\psi^*(z)/6\big)^{-1/3}$, we arrive at

$$A \le \frac{1}{\pi} \Big(\frac{4}{3}\Big)^{4/3} \int \rho_\psi^*(z)^{4/3} \mathrm{d}z \tag{4.23}$$

Finally, using the inequality $\|f^*\|_p \le A_p \|f\|_p$, in which $1 < p \le \infty$ and the constant A_p is given by $A_p = 2\big(5^3 p/(p-1)\big)^{1/p}$, we arrive at (4.15). □

DISCUSSION. To understand the error that such an inequality sacrifices, take a ball of radius $R \sim C|Z|^{-2/3}$, C large, which is expected to contain a fixed, large number of electrons. The term dropped, $(\mathcal{N} - \bar{\mathcal{N}})^2$ is approximately the variance of $\mathcal{N}$, which, if electrons were independent, would be like $\bar{\mathcal{N}} \sim$ large constant. For this ball, $dz\, dr/r^5$ contributes approximately R^{-1}, this giving an error of about $Z^{2/3}$. Since there are about Z of these balls, the total error is like $Z^{5/3}$.

Lower bound in (4.6). Now we prove the lower bound from (4.6). To estimate the last term in (4.15) we use first the Hölder inequality

$$\begin{aligned} \int \rho_\psi^{4/3} &\le \Big(\int \rho_\psi^{5/3} \Big)^{1/2} \Big(\int \rho_\psi \Big)^{1/2} \\ &= N^{1/r2} \Big(\int \rho_\psi^{5/3} \Big)^{1/2}. \end{aligned}$$

We assume now that ψ is s.t.

$$\int \rho_\psi^{5/3} \le \mathrm{C} N^{7/3} \tag{4.24}$$

with the constant C independent of N and M. Then

$$\int \rho_\psi^{4/3} \le \mathrm{C} N^{5/3}. \tag{4.25}$$

This together with (4.15) yields

$$\Big\langle \psi, \sum_{i<j} |x_i - x_j|^{-1} \psi \Big\rangle \ge \frac{1}{2} \iint \frac{\rho_\psi \rho_\psi}{|x-y|} - \mathrm{C} N^{5/3} \tag{4.26}$$

with the constant independent of N and M. Now we transform

$$\begin{aligned} \frac{1}{2} \iint \frac{\rho_\psi \rho_\psi}{|x-y|} &= \frac{1}{2} \iint \frac{(\rho_\psi - \rho_{\mathrm{TF}})(\rho_\psi - \rho_{\mathrm{TF}})}{|x-y|} \\ &\quad + \iint \frac{\rho_{\mathrm{TF}} \rho_\psi}{|x-y|} - \frac{1}{2} \iint \frac{\rho_{\mathrm{TF}} \rho_{\mathrm{TF}}}{|x-y|}, \end{aligned} \tag{4.27}$$

where, remember, ρ_{TF} is the Thomas-Fermi density. Combining the last two relations, we derive

$$\left\langle \psi, \sum_{i<j} |x_i - x_j|^{-1} \psi \right\rangle \geq \iint \frac{\rho_{\text{TF}} \rho_\psi}{|x-y|} - \mathrm{D}_{\text{TF}} + \|\rho_\psi - \rho_{\text{TF}}\|^2_{\text{Coul}} - \mathrm{C}N^{5/3}. \tag{4.28}$$

Note that the first term on the r.h.s. can be written also as

$$\iint \frac{\rho_{\text{TF}} \rho_\psi}{|x-y|} = \left\langle \psi, \left(\sum_{i=1}^{N} |x_i|^{-1} * \rho_{\text{TF}} \right) \psi \right\rangle. \tag{4.29}$$

The last two relations yield the lower bound on in (4.6).

In order to complete the proof of the lower bound it remains to demonstrate (4.24) which is done in

LEMMA 4.4. *Let* $\psi \in \wedge_{i=1}^N \mathrm{L}^2(\mathbb{R}^3 \times \mathbb{Z}_2) \cap \mathrm{D}(\Sigma\Delta_{x_i})$ *and let* $\langle H \rangle_\psi \leq 0$. *Then*

$$\int \rho_\psi^{5/3} \leq CN^{1/3}|Z|^2 \tag{4.30}$$

for a numerical constant C.

PROOF [**34**]. We split the Coulomb potential (into its "large and small" parts) as

$$|x|^{-1} = V_1 + V_2, \tag{4.31}$$

where $V_1 = \min(|x|^{-1}, \lambda)$ and $V_2 = (|x|^{-1} - \lambda)_+$. Then

$$\|V_1\|_\infty = \lambda \quad \text{and} \quad \|V_2\|_{5/2} = C\lambda^{1/5}$$

and therefore

$$\int |x|^{-1} \rho \leq \lambda \|\rho\|_1 + C\lambda^{-1/5} \|\rho\|_{5/3}$$

for any positive ρ from $\mathrm{L}^1 \cap \mathrm{L}^{5/3}$. Picking $\lambda = \left(\|\rho\|_{5/3}/\|\rho\|_1\right)^{5/6}$, we find

$$\int |x|^{-1} \rho \leq C_1 \|\rho\|_1^{1/6} \|\rho\|_{5/3}^{5/6} \tag{4.32}$$

and therefore

$$\sum_{j=1}^{M} Z_j \int |x - R_j|^{-1} \rho \leq C_1 |Z| \, \|\rho\|_1^{1/6} \|\rho\|_{5/3}^{5/6}. \tag{4.33}$$

Now return to our Hamiltonian:

$$0 \geq \langle H \rangle_\psi \geq \left\langle -\sum \frac{1}{2} \Delta_{x_i} - \sum Z_j |x_i - R_j|^{-1} \right\rangle_\psi. \tag{4.34}$$

Applying the Lieb-Thirring inequality [**58**]

$$\left\langle -\frac{1}{2} \sum \Delta_{x_i} \right\rangle_\psi \geq C_2 \int \rho_\psi^{5/3} \tag{4.35}$$

to the first sum on the r.h.s. of (4.34) and the inequality (4.33) to the second sum,

$$\left\langle -\sum Z_j |x_1 - R_j|^{-1} \right\rangle_\psi = -\sum Z_j \int |x - R_j|^{-1} \rho_\psi, \tag{4.36}$$

and using that $\int \rho_\psi = N$, we arrive at

$$0 \geq C_2 \|\rho_\psi\|_{5/3}^{5/3} - C_1 |Z| N^{1/6} \|\rho_\psi\|_{5/3}^{5/6}$$

with the numerical constants coming from (4.32) and (4.35). This implies (4.30). □

This completes the proof of the lower bound in (4.6). □

Since the upper bound was proven, this completes the proof of Theorem 4.2. □

The End of the proof of Theorem 4.1. Rewriting $\mu = \rho_{\mathrm{MF}} - \rho_{\mathrm{TF}}$ with the help of (2.10) and then using (2.9), estimate $E_N = \mathrm{O}(|Z|^{8/9})$, Theorem 2.3, relation $\int \phi_{\mathrm{TF}}^{3/5} \leq C|Z|^{17/15}$ and elementary L^p estimates, we derive

$$\iint \frac{\mu\mu}{|x-y|} \leq \mathrm{C}|Z|^{5/3}. \tag{4.37}$$

Theorem 4.2 and eq. (4.37) yield equation (4.1). Theorem 4.1 is proven. □

Proof of Theorem 0.1. This theorem follows from eq. (2.4) and Theorems 2.1 and 4.1. □

5. Estimates of the One-Particle Density and One-Particle Denisity Matrix

In this section we derive estimates on the one-particle density ρ_ψ, and on the one-particle density matrix, γ_ψ, for approximate ground states ψ. In both cases the estimates are derived by comparing lower and upper bounds on certain ground state energies. The first of these estimates follows directly from the upper and lower bounds on the ground state energy derived in the previous section, while to derive the second estimate we will be aided by the Feynman-Hellmann method.

We begin with the one-particle density. eqs. (4.6) and (4.37) imply

COROLLARY 5.1. *Under the conditions of Theorem* 0.1,

$$\|\rho_\psi - \rho_{\mathrm{TF}}\|_{\mathrm{Coul}} \leq C|Z|^{5/6} \tag{5.1}$$

with C independent of Z and R.

Now we turn to the one-particle density matricies. The one-particle density matrix, γ_ψ, for a state ψ is a trace class operator on the one-particle space $L^2(\mathbb{R}^3 \times \mathbb{Z}_2)$, given by the kernel

$$\gamma_\psi(x,\sigma;x',\sigma') = N \sum_{\sigma_2 \cdots \sigma_N} \int \psi(x,\sigma,x_2,\sigma_2,\ldots,x_N,\sigma_N) \times \psi(x',\sigma',x_2,\sigma_2,\ldots,x_N,\sigma_N)\,\mathrm{d}x_2 \ldots \mathrm{d}x_N$$

i.e. γ_ψ is a multiple of a partial trace of the rank one projection $|\psi\rangle\langle\psi|$. Clearly, $0 \leq \gamma_\psi \leq N$. A more subtle property which is due to the fact that the electrons are Fermions is

$$0 \leq \gamma_\psi \leq id. \tag{5.2}$$

Another important fact, the one which served as a motivation for the approach described here is

$$\gamma_\psi \text{ is a projection } \leftrightarrow \psi \text{ is a Slater determinant.}$$

Thus $\gamma_\psi - \gamma_\psi^2$ can serve as a measure of how much ψ differs from a Slater determinant. With this motivation the second main results of this section is

THEOREM 5.2 [**5**]. *Let* ψ *be* $|Z|^2$*-approximate ground state of* $\mathrm{H}(Z,R)$, *i.e.* $\langle \mathrm{H}(Z,R)\rangle_\psi \le \mathrm{E}_{\mathrm{TF}}(Z,R)+O(|Z|^2)$. *Then*

$$\mathrm{tr}(\gamma_\psi-\gamma_\psi^2)=O(|Z|^{6/7}) \tag{5.3}$$

PROOF [Essentially [**34**]]. We observe first that for any operator π satisfying $\pi \le \mathrm{id}$ and $\mathrm{tr}\,\pi \le \mathrm{tr}\,\gamma_\psi$

$$\mathrm{tr}(\gamma_\psi-\gamma_\psi^2)\le 2\,\mathrm{tr}\,\gamma_\psi(1-\pi). \tag{5.4}$$

Indeed, we have for $\gamma=\gamma_\psi$

$$\mathrm{tr}(\gamma-\gamma^2)\le \mathrm{tr}(\gamma-\gamma\pi\gamma)=\mathrm{tr}\,\gamma(1-\pi)+\mathrm{tr}\,\gamma\pi(1-\gamma).$$

Since $\mathrm{tr}\,\gamma\pi(1-\gamma)\le \mathrm{tr}\,\pi-\mathrm{tr}\,\pi\gamma\le \mathrm{tr}\,\gamma-\mathrm{tr}\,\pi\gamma$, inequality (5.19) follows. Note that $\delta(\gamma,\pi)\equiv \mathrm{tr}\,\gamma(1-\pi)$ for π a projection can be interpreted as the degree of condensation of a general state with a one particle density matrix γ into the fermionic state given by the projection π.

We choose $\pi=\mathrm{E_P}(E_N)$ ($=\gamma_{\mathrm{MF}}$, see eq. (4.4)) so that $\mathrm{tr}\,\pi\le N=\mathrm{tr}\,\gamma_\psi$. In order to esimate $\delta(\gamma_\psi,\pi)$ we use the Feynman-Hellmann technique which we describe briefly here. First we show that for π as above and for any one particle density matrix γ (i.e. $0\le\gamma\le\mathrm{id}$ and $\mathrm{tr}\,\gamma=N$)

$$\mathrm{tr}(\mathrm{P}\gamma)\ge E_C+C\delta(\gamma,\pi)^{7/3} \tag{5.5}$$

where, motivated by (2.11), we denoted $E_C=\mathrm{E}_{\mathrm{TF}}(Z,R)+D_{\mathrm{TF}}$, the classical energy. To prove this inequality we write (with $\mu=E_N$)

$$\begin{aligned}\mathrm{tr}(\mathrm{P}\gamma)-\alpha\delta(\gamma,\pi)&=\mathrm{tr}\,\big[P-\alpha(1-\pi)\big]\gamma\\&=\mathrm{tr}(\mathrm{P}-\mu+\alpha\pi)\gamma-(-\mu+\alpha)\,\mathrm{tr}\,\gamma.\end{aligned}$$

Next, using that due to Theorem 2.2 $\mathrm{tr}\,\gamma=\mathrm{tr}\,\pi+\mathrm{O}(|Z|^{2/3})$, and that

$$\chi_{\mathrm{P}-\mu+\alpha\pi\le 0}=\chi_{\mathrm{P}-\mu+\alpha\le 0},$$

we find

$$\begin{aligned}\mathrm{tr}(\mathrm{P}\gamma)-\alpha\delta(\gamma,\pi)&\ge \mathrm{tr}(\mathrm{P}-\mu+\alpha\pi)\chi_{\mathrm{P}-\mu+\alpha\pi\le 0}\\&\qquad-(-\mu+\alpha)\,\mathrm{tr}\,\pi-C\alpha|Z|^{2/3}\\&=\mathrm{tr}(\mathrm{P}-\mu+\alpha)\chi_{\mathrm{P}-\mu+\alpha\le 0}-\mathrm{tr}(\mathrm{P}-\mu+\alpha)\pi\\&\qquad+\mathrm{tr}(\mathrm{P}\pi)-C\alpha|Z|^{2/3}.\end{aligned} \tag{5.6}$$

A quasiclassical result of the type of (2.5) and elementary estimates (see [**34**] for the latter) show that

$$\mathrm{tr}(\mathrm{P}\pi)=E_C+\mathrm{O}(|Z|^2)$$

and

$$\mathrm{tr}(\mathrm{P}-\mu+\alpha)\chi_{0\le P-\mu+\alpha\le\alpha}\le C\alpha^{7/4}+C|Z|^2.$$

Using the last two estimates for the r.h.s. of (5.21) and optimizing in α by choosing $\alpha=\delta(\gamma,\pi)^{4/3}$ (we assume here that $\delta\gg|Z|^{2/3}$, otherwise the required estimate follows immediately), we arrive at (5.5).

Next we prove that if ψ is the ground state (or $\mathrm{O}(|Z|^2)$- close to it in energy), then

$$\mathrm{tr}(\mathrm{P}\gamma_\psi)\le E_C+C|Z|^2. \tag{5.7}$$

Indeed, write

$$\operatorname{tr}(\mathrm{P}\gamma_\psi) = \big\langle \mathrm{H}(Z,R)\big\rangle_\psi - I_\psi + \left\langle \sum_{i=1}^N |x_i|^{-1} * \rho_{\mathrm{TF}} \right\rangle_\psi .$$

Using the upper bound $\langle \mathrm{H}(Z,R)\rangle_\psi \leq \mathrm{E}_{\mathrm{TF}}(Z,R) + C|Z|^2$ from eq. (0.4), and using Lieb's inequality (4.15), we obtain

$$\operatorname{tr}(\mathrm{P}\gamma_\psi) \leq \mathrm{E}_{\mathrm{TF}}(Z,R) + C|Z|^2 + C\int \rho_\psi^{4/3} - \frac{1}{2}\iint \frac{\rho_\psi\rho_\psi}{|x-y|} + \iint \frac{\rho_\psi\rho_{\mathrm{TF}}}{|x-y|},$$

which due to (4.25) and (4.27) yields (5.7).

Now combining inequalities (5.5) and (5.7), we arrive at

$$\delta(\gamma,\pi) \leq C|Z|^{6/7}, \tag{5.8}$$

which together with (5.4) yields (5.3). □

6. Improved Correlation Estimate

The correlation energy in a state ψ can be defined as $I_\psi - 1/2\iint(\rho_\psi\rho_\psi)/(|x-y|)$, where, recall, $I_\psi = \langle 1/2\sum_{i\neq j}|x_i - x_j|^{-1}\rangle_\psi$, i.e. the Coulomb energy of electron-electron interaction in ψ minus the Coulomb energy of the one-electron density ρ_ψ. The latter is called the direct interaction and denoted by $D(\rho_\psi,\rho_\psi)$. It is a key problem in the theory of Coulomb systems to estimate the correlation energy in states close to the ground state. It was shown in Section 4 that for $\psi \in \bigwedge_{i=1}^N L^2(\mathbb{R}^3\times\mathbb{Z}_2)\cap D(\sum_{i=1}^N \Delta_{x_i})$ s.t. $\langle \mathrm{H}(Z,R)\rangle_\psi \leq 0$, the correlation energy is bounded below by $\mathrm{O}(|Z|^{5/3})$. The purpose of this section is to isolate the $\mathrm{O}(|Z|^{5/3})$ contribution to the correlations. This was first done in [**25–31**]. A stronger result by a different and simpler method was proven in [**5**], whose proof was further simplified and generalized in [**34**]. In our exposition we follow the latter work.

It is crucial in Bach's approach to pass from densities ρ_ψ to densitiy matricies γ_ψ. Now we can formulate the main result of this section

THEOREM 6.1. *Let $\psi \in \bigwedge_{i=1}^N L^2(\mathbb{R}^3\times\mathbb{Z}_2)\cap D(-\sum_{i=1}^N \Delta_{x_i})$ and satisfy the inequality $\langle (Z,R)\rangle_\psi \leq \mathrm{E}_{\mathrm{TF}}(Z,R) + \mathrm{O}(|Z|^2)$. Then*

$$I_\psi \geq \mathrm{Direct}_\psi + \mathrm{Exch}_\psi + O(|Z|^{5/3-\delta}), \tag{6.1}$$

where

$$\mathrm{Direct}_\psi = \frac{1}{2}\iint \frac{\rho_\psi\rho_\psi}{|x-y|}$$

and

$$\mathrm{Exch}_\psi = -\frac{1}{2}\sum_\sigma \iint \frac{|\gamma_\psi(x,\sigma;y,\sigma)|^2}{|x-y|}.$$

DISCUSSION. 1) As the notation suggests the main contribution comes from the exchange term. This term is due to the permutational symmetry of ψ (i.e. the symmetry w.r. to exchange of the particles) as can be discerned from considering Slater determinants:

$$\mathrm{slater} = \frac{1}{\sqrt{N!}}\psi_1\wedge\cdots\wedge\psi_N.$$

Indeed, in this case we have exactly

$$I_{\text{slater}} = \text{Direct}_{\text{slater}} + \text{Exch}_{\text{slater}}.$$

On the other hand, if $\psi = \psi_1 \otimes \cdots \otimes \psi_N$, then

$$I_\psi = \text{Direct}_\psi - SE_\psi,$$

where SE_ψ, the self-energy in ψ is given by

$$SE_\psi = \sum_{i=1}^{N} \int \frac{|\psi_i|^2 |\psi_i|^2}{|x-y|}.$$

Thus the difference, $\text{Exch}_{\text{slater}} + SE_{\text{slater}}$, is due to the antisymmetrization of $\psi_1 \otimes \cdots \otimes \psi_N$.

2) If ψ is close to the ground state of $\text{H}(Z, R)$, e.g. if it satisfies the conditions of Theorem 6.1, then

$$\frac{1}{2} \iint \frac{|\gamma_\psi(x,y)|^2}{|x-y|} = c_D \int \rho_{\text{TF}}^{4/3} + \left(|Z|^{5/3-\delta}\right) \tag{6.2}$$

for some $\delta > 0$, where $c_D = 3/4\pi\left((6\pi^2)/2\right)^{1/3}$, the coefficient computed by Dirac [**18**] in the model of the Fermi gas on the torus (in a box with periodic boundary conditions). Note that by the scaling property of ρ_{TF}

$$\int \rho_{\text{TF}}^{4/3} = C|Z|^{5/3} \tag{6.3}$$

with C depending only on $\lambda = Z|Z|^{-1}$ and $y = R|Z|^{1/3}$.

3) Dirac's idea was a starting point of the Fefferman-Seco proof. One can think about an atom, or molecule with internuclear distances of $\left(|Z|^{-1/3}\right)$, being placed in a box of size $\left(|Z|^{-1/3+\delta_1}\right)$ for some $\delta_1 > 0$. The exterior of the box does not contribute much into the ground state. Now divide this box into much smaller boxes of the size $\left(|Z|^{-2/3+\delta_2}\right)$ with $\delta_2 > \delta_1$ which are still very large compared to the interelectron distances which are $\text{O}\left(\left[\left(|Z|^{-1/3+\delta_1}\right)^3 |Z|^{-1}\right]^{1/3}\right) = O\left(|Z|^{-2/3+\delta_1}\right)$. Consider a state, ψ, which is sufficiently close to the ground state in energy. Such a state should be well approximated by a state, ψ_{TF}, which on most of smaller boxes looks like the Fermi gas with the density given by the Thomas-Fermi density, say, in the center of the box. Otherwise the energy of ψ would be different from the ground state energy in the leading order, which is given by the Thomas-Fermi energy. In particular, one shows that the correlation energy of ψ is given to the leading order by the correlation energy of ψ_{TF}. The latter is computed to the leading order to be $-c_D \int \rho_{\text{TF}}^{4/3}$.

4) In order to demonstrate the exchange effect in the general case, we define the operator Ex on $L^2(\mathbb{R}^3 \times \mathbb{Z}_2) \otimes L^2(\mathbb{R}^3 \times \mathbb{Z}_2)$ by extending the map $u \otimes v \to v \otimes u$ linearly and continuously. Thus the direct and exchange terms can be written as

$$\text{Direct}_\psi = \text{Tr}\left[V(\gamma_\psi \otimes \gamma_\psi)\right] \tag{6.4}$$

and

$$\text{Exch}_\psi = \text{Tr}\left[V\,\text{Ex}(\gamma_\psi \otimes \gamma_\psi)\right], \tag{6.5}$$

where V acts on $L^2(\mathbb{R}^3 \times \mathbb{Z}_2) \otimes L^2(\mathbb{R}^3 \times \mathbb{Z}_2)$ according to the following relation $(V\psi)(x,\sigma,x',\sigma') = V(x-x')\psi(x,\sigma;x',\sigma')$ and Tr stands for the trace on the product (two particle) space.

5) It is instructive to introduce also the two particle density matrix, Γ_ψ, which is defined in the same way as γ_ψ but leaving two sets of variables out of the integration and replacing the factor N by $N(N-1)$. Taking a partial-one particle-trace of Γ_ψ we obtain

$$\operatorname{tr}\Gamma_\psi = (N-1)\gamma_\psi,$$

where tr stands for the trace (on the one particle space) $L^2(\mathbb{R}^3 \times \mathbb{Z}_2)$, while the two particle trace gives

$$\operatorname{Tr}\Gamma_\psi = N(N-1), \tag{6.6}$$

the number of all pairs. We have $0 \le \Gamma_\psi \le N(N-1)$. One can show that (see [**14**])

$$\gamma_\psi^2 = \gamma_\psi \leftrightarrow \Gamma_\psi = (1-\mathrm{Ex})(\gamma_\psi \otimes \gamma_\psi). \tag{6.7}$$

This property motivates one to introduce the truncated two particle density matrix

$$\Gamma_\psi^T = \Gamma_\psi - (1-\mathrm{Ex})(\gamma_\psi \otimes \gamma_\psi), \tag{6.8}$$

which measures, in a sense, how close ψ is to a Slater determinant. In particular, we can write

$$I_\psi - \mathrm{Direct}_\psi - \mathrm{Exch}_\psi = \operatorname{Tr}(V\Gamma_\psi^T). \tag{6.9}$$

Note that by Lieb's variational principle (see [**51**] and also [**5**]), $\operatorname{tr}(\Gamma_\psi^T V) \le 0$. Thus our task can be formulated as estimating $\operatorname{Tr}(V\Gamma_\psi^T)$ below. Relation (6.7) suggests that we should try to estimate it in terms of $\gamma_\psi - \gamma_\psi^2$. Indeed, while Γ_ψ^T is inaccessible directly, $\gamma_\psi - \gamma_\psi^2$ can be estimated directly.

We will not give the proof of Theorem 6.1 here. This can be found in [**5, 34**]. Instead we list main ideas, with some comments, entering it.

IDEA 1 [DECOUPLING]. The operator V couples two one-particle (factor) spaces. We reduce it to the product of two operators acting on individual factors by using the Fefferman-de la Llave formula

$$|x-y|^{-1} = \frac{1}{\pi}\iint X_{z,r}(x)X_{z,r}(y)\frac{dz\,dr}{r^5}, \tag{6.10}$$

where $X_{z,r} = \chi_{B(z,r)}$. Then estimating $\operatorname{Tr}(V\Gamma^T)$ is reduced to estimating $\operatorname{Tr}(X_{z,r} \otimes X_{z,r})\Gamma^T$. In particular, one would like to prove Bach's inequality

$$\operatorname{Tr}(X \otimes X)\Gamma^T \ge -\operatorname{tr}(X\gamma)\min\left\{1, \sqrt{\frac{77}{2}}\left[\operatorname{tr}X(\gamma-\gamma^2)\right]^{1/2}\right\}, \tag{6.11}$$

where X is a projection and where we dropped the subindex ψ at Γ^T and γ. Thus the term containing two-electron correlations is estimated in terms of one-particle density matrix, γ, and, in particular, how close it is to a projection.

Note that for $X \otimes X$ the direct and exchange terms have a simple structure

$$\operatorname{Tr}(X \otimes X)(\gamma \otimes \gamma) = (\operatorname{tr}X\gamma)^2$$

and

$$\operatorname{tr}(X \otimes X)\mathrm{Ex}(\gamma \otimes \gamma) = \operatorname{tr}(X\gamma X\gamma).$$

It is convenient to go back to a more detailed form of $\mathrm{Tr}(X \otimes X)\Gamma$, namely

$$\mathrm{tr}(X \otimes X)\Gamma_\psi = \left\langle \sum_{i \neq j} X_i X_j \right\rangle_\psi, \tag{6.12}$$

where X_i denotes the operator X acing on the i-th factor in the product space $\bigwedge_{i=1}^N L^2(\mathbb{R}^3 \times \mathbb{Z}_2)$.

Now we sketch the proof (following [**34**]) of Bach's inequality (6.11).

IDEA 2 [COMPLETING THE SQUARE]. As in the proof of Lieb's inequality we want to complete the square so that the 2-particle terms appear in the form $|\sum Y_i|^2$ for some one-particle operators Y_i's, e.g.

$$\sum_{i \neq j} Y_i^* Y_j = \left|\sum Y_i\right|^2 - \sum |Y_i|^2. \tag{6.13}$$

This allows us to get rid of the 2-particle terms by expanding the square around its the mean value of $\sum Y_i$:

$$\left|\sum Y_i\right|^2 = \alpha \sum (Y_i + Y_i^*) - \alpha^2 + \left|\sum Y_i - \alpha\right|^2$$

and dropping the positive term (fluctuations) $\left|\sum Y_i - \alpha\right|^2$ which contains the 2-particle interactions.

IDEA 3 [MANAGING THE WASTE]. We could have completed the square in (6.12); however the waste $\left|\sum X_i - \alpha\right|^2$ would be too large no matter what α we chose. To minimize the waste we first split X into smaller pieces as

$$X = (P + Q)X(P + Q), \tag{6.14}$$

where $0 \leq P \leq \mathrm{id}$ and $Q = \mathrm{id} - P$, sort out different terms and then complete the square. (Note that there are always two ways to complete the square, e.g. as (6.13) or as $\sum_{i \neq j} Y_i^* Y_j = \sum_{i \neq j} Y_i Y_j^* = \left|\sum Y_i^*\right|^2 - \sum |Y_i^*|^2$. Usually, only one of the choices leads to the desired result.) The choice $P = \gamma$ leads to Bach's inequality, but there are other possibilities conveinent for obtaining the final result.

IDEA 4 [FERMIONS]. The fact that our Hilbert space is Fermionic is used in the inequality:

$$A \geq 0 \implies \sum_{i=1}^N A_i \leq \mathrm{tr}\, A. \tag{6.15}$$

Its proof is straightforward. Another way to see this inequality is to observe that $\left\langle \sum_{i=1}^N A_i \right\rangle_\psi = \mathrm{tr}(A\gamma_\psi)$, then (6.15) follows from the fact $\gamma_\psi \leq \mathrm{id}$, which characterizes the Fermion Hilbert space. This is the only place where the fact that the electrons are Fermions enters into the proof.

This completes our comments on the proof of inequality (6.11). Now we sketch the derivation of (6.1) from this inequality. First, combining inequality (6.11) with formula (6.10) and proceeding with the use of maximal functions similarly as in the proof of Lemma 4.3, we obtain

$$I_\psi \geq \mathrm{Direct}_\psi - \mathrm{Exch}_\psi - \mathrm{Rem}_\psi, \tag{6.16}$$

where the remainder is

$$\mathrm{Rem}_\psi = C\|\rho_\psi\|_{5/3}^{5/6}\|\rho_\psi\|_1^{1/6+\varepsilon}\big[\mathrm{tr}(\gamma_\psi - \gamma_\psi^2)\big]^{1/3-\varepsilon} \tag{6.17}$$

By the definition $\|\rho_\psi\|_1 = N = |Z|$, and, by (4.24), $\|\rho_\psi\|_{5/3}^{5/6} \le C|Z|^{7/3}$. Thus it remains to estimate the last factor on the r.h.s. of (6.17). This is done in the previous section.

Now combining estimates (6.16), (6.17) with (5.3) and remembering that $\|\rho_\psi\|_1 = |Z|$ and, by (4.24), $\|\rho_\psi\|_{5/3}^{5/6} \le C|Z|^{7/6}$, we arrive at (6.1). □

Proof of (6.2).
To prove (6.2) one can use another estimate of Bach, valid for any $\varepsilon \in (0, 1/6)$,

$$\mathrm{Tr}\, V\, \mathrm{Ex}(\gamma\otimes\gamma - \pi\otimes\pi) \le C\varepsilon^{-1/2}\big[\mathrm{tr}\,\gamma(1-\pi)\big]^{1/2-\varepsilon}N^{1/6-\varepsilon}|Z| \tag{6.18}$$

together with (5.8) and the one-body quasiclassical estimate

$$\left|\frac{1}{2}\int\frac{|\pi(x,y)|^2}{|x-y|} - c_D\int\rho_{\mathrm{TF}}^{4/3}\right| \le C|Z|^{5/3-\delta} \tag{6.19}$$

for some δ, whose proof we skip. Note that evoking (6.18) can be avoided if, following [**34**], we chose $\mathrm{P} = \pi$, or another suitable projection, in (6.14) to begin with and prove directly a variant of (6.11) in which $\mathrm{Tr}\big[(X\otimes X)\,\mathrm{Ex}(\gamma\otimes\gamma)\big]$ is replaced by $\mathrm{Tr}\big[(X\otimes X)\,\mathrm{Ex}(\pi\otimes\pi)\big]$.

The Dirac-Schwinger Hamiltonian. Finally, we modify $\mathrm{H}^{\mathrm{ind}}(Z,R)$ by subtracting from it the (Z-dependent) constant $c_D\int\rho_{\mathrm{TF}}^{4/3}$ responsible the exchange correlation:

$$\mathrm{H}^{\mathrm{DS}}(Z,R) = \mathrm{H}^{\mathrm{ind}}(Z,R) - c_D\int\rho_{\mathrm{TF}}^{4/3}, \tag{6.20}$$

the Dirac-Schwinger Hamiltonian. Let $\mathrm{E}_{\mathrm{DS}}(Z,R) = \inf \mathrm{H}^{\mathrm{DS}}(Z,R)$. We note that eqs. (6.1) and (6.2) together with an appropriate upper bound (using Slater determinants and Hartree-Fock functionals as in Section 4) yield

THEOREM 6.2 [Fefferman and Seco 1993 and Bach 1993]. *There are δ and C independent of Z, R and M so that*

$$\begin{aligned} -CM^2|Z|^{5/3-\delta} &\le \mathrm{E}(Z,R) - \mathrm{E}_{\mathrm{DS}}(Z,R) \\ &\le CM^2|Z|^{5/3-\delta} + \|\rho_{\mathrm{MF}} - \rho_{\mathrm{TF}}\|_{\mathrm{Coul}}^2, \end{aligned} \tag{6.21}$$

where $\rho_{\mathrm{MF}}(x) = \pi(x,x)$ (see (4.3)).

7. Binding

Binding is the property of matter to form stable compounds, such as atoms, molecules, nuclei, etc. It is measured by a gap between infspec H and infcontspec H. The gap is called the *ionization energy*,

$$\mathrm{IE} = \inf \mathrm{contspec}\, H - \inf \mathrm{spec} H.$$

In fact, this is the energy needed to destroy the atom in the ground state, the one corresponding to $\inf \mathrm{spec}\, H$.

Understanding how atoms (and molecules) bind is one of the central problems. We will not discuss this problem in its full extent here since it departs somewhat from the subject matter of these lectures. We will mention one important aspect of it which uses the main result and techniques of these lectures. It follows from Theorem 0.1 that

$$-\inf \operatorname{spec} = \mathrm{O}(|Z|^{7/3}).$$

On the other hand it is believed that IE $= \mathrm{O}(1)$.. Thus the existence of matter as we know it is due to a subtle phenomenon, indeed. It was shown in [**73**] for atoms and in [**71, 84**] for molecules that

THEOREM 7.1.

$$\mathrm{IE} \leq \mathrm{const}|Z|^{20/21}$$

The proof of this theorem is rather instructive. It uses Theorem 0.1 as well as the tools described in Section 4 and more.

Consider now an atom. Another way to measure binding is by the maximal number of electrons a nucleus of charge z can hold together, $N(z)$. It was shown in [**68, 78**] that $N(z)$ is finite. Moreover one has

THEOREM 7.2.

$$N(z) = z + \mathrm{O}(z^{5/7}).$$

It was shown in [**11, 84**] that if the electrons were bosons then

$$N^{\mathrm{boson}}(z) = 1.21z + o(z)$$

(the numerical value of the coefficient in front of z was computed in [**8**]). [**5**] proves that

$$\mathrm{IE}^{\mathrm{boson}} = \mathrm{O}(z^2).$$

Thus Pauli principle plays a crucial role in the formation of atoms.

The asymptotic behaviour of $N(z)$ was established in [**55**] and the remainder estimate was derived in [**23**]. A simpler proof is given in [**73**]. Similar results were established also for molecules [**69, 71, 85**].

We explain the main steps in the proof of Theorems 7.1 and 7.2 and the relation of this proof to the ground state problems. Given an approximate ground state, ψ, we define, besides of the one-electron density ρ_ψ, the two-electron density σ_ψ by

$$\sigma_\psi(x_1, x_2) = N(N-1) \sum_{\sigma_1, \dots, \sigma_N} \int |\psi(x_1, \sigma_1, \dots, x_N, \sigma_N)|^2 \, dx_3 \dots dx_N.$$

Now we define the unscreened portion of the nuclear charge at the distance R from the nucleus by

$$\nu(R) = Z - \int_{|x| \geq R} Q \Big/ \int_{|x| \geq R} \rho,$$

where $Q(x)$ is, roughly, a square of the charge distribution:

$$Q(x) = |x| \int \frac{\sigma_\psi(x, y)}{|x - y|} \, dy.$$

[**73**] proves the following bound

$$IE \leq \nu(R)/R + C/R^2$$

with C independent of Z and R. The first term on the r.h.s. is the work required to pull one electron from the distance R from the nucleus to infinity, while the last term on the r.h.s. is an error term.

Now it remains to estimate $\nu(R)$, which is done by using an estimate on $\rho-\rho_{\mathrm{TF}}$ implied by eq. (5.1) and the following correlation estimate

$$\left|\iint_{|x|\geq R,|y|\leq K}(\sigma_\psi(x,y)-\rho_\psi(x)\rho_{\mathrm{TF}}(y))\right|\leq CK^{1/2}\Delta(Z)\int_{|x|\geq R}\rho_\psi,$$

where $\Delta(Z)$ is the absolute value of the difference between the ground state energies of the atomic Hamiltonian and of the corresponding mean field Hamiltonian. Theorem 4.1 then yields that $\Delta(Z)\leq \mathrm{const}Z^{5/3}$ which implies the desired bound on σ_ψ and with it the desired bound on $\nu(R)$.

8. Open Problems

I. The next term. The problems of finding the $Z^{4/3}$-term for atoms and the $|Z|^{5/3}$-term for molecules are open. A preliminary result on the former problem is obtained in [**17**]. We discuss here the latter one. Theorem 5.3 reduces this problem to a one-body one, namely about the sum of negative eigenvalues and about the (one particle) density for the operator P (with $|Z|\to\infty$ or for K_β with $\beta\to 0$). To isolate the $|Z|^{5/3}$- (or β^{2-d}-) term one has to control the periodic orbits for the corresponding Hamiltonian flow. This is a new element relative to the analysis described in Section 3. It consists of two independent problems:

(a) Show that the set of periodic orbits in a small, Z-dependent neighbourhood of the zero energy is of measure zero.

(b) Assuming (a), obtain the two term (not counting Scott) expansion in (2.5) with a remainder $o(|Z|^{5/3})$.

The first problem belongs to Classical Mechanics while the second to PDE. [**25–31**] solve both problems in the context of atoms. Their analysis relies on the fact that the potential in P in this case is spherically symmetric. Indeed, the spherically symmetric case is the only case in which Classical Mechanics yields reasonable information about periodic orbits. In fact, it is only the case when it yields any information. Furthermore, the quasiclassical, or WKB, techniques in one dimension use extensively explicit computations.

The first step in analysis of problems (a) and (b) is to realize that they are wrongly and/or badly formulated. We start with the problem (a). First of all, because of uncertainty principle, QM does not distinguish between closed and "almost" closed orbit (remember, we are talking about the *phase-space* orbits). More precisely, remember that each point x has the length $\ell(x)$, momemtum $f(x)$, and action $\beta=|Z|^{-1/3}$, scales. Thus its dimensionless quasiclassical parameter is

$$\alpha(x)=\frac{|Z|^{-1/3}}{f(x)\ell(x)}$$

Moreover, one has to remember also the time scale

$$T(x)=\ell(x)/f(x).$$

Secondly, we do not care at all whether the orbit is periodic or not as long as it does not revisit a neighbourhood of x of interest for sufficiently long time, say $O(|\ln\alpha(x)|)$, on the time scale $T(x)$, or course. We also have to be sure that the orbits issuing from our neighbourhood stay away from an appropriate neighbourhood of singularities for similarly long time. This discussion makes it clear that problem (a) should be reformulated as follows:

(a′) Consider a ball $B(z,\ell(z))$ outside of an appropriate (say, $|Z|^{-2/3}$-) neighbourhood of singularities. Show that there is a time direction (depending on z) s.t. the classical orbits issuing from $\mathrm{B}(z,\ell(z))$ do not approach closely the singularities and do not return to $\mathrm{B}(z,\ell(z))$ for the time $\tau(z)$ s.t. $\tau(z)/T(z)\to\infty$ as $|Z|\to\infty$.

Now we turn to the problem (b). Remember that the philosophy of Section 3 is that we do analysis only for a standard problem on the unit ball, $\mathrm{B}(0,1)$. Thus problem (b) should be first localized to an arbitrary ball, $\mathrm{B}(z,\ell(z))$, outside the singularities and then rescaled to a unit ball. Taking into account (a′), we arrive at

(b′) Assume the trajectories for the Classical Hamiltonian $h(x,\xi)=|\xi|^2/2-W(x)$ starting in $\mathrm{B}(0,1)\times\mathrm{B}(0,K)$, where $K=2\big(\sup_{\mathrm{B}(0,2)}W(x)\big)_+^{1/2}+1$, exit and do not revisit this set for the time $T(\alpha)$ s.t. $T(\alpha)\to\infty$ as $\alpha\to 0$. Then $\operatorname{tr}\psi g_{s=1}(H_\alpha)$ (see Theorem 3.1) admits the two term asymptotics (the α^{-d}- and α^{2-d}-terms) with the remainder $o(\alpha^{2-d})$, at least for $d=3$.

In the absence of spherical symmetry, problem (a′) is a tough one. Problem (b′) is better understood (see [**12, 19, 42, 64, 66**] for various related results). In particular, results of Section 3 (supplemented by some estimates concerning propagation of singularities) can be used to tackle this problem.

II. Correlation estimate. Isolate the next $\big(|Z|^{4/3}\big)$ term in the correlation inequality (5.1). Besides of the formal perturbation theory around the Hartree-Fock wave function which yields a term of the order $|Z|$ there is no physical argument to what such a term should be and what its meaning is. One can see already on the simplified Jellium-case how hard this problem is.

III. Relativistic corrections. For heavy atoms the relativistic correction becomes more significant. A non-rigorous and rough estimate puts them at $5\cdot 10^{-4}Z^{13/6}\%$ of the ground state energy for the atoms (see [**74**]), which gives $\approx 0.06\%$ for Sodium ($Z=11$) and $\approx 4.5\%$ for Uranium ($Z=92$). The problem here is how to understand these corrections without going to the full-fledged Quantum Electrodynamics. One can try the caricature Hamiltonian

$$H(N,V)=\sum_{i=1}^{N}\big(\sqrt{-\Delta_{x_i}+m^2}+V(x_i)\big)+\frac{1}{2}\sum_{i\neq j}|x_i-x_j|^{-1} \tag{8.1}$$

on $\bigwedge_{i=1}^N L^2(\mathbb{R}^3\times\mathbb{Z}_2)$, where $V(x_i)$ is the potential of the interaction of the i-th electron with the nuclei. Here we replaced the non-relativistic electron kinetic energy $-\Delta_x/(2m)$ by the relativistic one $\sqrt{-\Delta_x+m^2}$, but otherwise did not change the Hamiltonian. We will not discuss the merits of this model. It is often used as a testing ground to see the relativistic effects in a quantum mechanical picture. For the question of stability for this system see [**16, 21, 60–61**], a review of the subject can be found in [**52**] (many of the papers mentioned above are included in [**53**]). The problem here is that for "atoms" with $Z>2/\pi$ this Hamiltonian is unbounded below (the electrons fall onto the nucleus) so that the large Z asymptotics as was understood before does not make sense. However, I expect that this Hamiltonian has a *resonance* near the ground state energy of its non-relativistic counterpart. Indeed, the electrons initially localized away from the nucleus would think for a long time that they are non-relativistic until they tunnel into the inner core of the atom where the velocities are appreciably relativistic. To prove (or disprove) this and find the asymptotics of the resonance energy and life-time is an interesting problem. A relativistic TF Theory?

IV. From large-Z asymptotics to stability of matter (W. Thirring, B. Helffer). Remove or weaken the condition $\min Z_j \geq \delta_1 |Z|$ in Theorem 0.1 keeping at the same time a reasonable estimate of the remainder. Uniformity in $\lambda = Z|Z|^{-1}$ is a delicate point since some of the λ_j's might vanish as $|Z| \to \infty$. Optimally, the remainder should be such that for fixed Z_j's but with M arbitrary the estimate replacing (0.3) should turn into stability of matter estimate.

V. Binding. Show that the ionization energy for atoms and molecules is $I = \mathrm{O}(1)$ as $Z \to \infty$ and that the maximal number, $N(Z)$, of bound electrons is $N(Z) = |Z| + \mathrm{O}(1)$. There is no physical argument justifying these conjectures, only the experimental evidence for real atoms. Note that the latter does not contradict any other power law, provided that the coefficient is sufficiently small.

VI. Rigidity of molecules. Estimate the constant in the remainder in inequality (4.1) or in (6.21). Understand why such a complex object as a molecule is so well approximated in the ground state by a simple mean field theory as one given by (2.1) or by (6.20). Is the rest of the spectrum of $\mathrm{H}^{\mathrm{ind}}(Z, R)$, or of $\mathrm{H}^{\mathrm{DS}}(Z, R)$, relevant for the molecular Hamiltonian $\mathrm{H}(Z, R)$? Note that in the latter case one might want to replace in $\mathrm{H}^{\mathrm{ind}}(Z, R)$ (i.e. in P as well) the Thomas-Fermi potential for the neutral system of $N = |Z|$ electrons by the Thomas-Fermi potential for an ion with $|Z| - 1$ electrons.

Acknowledgement

These notes present the slightly expanded material of three lectures given at the Vancouver conference on Field Theory and Quantum Mechanics. The bulk of the notes was prepared while the author was visiting Institut der Theoretische Physik, ETH-Zürich. The author is grateful to J. Feldman, R. Froese and L. Rosen for organizing a stimulating conference, and to J. Fröhlich and W. Hunziker for hospitality at Zürich. The author is indebted to G.-M. Graf for many instructive discussions and, in particular, for teaching him several proofs incorporated in these notes, and to V. Bach and G.-M. Graf for reading the manuscript and making numerous useful comments. The birthday anniversary gave the author and an opportunity to acknowledge his debt to and admiration for his two colleagues and friends whose contributions determined the subject.

References

1. C. Albanese, *On the accuracy of the Schwinger-Englert harmonic approximation in the semi-classical theory of atoms and molecules*, 1993.
2. V. Arnold, *Mathematical methods of classical mechanics*, 2nd edition, Springer, 1989.
3. K. Asada and D. Fujiwara, *On some oscillatory integral transformations in* $L^2(\mathbb{R}^n)$, Japan J. Math. **4** (1978), 299–361.
4. V. Bach, *Ionization energies of bosonic Coulomb systems.*, Lett. Math. Phys. **21** (1991), 131–149.

5. ——, *Error bound for the Hartree-Fock energy of atoms and molecules*, Comm. Math. Phys. **147** (1992), 527–548.
6. ——, *Accuracy of mean field theory approximation for atoms and molecules*, Comm. Math. Phys. (1993).
7. V. Bach, R. Lewis, E. H. Lieb, and H. Siedentop, *On the number of bound states of a bosonic Coulomb systesm*, Math. Z. (1993) (to appear).
8. B. Baumgartner, *On the Thomas-Fermi-von-Weizsäcker and Hartree energies as functions of the degree of ionization*, J. Phys. A **17** (1984), 1593–1602.
9. R. Beals, *A general calculus of pseudo-differential operators*, Duke Math. J. **42** (1975), 1–42.
10. R. Beals and C. Fefferman, *Spatially inhomogenous pseudodifferential operators*, Pure Appl. Math. **27** (1974), 1–24.
11. R. Benguria and E. H. Lieb, *Proof of the stability of highly negative ions in the absence of the Pauli principle*, Phys. Rev. Lett. **50** (1983), 1771–1774.
12. J. Chazarain, *Formules de Poisson pour les variétés riemaniennes*, Invent. Math. **24** (1974), 65–87.
13. ——, *Spectre d'un hamiltonien quantique et méchanique classique*, Comm. Partial Differential Equations **5** (1980), no. 6, 595-644.
14. A. J. Coleman, *Structure of fermion density matricies*, Rev. Modern Phys. **35** (1963), no. 3, 668–689.
15. J. Conlon, *Semiclassical limit theorems for Hartree-Fock theory*, Comm. Math. Phys. **88** (1983), no. 1.
16. ——, *The ground state energy of a classical gas*, Comm. Math. Phys. **94** (1984), 439.
17. A. Cordoba, Ch. Fefferman, and L. Seco, *Weyl sum and atomic energy oscillations*, Rev. Mat. Iberoamericana (to appear in 1995).
18. P. A. M. Dirac, *Note on exchange phenomena in the Thomas-Fermi atom*, Proc. Cambridge Phil. Soc. **26** (1930), 376–385.
19. J. J. Duistermaat and V. W. Guillemin, *The spectrum of positive elliptic operators and periodic bicharacteristics*, Invent. Math. **29** (1975), 39–79.
20. B.-G.Englert and J. Schwinger, *Thomas-Fermi revisited: The outer regions of the atom*, Phys. Rev. A **26** (1982), no. 3, 2322–2329.
21. C. Fefferman and R. de la Llave, *Relativistic stability of matter* I, Rev. Mat. Iberoamericana **2** (1986), 119–213.
22. C. Fefferman, V. Ivrii, L. Seco, and I. M. Sigal, *The energy asymptotics of large Coulomb systems*, Schrödinger Operators (E. Balslev, ed.), Springer-Verlag, 1992.
23. C. Fefferman and L. Seco, *Asymptotic neutrality of large ions*, Comm. Math. Phys. **128** (1990), 109–130.
24. ——, *On the energy of a large atom*, Bull. Amer. Math. Soc. **23** (1990), 525–530.
25. ——, *On the Dirac and Schwinger corrections to the ground-state energy of an atom*, Adv. in Math. **107** (1994), 1–185.
26. ——, *Eigenvalues and eigenfunctions of ordinary differential operators*, Adv. in Math. **95** (1993), no. 2, 145–305.
27. ——, *The density in a one-dimensional potential*, Adv. in Math. **107** (1993), 187–364.
28. ——, *The eigenvalue sum for a one-dimensional potential*, Adv. in Math. (1993) (to appear).
29. ——, *The density in a three-dimensional radial potential*, Adv. in Math. (1993) (to appear).
30. ——, *The egenvalue sum in a three-dimensional potential*, Adv. in Math. (1993) (to appear).
31. ——, *Aperiodicity of the Hamiltonian flow in the Thomas-Fermi potential*, Rev. Mat. Iberoamericana **9** (1993), no. 3, 407–530.
32. E. Fermi, *Un metodo statistico per la determinazione di alcune priorieta dell'atomo*, Rend. Accad. Naz. Lincei **6** (1927), 602–607.
33. J. Fröhlich, G.-M. Graf, J. M. Richard, and M. Siefert, *Proof of stability of the hydrogen molecule* (1993) (preprint).
34. G.-M. Graf and J.-P. Solovej, *A Correlation estimate with applications to quantum systems with Coulomb interactions* (1993) (preprint).
35. B. Hellfer, A. Knauf, H. Siedentop, and R. Weikard, *On the absence of a first order correction for the number of bound states of a Schrödinger operator with Coulomb singularity*, Comm. Partial Differential Equations **17** (1992), no. 3, 4, 615–639.
36. B. Helffer and D. Robert, *Riesz means of bound states and semiclassical limit* ... I, Asymptotic Anal. **3** (1990), 91–103.

37. L. Hörmander, *The spectral function of an elliptic operator*, Acta Math. **121** (1968), 193–218.
38. ——, *The analysis of linear partial differential operators* I–IV, Springer-Verlag, 1983–1985.
39. W. Hughes, *An atomic energy lower bound that agrees with Scott's correction*, Adv. in Math. **79** (1990), 213–270.
40. W. Hunziker, *On the asymptotic behaviour of atomic spectra near the continuum*, Helv. Phys. Acta **48** (1975), 145–150.
41. W. Hunziker and C. Günther, *Bound states in dipole field and continuity properties of electronic spectra*, Helv. Phys. Acta **53** (1980), 201–208.
42. V. Ivrii, *On the second term in the spectral asymptotics for the Laplace-Bettrami operator on a manifold with boundary*, Functional Anal. Appl. **14** (1980), 98–106.
43. ——, *Estimations pour le nombre de valeurs propres négatives ...*, C. R. Acad. Sci. Paris Sér. I Math. **302** (1986), 467–470; 491–494.
44. ——, *Semiclassical microlocal analysis and precise spectral asymptotics* I–IX (1990–93) (preprints).
45. V. Ivrii and I. M. Sigal, *Asymptotics of the ground state energies of large Coulomb systems*, Ann. of Math. (2) **138** (1993), 143–335.
46. J. B. Keller, *Corrected Bohr-Sommerfeld quantum conditions for non-separable systems*, Ann. Physics **4** (1958), 180–188.
47. H. Kumano-go, *Pseudo-differential operators*, The MIT Press, 1981.
48. P. D. Lax, *Asymptotic solutions of oscillatory initial value problems*, Duke Math. J. **24** (1957), 627–646.
49. E. H. Lieb, *A lower bound for Coulomb energies.*, Phys. Lett. A **70** (1979), 444–446.
50. ——, *Thomas-Fermi and related theories of atoms and molecules*, Rev. Modern Phys. **53** (1981).
51. ——, *Variational principle for many-fermion systems*, Phys. Rev. Lett. **46** (1981), no. 7, 457–459.
52. ——, *Stability of matter: From atoms to stars*, Bull. Amer. Math. Soc. (N.S.) **22** (1990), 1–49.
53. ——, *The Stability of Matter: From Atoms to Stars, Selecta of Elliott H. Lieb* (W. Thirring, ed.), Springer-Verlag, 1991.
54. E. H. Lieb and S. Oxford, *An improved lower bound on the indirect Coulomb energy*, Internat. J. Quantum Chem. **19** (1981), 427–439.
55. E. H. Lieb, I. M. Sigal, B. Simon, and W. Thirring, *Approximate neutrality of large Z-ions*, Comm. Math. Phys. **116** (1988), 635–644.
56. E. H. Lieb and B. Simon, *Thomas-Fermi theory of atoms, molecules and solids,*, Adv. in Math. **23** (1977), 22–116.
57. E. H. Lieb, J.-P. Solovej, and J. Yngvason, *Asymptotics of heavy atoms in high magnetic fields* (1993) (preprint).
58. E. H. Lieb and W. Thirring, *Bound for the kinetic energy of fermions which proves the stability of matter*, Phys. Rev. **35** (1975), 687–689.
59. ——, *Universal nature of the van der Waals forces for Coulomb systems*, Phys. Rev. A **34** (1990), 40–46.
60. E. H. Lieb and H.-T. Yau, *The stability and instability of relativistic matter*, Comm. Math. Phys. **118** (1988), 177–213.
61. ——, *Many-body stability implies a bound on the find structure constant*, Phys. Rev. Lett. **61** (1988), 1695–1697.
62. N. M. March, *Self-consistent field in atoms*, Perganon Press, 1975.
63. V. P. Maslov, *Theory of Perturbations and Asymptotic Methods*, Moscow, 1965. (Russian)
64. R. B. Melrose, *Weyl's conjecture for manifold with concave boundary*, Proc. Sympos. Pure Math., vol. 36, Amer. Math. Soc., Providence, 1980.
65. N. Narnhofer and W. Thirring, *Asymptotic exactness of finite temperature Thomas-Fermi theory*, Ann. Physics **134** (1981), 128–140.
66. V. Petkov and D. Robert, *Asymptotique semi-classique du spectre d'hamiltoniens quantiques trajectoires classiques p'eriodiques*, Comm. Partial Differential Equations **10** (1985), no. 4, 365–390.
67. D. Robert, *Autour de l'approximation semi-classique*, Birkhäuser, 1987.
68. M. B. Ruskai, *Absence of discrete spectrum in highly negative ions* I, Comm. Math. Phys. **82** (1982), 457–469; II **85**, 325–327.

69. ____, *Limit on the excess negative charge of a dynamic diatomic molecule*, Ann. Inst. H. Poincaré. Phys. Théor. **52** (1990), 397–414.
70. ____, *Improved estimate on the number of bound states of negatively charged bosonic atoms* (1993) (preprint).
71. M. B. Ruskai and J.-P. Solovej, *Asymptotic neutrality of polyatomic molecules* (1993) (in preparation).
72. J. Schwinger, *Thomas-Fermi model: The second correction*, Phys. Rev. A **24** (1981), no. 5, 2353–2361.
73. L. Seco, I. M. Sigal, and J.-P. Solovej, *Bound on the ionization energy of large atoms*, Comm. Math. Phys. **131** (1990), 307–315.
74. J. M. C. Scott, *The binding energy of the Thomas-Fermi atom*, Phil. Mag. **43** (1952), 859–867.
75. H. Siedentop and R. Weikard, *On the leading energy correction for the statistical model of the atom: Interacting case*, Comm. Math. Phys. **112** (1987), 471–490.
76. ____, *On the leading correction of the Thomas-Fermi model: Lower bound*, Invent. Math. **97** (1989), 159–193.
77. ____, *A lower bound of Scott type by a new microlocalization technique*, Ann. Sci. École Norm. Sup. (4) (1990).
78. I. M. Sigal, *Geometric methods in the quantum many-body problem. Nonexistence of very negative ions*, Comm. Math. Phys. **85** (1982), 309–324.
79. ____, *Quantum mechanics of many-particle systems*, Proceedings of the International Congress of Mathematicians (Kyoto 1990), Springer-Verlag, 1991, pp. 1397–1407.
80. I. M. Sigal and A. Soffer, *Local decay and velocity bounds for time-dependent and time-independent Hamiltonians*, Princeton (1988) (preprint).
81. B. Simon, *Fifteen Problems in Mathematical Physics*, Anniversary of Oberwolfach 1984, Perspect. Math., pp. 423–454.
82. M. Shubin, *Spectral theory of pseudodifferential operators*, Springer-Verlag.
83. A. Sobolev (1993) (preprint).
84. J.-P. Solovej, *Asymptotics for bosonic atoms*, Lett. Math. Phys. **20** (1990), 165–172.
85. ____, *Asymptotic neutrality of diatomic molecules*, Comm. Math. Phys. **130** (1990), 185–204.
86. ____, *Proof of the ionization conjecture in a reduced Hartree-Fock model*, Invent. Math. **104** (1991), 291–311.
87. ____, *Atomic and molecular structure—A renormalized picture.*
88. E. Stein, *Singular integrals and differentiability properties of functions*, Princeton Univ. Press, Princeton, 1970.
89. H. Tamura, *The asymptotic formulas for the number of bound states in the strong coupling limit*, J. Math. Soc. Japan **36** (1984), 355–374.
90. W. Thirring, *A lower bound with the best possible constant ...*, Comm. Math. Phys. **79** (1981), 1–7.
91. L. H. Thomas, *The calculation of atomic fields*, Math. Proc. Cambridge Philos. Soc. **23** (1927), 542–548.
92. S. A. Vugalter and G. M. Zhislin, *Asymptotics of the discrete spectrum of Hamiltonians of multiparticle quantum systems in a homogeneous magnetic field*, Algebra i Analiz **3** (1991), 1313–1349.
93. ____, *On the asymptotics of the discrete spectrum of a given symmetry of multi-particle Hamiltonians,*, Trudy Moskov. Mat. Obshch. (1993) (to appear); English transl., Trans. Moscow Math. Soc. (to appear).
94. J. Yngvason, *Thomas-Fermi theory for matter in a magnetic field as a limit of quantum mechanics,*, Lett. Math. Phys. **22** (1991), 107–117.
95. G. Zhislin, *Discussion of the spectrum of the Schrödinger operator for systems of many particles*, Trudy Moskov. Mat. Obshch. **9** (1960), 81–128.

Department of Mathematics, University of Toronto, Toronto, ON, Canada M5S 1A1

E-mail address: sigal@math.toronto.edu

Centre de Recherches Mathématiques
CRM Proceedings and Lecture Notes
Volume 8, 1995

Spectral Analysis of Rank One Perturbations and Applications

Barry Simon

ABSTRACT. A review of the general theory of self-adjoint operators of the form $A + \alpha B$ where B is rank one is presented. Applications include proofs of localization for Schrödinger operators, results on inverse spectral theory, and examples of operators with singular continuous spectrum.

1. Introduction, Borel Transforms, the Krein Spectral Shift, and All That

Introduction. Our goal here is to review the spectral theory (with applications) of rank one perturbations of positive self-adjoint operators on a separable complex Hilbert space, $\mathcal{H}$:

$$A_\alpha = A + \alpha B \quad B = (\varphi, \cdot)\varphi. \tag{1.1}$$

The cynic might feel that I have finally sunk to my proper level. I started with quantum field theory, analysis in infinitely many variables. That was too hard so I switched to the N-body Schrödinger equation; but that was too hard so I switched to one-body, then one-dimensional, then discrete one-dimensional. Finally to rank one perturbations—maybe something so easy that I can say something useful! Alas, we'll see even this is hard and exceedingly rich.

I should warn you that (1.1) is somewhat more general than you might think. First, I'll bet you thought that φ was a unit vector. It need not be. Big deal, you think—normalize φ and renormalize α. But φ need not be a normalizable vector. We'll consider B's which are rank one but only form bounded perturbations of A. In one fell swoop, we've absorbed the theory of variation of boundary conditions for Sturm-Liouville problems on a half-line!

Moreover, as I'll discuss in §1.5, α may be infinite.

In some ways, my recent relationship with rank one perturbations reminds me of my relationship with trace ideals fifteen years ago. Then it turned out three distinct problems: Yukawa field theories, scattering theory, and semi-classical bounds on the number of bound states led me to similar mathematics, viz estimates on the trace ideal properties of certain pseudo-differential operators.

1991 *Mathematics Subject Classification.* Primary: 47B15, 34L05; Secondary: 81Q10, 34F05.

Lecture given at the Vancouver Summer School in Mathematical Physics, August 10–14, 1993. This material is based upon work supported by the U. S. National Science Foundation under Grant No. DMS-9101715.

This is the final form of the paper.

This past year, three distinct research projects:

1. my work, in part with del Rio, Jitomirskaya, and Makarov, on singular spectrum;
2. my trying to understand the work of Aizenman and Molchanov;
3. my work with Gesztesy on trace formulas

all led to rank one perturbations and enriched my own knowledge of the subject. 1–3 are the themes of Sections 2–4 of these notes.

In this section I'll discuss the general theory in a new systematic way; many of the ideas were developed in discussions with Fritz Gesztesy, to whom I am grateful. A key role will be played by

$$F_\alpha(z) = (\varphi, (A_\alpha - z)^{-1}\varphi). \tag{1.2}$$

In terms of a suitable spectral measure $d\mu_\alpha$:

$$F_\alpha(z) = \int \frac{d\mu_\alpha(\lambda)}{\lambda - z} \tag{1.3}$$

so we'll begin with an exhaustive study of the Borel transform of a measure and its relation to the measure in §1.1. That section and §1.2 are the technical core behind most of what follows.

1.1. Borel Transforms of Measures. Throughout, we let $d\mu$ be a (positive) measure on $[a, \infty)$ for some $a > -\infty$ with

$$\int \frac{d\mu(\lambda)}{|\lambda| + 1} < \infty. \tag{1.4}$$

REMARKS. 1. We'll suppose that $\operatorname{supp}(\mu)$ is bounded from below and the operator A in (1.1) is positive. Much of our theory works without those restrictions but the details are a little simpler with them and they hold in the applications I want to make, so I've made them.

2. For some results when one only knows that $\int d\mu(\lambda)/(1+\lambda^2) < \infty$, see Aronszajn-Donoghue [**4**].

When (1.4) holds, we can define for $z \in \mathbb{C} \setminus (-\infty, a)$:

$$F(z) = \int \frac{d\mu(\lambda)}{\lambda - z} \tag{1.5}$$

the *Borel transform* of μ (also called the Stieltjes transform or the Borel-Stieltjes transform). The key issues concern boundary values of F as $z = x + i\epsilon \downarrow x \in \operatorname{supp}(d\mu)$. We begin with the simple stuff away from $\operatorname{supp}(d\mu)$.

THEOREM 1.1. (i) *F is an analytic function on $\mathbb{C} \setminus (a, \infty)$.*

(ii) *F is positive on $(-\infty, a)$ and* $\operatorname{Im} F > 0$ *if* $\operatorname{Im} z > 0$.

(iii) $\lim_{\gamma\to\infty} F(-\gamma) = 0$.

(iv) $\lim_{\gamma\to\infty} F(-\gamma)(\gamma) = \int d\mu(\lambda)$ *where $\int d\mu = \infty$ is allowed. Similarly,*

$$\lim_{\gamma\to\infty} F'(-\gamma)(\gamma)^2 = \int d\mu(\lambda).$$

(v) $\lim_{\gamma\to\infty} F'(-\gamma)/F^2(-\gamma) = \left(\int d\mu(\lambda)\right)^{-1}$ *where $\int d\mu = \infty$ is allowed with* $\infty^{-1} = 0$.

PROOF. (i) and (ii) are elementary. (iii) uses:

$$\operatorname{Im} F(x+i\beta) = \int \frac{\beta}{(x-y)^2+\beta^2}\, d\mu(y). \tag{1.6}$$

(iii) and (iv) follow from the monotone convergence theorem. Actually, by a slightly more involved argument if $|z| \to \infty$ with $|\arg(-z)| < \epsilon$ with $\epsilon > 0$ and small, it is not hard to see that $(-z)F(z) \to \int d\mu$.

If $\int d\mu < \infty$, a similar use of the monotone convergence theorem shows that

$$\gamma^2 F'(-\gamma) = \gamma^2 \int \frac{d\mu(\lambda)}{(\lambda+\gamma)^2} \longrightarrow \int d\mu(\lambda)$$

and (iv) follows. If $\int d\mu = \infty$, we use the fact that $\left|1/[zF(z)]\right| \to 0$ as $|z| \to \infty$ with $\operatorname{Re} z < 0$. Thus, writing

$$\left(\frac{1}{F}\right)'(z) = (2\pi i)^{-1} \int_{|w-z|=|z|/2} (w-z)^{-2} \frac{1}{F(w)}\, dw$$

we see that $(1/F)' \to 0$ but $(1/F)' = -F'/F^2$ implying (v). $\square$

Next we talk about boundary values of F. The function

$$G(x) = \int \frac{d\mu(y)}{(x-y)^2}$$

which is defined for $x \in (-\infty, \infty)$ and takes values in $(0, \infty]$ (note ∞ is allowed), will play an important role.

THEOREM 1.2. (i) *$G(x) < \infty$ implies that*

$$\int \frac{d\mu(y)}{|x-y|} < \infty. \tag{1.7}$$

(ii) **[12, 27]** *$\{x \mid G(x) = \infty\}$ is a dense G_δ in* $\operatorname{supp}(d\mu)$.

(iii) *If (1.7) holds, and, in particular if $G(x) < \infty$, then $\lim_{\epsilon \downarrow 0} F(x+i\epsilon)$ exists and is real.*

(iv) $\lim_{\epsilon \downarrow 0} \epsilon^{-1} \operatorname{Im} F(x+i\epsilon) \to G(x)$.

(v) *If $G(x) < \infty$, then $\lim_{\epsilon \downarrow 0} (i\epsilon)^{-1}[F(x+i\epsilon) - F(x+i0)] = G(x)$.*

REMARK. A G_δ is a countable intersection of open sets.

PROOF. (i) is immediate if one notes that

$$\frac{1}{|x-y|} \le \frac{1}{|x-y|^2} + \frac{|x|+2}{|y|+1}$$

since

$$|y| + 1 \le |x| + |x-y| + 1 \le (|x-y|)(|x|+2)$$

if $|x-y| \ge 1$.

(ii) By the lemma below, if $\mu([0,1]) \ge 1$, there is an $x_0 \in [0,1]$ with

$$\int d\mu(y)/|x_0 - y| = \infty.$$

So, by scaling, if $\mu([a,b]) > 0$, there is an $x_0 \in [a,b]$ with $\int d\mu(y)/|x_0-y| = \infty$. It follows that if $a_0 \in \operatorname{supp}(d\mu)$, then there exists $a_n \to a_0$ with $\int d\mu(y)/|a_n - y| = \infty$.

So, by (i), $\{x \mid G(x) = \infty\}$ is dense in $\operatorname{supp}(d\mu)$. It is a G_δ because if we define $G_n(x) = \int (|x-y|^2 + n^{-1})^{-1}\, d\mu(y)$, then G_n is continuous and

$$\{x \mid G(x) = \infty\} = \bigcap_m \bigcup_n \{x \mid G_n(x) > m\}.$$

(iii) By (1.6), for any δ

$$\operatorname{Im} F(x+i\beta) \le \beta \int_{|x-y|>\delta} |x-y|^{-2}\, d\mu(y) + \int_{|x-y|\le\delta} |x-y|^{-1}\, d\mu(y)$$

so

$$\overline{\lim_{\beta\downarrow 0}} \operatorname{Im} F(x+i\beta) \le \overline{\lim_{\delta\downarrow 0}} \int_{|x-y|\le\delta} |x-y|^{-1}\, d\mu(y) = 0$$

if (1.7) holds. (1.7) also implies

$$\lim_{\beta\downarrow 0} \operatorname{Re} F(x+i\beta) = \int (y-x)^{-1}\, d\mu(y)$$

by the dominated convergence theorem.

(iv) follows from (1.6) and the monotone convergence theorem.

(v) If $G(x) < \infty$, then, by the proof of (iii)

$$F(x+i0) = \int \frac{d\mu(y)}{y-x}$$

so

$$(i\epsilon)^{-1}[F(x+i\epsilon) - F(x-i0)] = \int \frac{d\mu(y)}{(y-x)(y-x-i\epsilon)} \to G(x)$$

by the dominated convergence theorem if $G(x) < \infty$. $\square$

LEMMA 1.3. *Let $\mu([0,1]) \ge 1$. Then, there exists $x_0 \in [0,1]$ with*

$$\int d\mu(y)/|x_0 - y| = \infty.$$

PROOF. Define I_n inductively as follows. Let I_1 be one of $[0,1/2]$, $[1/2,1]$ with $\mu(I_1) \ge 1/2$. Now subdivide I_1 in half and choose I_2 to be one of those halves with $\mu(I_2) \ge 1/4$. By induction

$$I_1 \supset I_2 \supset \cdots \quad \text{all closed;} \quad \mu(I_n) \ge 2^{-n}; \quad |I_n| = 2^{-n}.$$

By compactness, $\cap I_n = \{x_0\}$ for some x_0. Obviously

$$\int_{I_n} |x - x_0|^{-1}\, d\mu(x) \ge (2^{-n})^{-1} 2^{-n} = 1.$$

But if $\int |x-x_0|^{-1}\, d\mu(x) < \infty$, $\int_{I_n} |x-x_0|^{-1}\, d\mu(x) \downarrow 0$ by monotone convergence, so we must have $\int |x-x_0|^{-1}\, d\mu(x) = \infty$. $\square$

THEOREM 1.4. *$\lim_{\beta\downarrow 0} F(x+i\beta)$ exists and is finite for a.e. x.*

PROOF. This is a standard harmonic analysis result on non-tangential boundary values of analytic maps, see [**31**], so we'll only sketch the details. Let $g(z) = (z-i)/(z+i)$, the fractional linear map of the upper half plane to the unit disc. Thus $\widetilde{F} = g \circ F \circ g^{-1}$ is a map of the unit disc to itself. Such maps have boundary values by a standard maximal function argument, see Katznelson [**31**]. By the same argument as in the next theorem, $\{\theta \mid \lim_{r\uparrow 1} \widetilde{F}(re^{i\theta}) = -1\}$ has measure zero, so $\lim_{\beta\downarrow 0} F(x+i\beta) = \infty$ on a set of measure 0 and so F is finite a.e. $\square$

THEOREM 1.5. *Let F_1, F_2 be the Borel transforms of two (positive) measures (one may be zero) and fix $\alpha \in \mathbb{C}$. Then*

$$|\{x \mid F_1(x+i0) - F_2(x+i0) = \alpha\}| > 0$$

only if $\alpha = 0$ and $F_1 = F_2$.

PROOF. Let g be defined as in the proof of Theorem 1.4 and let

$$h = e^{iF_1 \circ g^{-1}} - (e^{iF_2 \circ g^{-1}})e^{i\alpha}.$$

h is a bounded analytic function on the disc and the theorem asserts that under $h \not\equiv 0$, $\{\theta \mid \lim_{r\uparrow 1} h(re^{i\theta}) = 0\}$ has Lebesgue measure zero. This is a standard theorem; see Katznelson [**31**]. $\square$

Let $d\mu_{\text{ac}}, d\mu_{\text{sc}}, d\mu_{\text{pp}}$ be the absolutely continuous, singular continuous and pure point parts of a measure $d\mu$. Let $d\mu_{\text{sing}} = d\mu_{\text{sc}} + d\mu_{\text{pp}}$.

The main theorem on Borel transforms is

THEOREM 1.6. *Let $F(z)$ be the Borel transform of a measure $d\mu$ obeying* (1.4). *Then*

(i) *$\pi^{-1} \operatorname{Im} F(E+i\epsilon)\, dE \to d\mu$ weakly in the sense that*

$$\int \pi^{-1} f(E) \operatorname{Im} F(E+i\epsilon)\, dE \longrightarrow \int f(E)\, d\mu(E) \tag{1.8}$$

for all continuous functions, f, of compact support.

(ii) *μ_{sing} is supported on $\{E \mid \lim_{\epsilon\downarrow 0} \operatorname{Im} F(E+i0) = \infty\}$.*

(iii) *$\mu(\{E_0\}) = \lim_{\epsilon\downarrow 0} \epsilon \operatorname{Im} F(E_0+i\epsilon)$. Moreover, for any $E_0 \in \mathbb{R}$:*

$$\lim_{\epsilon\downarrow 0} \epsilon \operatorname{Re} F(E_0+i0) = 0.$$

(iv) *$d\mu_{\text{ac}}(E) = \pi^{-1} \operatorname{Im} F(E+i0)\, dE$.*

PROOF. (i) It is a well-known calculation that for f, continuous of compact support:

$$\lim_{\epsilon\downarrow 0} \frac{1}{\pi} \int f(E') \frac{\epsilon}{(E-E')^2+\epsilon^2}\, dE' \to f(E).$$

This and the dominated convergence theorem easily imply the result.

(ii) We will prove the weaker fact that μ_{sing} is supported on $\{E \mid \overline{\lim} \operatorname{Im} F(E+i2^{-n}) = \infty\}$. The stronger fact (theorem of de Vallée Poussin) can be found in Saks [**40**], but for most applications, this weaker fact suffices. Since the stronger result is true and more elegant, we'll use it. For the theorem of Aronszajn on mutual singularity (Theorem 2.2(iv)), one does need the stronger fact.

We will prove that for any a

$$\Big\{E \Bigm| \sup_n \operatorname{Im} F(E+i2^{-n}) \le a\Big\} = A_a$$

has μ_{sing}-measure 0 from which the result follows. Notice that

$$\begin{aligned}\operatorname{Im} F(x+i\epsilon) &= \int \frac{\epsilon}{(x-y)^2+\epsilon^2}\, d\mu(y) \\ &\geq \frac{1}{2\epsilon}\int_{|x-y|\leq\epsilon} d\mu(y).\end{aligned}$$

So for fixed $x \in A_a$

$$\sup_n 2^n \mu\left([x-2^{-n}, x+2^{-n}]\right) \leq 2a$$

and so, since any $2^n \leq \epsilon^{-1} \leq 2^{n+1}$ for some n

$$\sup_n \epsilon^{-1}\mu\left([x-\epsilon, x+\epsilon]\right) \leq 4a$$

and thus looking at the two subintervals:

$$\sup_{x\in[\alpha,\beta]} |\beta-\alpha|^{-1}\mu([\alpha,\beta]) \leq 8a.$$

By regularity of measures, for any ϵ, we can find a closed set C and open O with $C \subset O$, so $\mu_{\text{sing}}(\mathbb{R}\setminus C) \leq \epsilon$ and $|O| \leq \epsilon$. We write $O = \bigcup_{n=1}^{\infty}(\alpha_n, \beta_n)$ as a disjoint union of intervals (union of connected components). Then

$$\begin{aligned}\mu_{\text{sing}}(C\cap A_a) &\leq \mu_{\text{sing}}\left(\bigcup_{\{n|(\alpha_n,\beta_n)\cap A_a\neq\varnothing\}} (\alpha_n,\beta_n)\right) \\ &\leq \sum_{\{n|(\alpha_n,\beta_n)\cap A_a\neq\varnothing\}} \mu(\alpha_n,\beta_n) \leq 8a\sum_n |\beta_n-\alpha_n| = 8a|O| \leq 8a\epsilon\end{aligned}$$

so

$$\mu_{\text{sing}}(A_a) \leq \mu_{\text{sing}}(C\cap A_a) + \mu_{\text{sing}}(\mathbb{R}\setminus C) \leq \epsilon + 8a\epsilon.$$

Since ϵ is arbitrary, $\mu_{\text{sing}}(A_a) = 0$ as was to be proven.

(iii) The first part is a trivial consequence of the monotone convergence theorem, if we note that $\epsilon^2/(x^2+\epsilon^2)$ converges monotonically to 0 (resp. 1) if $x \neq 0$, (resp. $x = 0$). For the second:

$$\epsilon\int_{|x-y|\geq 1} \frac{|x-y|}{|x-y|^2+\epsilon^2}\, d\mu(y) \leq \epsilon\int_{|x-y|\geq 1}\frac{1}{|x-y|}\, d\mu(y) \longrightarrow 0$$

while

$$\lim_{\epsilon\downarrow 0}\int_{|x-y|<1}\frac{\epsilon|x-y|}{(x-y)^2+\epsilon^2}\, d\mu(y) = 0$$

by the dominated convergence theorem since $\alpha\beta/(\alpha^2+\beta^2) \leq 1/2$.

(iv) Write $d\mu_{\text{ac}}(E) = h(E)\, dE$ and let $F_{\text{ac}}(z)$ be the Borel transform of $d\mu_{\text{ac}}$. By the Lebesgue theorem on the derivatives of the integral (that says if $h \in L^1$, for a.e. x, $h(x) = \lim_{a\to 0}(2a)^{-1}\int_{x-a}^{x+a} h(y)\, dy$; see Katznelson [**31**]), one sees that $1/\pi \operatorname{Im} F_{\text{ac}}(E+i0) = h(E)$ for a.e. E. By Fatou's lemma, for $f \geq 0$ and continuous:

$$\begin{aligned}\pi^{-1}\int [\operatorname{Im} F(E+i0)] f(E)\, dE &\leq \underline{\lim}\, \pi^{-1}\int \operatorname{Im} F(E+i\epsilon) f(E)\, dE \\ &= \int f(E)\, d\mu(E).\end{aligned}$$

Thus $\pi^{-1} \operatorname{Im} F(E+i0)\, dE \leq d\mu$ and so for a.e. E:

$$\frac{1}{\pi} \operatorname{Im} F(E+i0) \leq h(E) = \frac{1}{\pi} \operatorname{Im} F_{\rm ac}(E+i0) \leq \frac{1}{\pi} \operatorname{Im} F(E+i0)$$

since $d\mu \geq d\mu_{\rm ac}$ implies $\operatorname{Im} F \geq \operatorname{Im} F_{\rm ac}$. Thus

$$h(E) = \frac{1}{\pi} \operatorname{Im} F(E+i0)$$

as was to be proven. □

1.2. Rank One Perturbations: The Set-Up and Basic Formulas. Let $A \geq 0$ be a possibly unbounded self-adjoint operator on a Hilbert space $\mathcal{H}$. One defines the scale of spaces $\mathcal{H}_s(A)$ associated to A as follows. For $s \geq 0$, $\mathcal{H}_s(A)$ is $D(A^{s/2})$ with the norm

$$\|\varphi\|_s = \left\|(A+1)^{s/2}\varphi\right\| \tag{1.9}$$

in which $\mathcal{H}_s$ is easily seen to be complete. For $s < 0$, take $\mathcal{H}$ with the norm given by (1.9) and complete it. In a natural way, $\mathcal{H}_s$ and $\mathcal{H}_{-s}$ are duals in such a way that $\varphi \in \mathcal{H}$ is associated to the functional on $\mathcal{H}_s$ given by $\ell_\varphi(\eta) = (\varphi, \eta)$, so $|\ell_\varphi(\eta)| \leq \|\varphi\|_{-s}\|\eta\|_s$ and $\mathcal{H}_s \subset \mathcal{H}_t$ if $s > t$.

$(A+1)^{-t/2}$ is an isometry from $\mathcal{H}_s$ to $\mathcal{H}_{s+t}$ so for $s > 0$, if Δ is a bounded subset of $\mathbb{R}$, $E_\Delta(A)$, the spectral projection of A, maps $\mathcal{H}_{-s}$ to $\mathcal{H}_s$ since $(A+1)^{+s/2}E_0(A)(A+1)^{s/2}$ is bounded. In particular, we can define a spectral measure $d\mu_A^\varphi$ for any $\varphi \in \mathcal{H}_{-s}(A)$ by

$$\mu_A^\varphi(\Delta) = (\varphi, E_\Delta(A)\varphi).$$

It is easy to see that

$$\int \frac{d\mu_A^\varphi(x)}{(|x|+1)^s} < \infty \tag{1.10}$$

and, in particular, the spectral measures of $\varphi \in \mathcal{H}_{-1}(A)$ have Borel transforms.

Now let $\varphi \in \mathcal{H}_{-1}(A)$ and let b be the quadratic form on $\mathcal{H}_{+1}(A)$ given by

$$b_\varphi(\eta, \eta') = \overline{\ell_\varphi(\eta)}\, \ell_\varphi(\eta').$$

b_φ is positive but if $\varphi \notin \mathcal{H}$, it is neither closed nor closable.

See [**30, 38**] for a discussion of forms and their perturbation theory.

PROPOSITION 1.7. *For any φ, b_φ is a form bounded perturbation of A with relative bound zero.*

PROOF. Given ϵ, find $\varphi_\epsilon \in \mathcal{H}_{-1}(A)$ so $\|\varphi_\epsilon\|_{-1}^2 \leq 1/2\epsilon$ and $\psi_\epsilon \equiv \varphi - \varphi_\epsilon \in \mathcal{H}$. Then

$$b_\varphi = b_{\varphi_\epsilon + \psi_\epsilon} \leq b_{\varphi_\epsilon + \psi_\epsilon} + b_{\varphi_\epsilon - \psi_\epsilon} = 2b_{\varphi_\epsilon} + 2b_{\psi_\epsilon}.$$

So, for any $\eta \in \mathcal{H}_{+1}(A)$

$$\begin{aligned} b_\varphi(\eta, \eta) &\leq 2\|\varphi_\epsilon\|_{-1}^2\|\eta\|_{+1}^2 + 2\|\psi_\epsilon\|^2\|\eta\|^2 \\ &\leq \epsilon(\eta, A\eta) + (2\|\psi_\epsilon\|^2 + \epsilon)\|\eta\|^2 \end{aligned}$$

as was to be proven. □

The perturbation theory of forms [**30, 38**] thus implies that for any $\alpha \in \mathbb{R}$, $A + \alpha b$ is the quadratic form of a self-adjoint operator A_α with $\mathcal{H}_s(A_\alpha) = \mathcal{H}_s(A)$ if $|s| \leq 1$. We will write

$$A_\alpha = A + \alpha B = A + \alpha(\varphi, \cdot)\varphi \tag{1.11}$$

although there is no operator B or vector φ in the usual sense if $\varphi \notin \mathcal{H}$. We will occasionally write $\|\varphi\|^2$ which may be infinite if $\varphi \notin \mathcal{H}$.

Since $\mathcal{H}_{-1}(A_\alpha) = \mathcal{H}_{-1}(A)$ and so $(A_\alpha - z)^{-1}\varphi \in \mathcal{H}_{+1}(A_\alpha) = \mathcal{H}_{+1}(A) \subset \mathcal{H}$. The strong resolvent formula holds, viz

$$(A_\alpha - z)^{-1} - (A - z)^{-1} = -\alpha\big((A_\alpha - \bar{z})^{-1}\varphi, \cdot\big)(A - z)^{-1}\varphi. \tag{1.12}$$

Don't blink now because in just about one page we're going to present four of the five critical formulas in the subject (the fifth is (1.17) below).

Define

$$F_\alpha(z) = (\varphi, (A_\alpha - z)^{-1}\varphi) \equiv \int \frac{d\mu_\alpha(x)}{x - z}$$

where $d\mu_\alpha \equiv d\mu^\varphi_{A_\alpha}$ is the spectral measure for φ which has a Borel transform by (1.10) and $\mathcal{H}_{-1}(A_\alpha) = \mathcal{H}_{-1}(A)$. We define $F(z) \equiv F_{\alpha=0}(z)$.

Taking matrix elements of (1.12) with φ on both sides, we get $F_\alpha(z) - F(z) = -\alpha F_\alpha(z)F(z)$, or solving for F_α, we get the Aronszajn-Krein formula:

$$F_\alpha(z) = \frac{F(z)}{1 + \alpha F(z)}, \tag{1.13}$$

the first critical formula. Applying (1.12) to φ and using (1.13) to see that $1 - \alpha F_\alpha = [1 + \alpha F(z)]^{-1}$, we get the second key formula:

$$(A_\alpha - z)^{-1}\varphi = \big(1 + \alpha F(z)\big)^{-1}(A - z)^{-1}\varphi. \tag{1.14}$$

Plugging this into (1.12) we get the third key formula:

$$(A_\alpha - z)^{-1} = (A - z)^{-1} - \frac{\alpha}{1 + \alpha F(z)}\big((A - \bar{z})^{-1}\varphi, \cdot\big)(A - z)^{-1}\varphi. \tag{1.15}$$

Notice that (1.14) says for η, z fixed, $(\eta, (A_\alpha - z)^{-1}\varphi) = a/(\alpha + b)$, the key to the Aizenman-Molchanov theory.

For the final formula, (1.12) says that $(A_\alpha - z)^{-1} - (A - z)^{-1}$ is rank one, so trace class, and by (1.15) we have

$$\operatorname{Tr}[(A - z)^{-1} - (A_\alpha - z)^{-1}] = \frac{\alpha}{1 + \alpha F(z)}(\varphi, (A - z)^{-2}\varphi).$$

Notice that $(\varphi, (A - z)^{-2}\varphi) = dF(z)/dz$. Thus, we have what I'll call the trace formula, the key to our discussion of the Krein spectral shift in §1.4 below:

$$\operatorname{Tr}[(A - z)^{-1} - (A_\alpha - z)^{-1}] = \frac{d}{dz}\ln\big(1 + \alpha F(z)\big) \tag{1.16}$$

where we take the branch of the log which is positive for z real and very negative (recall by Theorem 1.1(iii), that even for $\alpha < 0$, $1 + \alpha F(z) > 0$ for z real and very negative).

1.3. The Integral Formula. The critical last formula of the general theory is

THEOREM 1.8.

$$\int_{-\infty}^{\infty} [d\mu_\alpha(E)]\, d\alpha = dE \tag{1.17}$$

in the sense that if f is in $L^1(\mathbb{R}, dE)$, then $f \in L^1(\mathbb{R}, d\mu_\alpha)$ for a.e. α, $\int f(E)\, d\mu_\alpha(E)$ $\in L^1(\mathbb{R}, d\alpha)$ and

$$\int \Big(\int f(E)\, d\mu_\alpha(E) \Big)\, d\alpha = \int f(E)\, dE. \tag{1.18}$$

PROOF. By an elementary argument, it suffices to prove this result for $f_z(E) = (E-z)^{-1} - (E+i)^{-1}$ for any $z \in \mathbb{C} \setminus \mathbb{R}$ (use analyticity in z and then Stone-Weierstrass approximation). By closing the contour in the upper half plane

$$\int_{-\infty}^{\infty} f_z(E)\, dE = \begin{cases} 0 & \text{if } \operatorname{Im} z < 0 \\ 2\pi i & \text{if } \operatorname{Im} z > 0. \end{cases}$$

On the other hand, by (1.13)

$$\begin{aligned} \int f_z(E)\, d\mu_\alpha(E) &= F_\alpha(z) - F_\alpha(-i) = \frac{1}{\alpha + F(z)^{-1}} - \frac{1}{\alpha + F(-i)^{-1}} \\ &\equiv h_z(\alpha). \end{aligned}$$

Now, if $\pm \operatorname{Im} z > 0$, then $\pm \operatorname{Im} F(z) > 0$ so $\pm \operatorname{Im} F(z)^{-1} < 0$. Thus, $h_z(\alpha)$ has either two poles in the lower half plane if $\operatorname{Im} z < 0$ or one in each half plane if $\operatorname{Im} z > 0$. The same contour integral as for E (but now for α!) implies that

$$\int_{-\infty}^{\infty} d\alpha\, h_z(\alpha) \begin{cases} = 0 & \text{if } \operatorname{Im} z < 0 \\ \equiv 2\pi i & \text{if } \operatorname{Im} z > 0 \end{cases}$$

so 1.18 is proven for $f_z(E)$. □

The above implies a regularity result for the density of states in n-dimensional Jacobi matrices (see Theorem 1.19 below) due to Wegner [**49**]. Wegner proved his result by tracking eigenvalues in finite-dimensional approximations, but his underlying result is similar to (1.17) so this is sometimes called a Wegner estimate. The Russian literature, however, has similar results much earlier in the case of boundary condition dependence; see Javrjan [**28**].

1.4. The Krein Spectral Shift. In this section, we make a presentation of the Krein spectral shift [**33**] due to Gesztesy and me. By the general theory of Borel transforms, $F(z)$ has a boundary value $F(\lambda + i0)$ for a.e. λ in $\mathbb{R}$. Moreover, by Theorem 1.5, if α is fixed, $F(\lambda + i0) \neq -\alpha^{-1}$ for a.e. α. Thus we can define $\operatorname{Arg}\big(1 + \alpha F(\lambda + i0)\big)$ for a.e. λ. Since $\operatorname{Im} F(\lambda + i0) \geq 0$, if $\alpha > 0$ (resp. $\alpha < 0$), this arg lies in $[0, \pi]$ (resp. $[-\pi, 0]$). If $1 + \alpha F(\lambda + i0) \in (-\infty, 0)$ that sets the argument to $(\operatorname{sgn} \alpha)\pi$.

DEFINITION. The Krein spectral shift, $\xi_\alpha(\lambda)$, is defined by

$$\xi_\alpha(\lambda) = \frac{1}{\pi} \operatorname{Arg}\big(1 + \alpha F(\lambda + i0)\big).$$

DEFINITION. $i(\alpha) \equiv \min\big(\inf \operatorname{spec}(A), \inf \operatorname{spec}(A_\alpha)\big)$.

THEOREM 1.9. (i) $0 \le \pm\xi_\alpha(\lambda) \le 1$ *if* $\pm\alpha > 0$; $\xi_\alpha(\lambda) = 0$ *if* $\lambda < i(\alpha)$.
(ii)

$$\operatorname{Tr}\big((A-z)^{-1} - (A_\alpha - z)^{-1}\big) = \int (\lambda - z)^{-2} \xi_\alpha(\lambda)\, d\lambda \tag{1.19}$$

if $z \in \mathbb{C} \setminus [i(\alpha), \infty)$.
(iii) *If* $\|\varphi\| < \infty$, *then* $\xi_\alpha \in L^1$ *and*

$$\int \xi_\alpha(\lambda)\, d\lambda = \alpha \|\varphi\|^2. \tag{1.20}$$

PROOF. (i) The first part follows from the discussion above. If $\alpha > 0$, $i(\alpha) = \inf \operatorname{spec}(A)$ and $F(\lambda + i0) > 0$ for $\lambda < i(\alpha)$ so $\xi_\alpha(\lambda) = 0$. If $\alpha < 0$, $i(\alpha)$ is either $\inf \operatorname{spec}(A)$ or the unique $\lambda < i(\alpha)$ with $1 + \alpha F(\lambda) = 0$. Either way, since $F(\lambda)$ is monotone increasing on $(-\infty, i(0))$, below $i(\alpha)$, we have that $1 + \alpha F(\lambda) > 0$.

(ii) Let $H_\alpha(z) = \ln\big(1 + \alpha F(z)\big)$ on $\big(-\infty, i(\alpha)\big)$ analytically continued to $\mathbb{C} \setminus [i(\alpha), \infty)$. Consider a contour, C, which goes from $i(\alpha)$ just above the real axis to R, then directly up to $R + iR$, then on a circle of radius $R\sqrt{2}$ to $R - iR$, then up to just above the real axis, then along the bottom of the real axis to $i(\alpha)$. Consider for $z_0 \in \big(-\infty, i(\alpha)\big)$:

$$\frac{1}{2\pi i} \oint_C \frac{H_\alpha(z)}{(z - z_0)^2}\, dz \equiv I_\alpha(z_0).$$

By Cauchy's formula, once $R\sqrt{2} > |z_0|$, the integral is just dH_α/dz, which is the left side of (1.19) by (1.16).

Since $|\operatorname{Im} H_\alpha| \le \pi$, and the real parts of the contour from R to $R + iR$ and from R to $R - iR$ cancel, and since $F(z) \to 0$ as $z \to \infty$ with $|\operatorname{Arg} z| > \pi/2$, the infinite part of the contour contributes 0 as $R \to \infty$. Since $|\operatorname{Im} H_\alpha| \le \pi$, dominated convergence lets us take the contour to the real axis and get $\operatorname{Im} \ln\big(1 + \alpha F(z)\big) = \operatorname{Arg}\big(1 + \alpha F(z)\big)$. Thus, we get the right side of (1.19).

(iii) By Theorem 1.1(iii) and (iv) and the fact that $\int d\mu_0(z) = \|\varphi\|^2$, we see that

$$\lim_{\gamma \to \infty} \gamma^2 \frac{d}{dz} H_\alpha(-\gamma) = \alpha \|\varphi\|^2$$

so, by (ii)

$$\lim_{\gamma \to \infty} \int \frac{\gamma^2}{(x + \gamma)^2} \xi_\alpha(x)\, dx = \alpha \|\varphi\|^2.$$

By monotone convergence we obtain (1.20). □

The next result extends (1.19) to more general functions than $f(x) = (x - z)^{-1}$. The formula we want is:

$$\operatorname{Tr}\big(f(A) - f(A_\alpha)\big) = -\int f'(x) \xi_\alpha(x)\, dx. \tag{1.21}$$

Formally this holds for any reasonable f since it holds for $f(x) = (x - z)^{-1}$ and those f's are "dense" in reasonable f. One issue is that we don't even know that $f(A) - f(A_\alpha)$ is trace class! Our hypotheses on f aren't optimal but include the critical example $f(x) = e^{-sx}$ (cutoff near $-\infty$).

THEOREM 1.10. *Let f be a C^2 on $\mathbb{R}$ with $(1+|x|)^2 f^{(j)} \in L^2(0,\infty)$ for $j=1,2$. Then for any α, $f(A) - f(A_\alpha)$ is trace class and* (1.21) *holds.*

PROOF. We only sketch the details supposing w.l.o.g. that $\alpha > 0$ and $A \geq 1$ (otherwise, reverse the roles and/or add a constant to A). And we write $B \equiv A_\alpha$ for notational simplicity.

1. $B^{-1} - A^{-1}$ is rank 1 and so trace class. It follows that $B^{-2}e^{isB} - A^{-2}e^{isA}$ is trace class with trace norm bounded by $c(1+|s|)$ since

$$B^{-2}e^{isB} - A^{-2}e^{isA} = B^{-1}e^{isB}(B^{-1} - A^{-1}) \\ - i\int_0^s e^{iuB}(B^{-1} - A^{-1})e^{i(s-u)A} + (B^{-1} - A^{-1})e^{isA}A^{-1}.$$

2. Let $\hat{f}$ be the Fourier transform of f. Under the hypotheses

$$\int (1+|k|)\left[\left|\frac{d\hat{f}}{dk}\right| + \left|\frac{d^2\hat{f}}{dk^2}\right|\right] dk < \infty \tag{1.22}$$

and that implies that we can integrate by parts twice to find that

$$f(B) - f(A) = -(2\pi)^{-1/2}\int \left[\frac{d^2\hat{f}}{dk^2}(k)\right][B^{-2}e^{ikB} - A^{-2}e^{ikA}]\, dk.$$

It follows by step 1 that $f(B) - f(A)$ is trace class.

3. From (1.19) one deduces (1.21) for $f(x) = x^{-2}e^{ixs}$ by writing

$$f(x) = \lim_{n\to\infty} [x^{-2}(1 - ixs/n)^{-n}]$$

for one can use analyticity to get (1.21) for $f(x) = (x-z)^{-n}$ and then for this special f.

4. (1.21) for general f follows from the formula in step 2 and the equality proven in step 3. $\square$

REMARK. This does not capture all f for which (1.21) holds. For example, $f(x) = (x+1)^{-\beta}$ with $0 < \beta < 1$ does not obey the hypotheses, but using

$$(A+1)^{-\beta} = \frac{\sin\pi\beta}{\pi}\int_0^\infty (A+1+x)^{-1}|x|^{-\beta}\, dx$$

and $\|(A+x)^{-1} - (A_\alpha + x)^{-1}\|_1 \leq C(|x|+1)^{-2}$ for x large, one can prove (1.21) for such f.

PROPOSITION 1.11. *Suppose that (a,b) is an interval disjoint from* $\operatorname{spec}(A) \cup \operatorname{spec}(A_\alpha)$. *Then*

(i) *$\xi_\alpha(x)$ is constant on (a,b).*

(ii) *Its value is either* 0 *or* 1.

(iii) *$P_{(-\infty,a)}(A_\alpha) - P_{(-\infty,a)}(A)$ is trace class and its trace is the common value of $\xi_\alpha(x)$ in (a,b).*

PROOF. (i) Let $f \in C_0^\infty(a,b)$. Then, $f(A) = f(A_\alpha) = 0$ so $\int f'(x)\xi_\alpha(x)\,dx = 0$ by Theorem 1.10. Thus the distributional derivative $d\xi_\alpha/dx = 0$ on (a,b) so ξ_α is constant there.

(iii) Let f be a function in $C^\infty(\mathbb{R})$ which is 1 on $(i(\alpha), a)$, 0 on $[b,\infty)$. Then $f(A) = P_{(-\infty,a)}(A)$, $f(A_\alpha) = P_{(-\infty,a)}(A_\alpha)$, and $-\int f'(x)\xi_\alpha(x) = -\int_a^b f'(x)\xi_\alpha(x)$ $= -\xi_\alpha(x_0)\int_a^b f'(x)\,dx = \xi_\alpha(x_0)$, the common value of $\xi_\alpha(x)$ in (a,b).

(ii) It is a general fact about the trace of a difference of projections that is trace class, that it is an integer [**16, 5**]. Since $0 \le \xi_\alpha(x) \le 1$, it must be 0 or 1. [There is a simple argument: Since $(a,b) \cap \operatorname{spec}(A) = \varnothing$, $F(x+i0)$ is real and continuous on (a,b). If $F(x_0+i0) = -\alpha^{-1}$ anywhere on (a,b), then $F_\alpha(z) = F(z)/[1+\alpha F(z)]$ has a pole at x_0 and so x_0 is an eigenvalue of A_α. Since $(a,b) \cap \operatorname{spec}(A_\alpha) = \varnothing$, this cannot happen; that is, $1+\alpha F(x+i0)$ has no zeros and so its argument is constant at either 0 or 1.] □

REMARK. Because of (iii), $\xi_\alpha(x)$ is called a spectral shift.

There is an interesting extension of Theorem 1.8, an abstraction of a result of Javrjan [**28**]:

THEOREM 1.12. *Let $\beta < \gamma$. Then*

$$\int_\beta^\gamma \big(d\mu_\alpha(E)\big)\, d\alpha = \big(\xi_\gamma(E) - \xi_\beta(E)\big)\, dE. \tag{1.23}$$

REMARK. If $c \neq 0$, $\lim_{\alpha\to\infty} \operatorname{Arg}(1+\alpha c) - \operatorname{Arg}(1-\alpha c) = \pi$ so $\lim_{\alpha\to\infty}[\xi_\alpha(E) - \xi_{-\alpha}(E)] = 1$ and (1.23) implies (1.17).

PROOF. By (1.13), if $\operatorname{Im} z \neq 0$:

$$\begin{aligned}\int_\beta^\gamma \left[\int \frac{d\mu_\alpha(E)}{(E-z)^2}\right] d\alpha &= \int_\beta^\gamma \frac{d}{dz}\left[\frac{F(z)}{1+\alpha F(z)}\right] d\alpha \\ &= \frac{d}{dz}\int_\beta^\gamma \frac{F(z)}{1+\alpha F(z)}\, d\alpha \\ &= \frac{d}{dz}\left[\ln\big(1+\gamma F(z)\big) - \ln\big(1+\beta F(z)\big)\right] \\ &= \int \frac{\xi_\gamma(E) - \xi_\beta(E)}{(E-z)^2}\, dE\end{aligned}$$

proving (1.23) smeared with $(E-z)^2$. Such functions have linear combinations which are dense in a big enough space to imply (1.23) as an equality of measures. □

Since $d\mu_\alpha$ obeys $\int d\mu_\alpha(\lambda)(|\lambda|+1)^{-1} < \infty$, with bounds uniform in α on compacts, this result implies a similar estimate for ξ_α. Here's another proof:

THEOREM 1.13. (i) *If $\alpha < \infty$, then*

$$\int |\xi_\alpha(E)|(1+|E|)^{-1}\, dE < \infty \tag{1.24}$$

and

$$\int \xi_\alpha(E)(E-z)^{-1}\, dE = \ln\big(1+\alpha F(z)\big). \tag{1.25}$$

(ii) $\int |\xi_\alpha(E)|\, dE \equiv \infty$ *if and only if* $\varphi \notin \mathcal{H}$.

PROOF. (i) Integrating

$$\frac{d}{dz}\ln\big(1+\alpha F(z)\big) = \int \xi_\alpha(y)(y-z)^{-2}\, dy.$$

From $-w$ to $-x$ for $x < w$ real, we get

$$\ln\big(1+\alpha F(-x)\big) - \ln\big(1+\alpha F(-w)\big) = \int \xi_\alpha(y) \frac{1}{y+x}\left(\frac{w-x}{y+w}\right) dy.$$

As $w \to \infty$, $F(-w) \to 0$ so $\ln\bigl(1 + \alpha F(-w)\bigr) \to 0$ while $(w-x)/(y+w) = \bigl(1 + (y+x)/(w-x)\bigr)^{-1}$ is monotone increasing in w so (1.24) is proven and (1.25) is proven for $z = -x$, x real. By analytic continuation, it holds for all $z \in \mathbb{C}\backslash[i(\alpha), \infty)$.

(ii) By (1.25) and the monotone convergence theorem

$$\begin{aligned}\int |\xi_\alpha(E)|\, dE &= (\operatorname{sgn}\alpha) \lim_{x\to\infty} \int \frac{x}{E+x}\, [\xi_\alpha(E)]\, dE \\ &= \operatorname{sgn}\alpha \lim_{x\to\infty} x \ln\bigl(1 + \alpha F(-x)\bigr) \\ &= |\alpha| \|\varphi\|^2\end{aligned}$$

by Theorem 1.1(iv). □

REMARK. It is not hard to show that $\int x^n |\xi_\alpha(x)|\, dx < \infty$ if and only if $\varphi \in \mathcal{H}_{n/2}(A)$ and in that case

$$\int x^n \xi_\alpha(x)\, dx = \operatorname{Tr}(A_\alpha^{n+1} - A^{n+1})/(n+1).$$

Appendix to §1.4. The Krein Spectral Shift for Trace Class Perturbations. Let $C - A = (\varphi, \cdot)\varphi$ be rank one with $\|\varphi\| < \infty$ and $C \geq A$. Then for nice f, we have

$$\operatorname{Tr}\bigl(f(C) - f(A)\bigr) = \int f'(x) \xi_{C,A}(x)\, dx \tag{1.26a}$$

where

$$\int \xi_{C,A}(x)\, dx = \operatorname{Tr}(C - A) \tag{1.26b}$$

and

$$\int |\xi_{C,A}(x)|\, dx = \operatorname{Tr}(|C - A|). \tag{1.26c}$$

If $C - A$ is finite rank, we can diagonalize it and get it as a sum of rank one perturbations and (1.26) continue to hold. By (1.26c), ξ has a limit in L^1 norm as $C - A$ approaches a general trace class of operator. This proves

THEOREM 1.14 (KREIN'S SPECTRAL SHIFT FORMULA). *If A is self-adjoint and bounded below, B is trace class, and $C = A + B$, then there exists an L^1 function, $\xi_{C,A}(x)$, on $\mathbb{R}$ vanishing for x near $-\infty$ so that* (1.26) *hold.*

REMARK. ξ is no longer given as $\operatorname{Arg}\bigl(1 + \alpha F(z)\bigr)$ with $F(z)$ as simple as in the rank one case. However, in the rank one case

$$\begin{aligned}\operatorname{Tr}\bigl(1 + \alpha F(z)\bigr) &= 1 + \operatorname{Tr}\bigl((C - A)(A - z)^{-1}\bigr) \\ &= \det(1 + (C - A)(A - z)^{-1}) \\ &= \det\bigl((C - z)(A - z)^{-1}\bigr)\end{aligned}$$

and the argument of this has the correct additive properties. The result is [**33**]

$$\xi_{C,A}(\lambda) = \lim_{\epsilon \downarrow 0} \arg[\det(1 + (C - A)(A - \lambda - i\epsilon)^{-1})]$$

where det is a Fredholm determinant (see [**42**]).

1.5. Infinite Coupling. In the rank one setting we have been describing, let P be the projection onto the space spanned by φ if $\|\varphi\| < \infty$ and $P = 0$ if $\|\varphi\| = \infty$.

Look at the formula (1.15) for $(A_\alpha - z)^{-1}$. It clearly shows the first part of

THEOREM 1.15. (i) *As* $\alpha \to \pm\infty$, $(A_\alpha - z)^{-1}$ *converges in norm to a limit we call* $(A_\infty - z)^{-1}$ *given by*

$$(A_\infty - z)^{-1} = (A - z)^{-1} - F(z)^{-1}\big((A - \bar{z})^{-1}\varphi, \cdot\big)(A - z)^{-1}\varphi. \tag{1.27}$$

(ii) *If* $\|\varphi\| = \infty$, *then there is a self-adjoint operator* A_∞ *on* $\mathcal{H}$ *whose resolvent is* $(A_\infty - z)^{-1}$. *If* $\|\varphi\| < \infty$, *then there is a self-adjoint operator*, $\tilde{A}_\infty$, *on* $(1-P)\mathcal{H}$ *so that* $(A_\infty - z)^{-1}\eta = (\tilde{A}_\infty - z)^{-1}(1 - P)\eta$. *(We'll call* $\tilde{A}_\infty$ *as just* A_∞ *with the understanding that* $(A_\infty - z)^{-1} = 0$ *on* $P\mathcal{H}$ *and we'll write* $\mathcal{H}(A_\infty) = (1 - P)\mathcal{H}$.)

(iii) $\mathcal{H}_{+1}(A_\infty) = \{\psi \in \mathcal{H}_{+1}(A) \mid (\varphi, \psi) = 0\}$ *and*

$$(\psi, A_\infty\psi) = (\psi, A\psi)$$

for $\psi \in \mathcal{H}_{+1}(A_\infty)$.

PROOF. We have proven (i). To prove (ii), (iii), one uses the following monotone convergence theorem for forms ([**30, 41**]):

Let C_n be a family of non-negative self-adjoint operators with $\mathcal{H}_{+1}(C_{n+1}) \subseteq \mathcal{H}_{+1}(C_n)$ and $(\psi, C_{n+1}\psi) \geq (\psi, C_n\psi)$ if $\psi \in \mathcal{H}_{+1}(C_{n+1})$. Let

$$\mathcal{H}_{+1}(C_\infty) = \left\{\psi \in \bigcap_n \mathcal{H}_{+1}(C_n) \,\middle|\, \sup_n(\psi, C_n\psi) < \infty\right\}.$$

Let c_∞ be the quadratic form on $\mathcal{H}_{+1}(C_\infty)$ given by $c_\infty(\psi, \psi) = \sup_n(\psi, c_n\psi)$. Then, there is a self-adjoint operator C_∞ on $\overline{\mathcal{H}_{+1}(C_\infty)}$ so that if $(C_\infty - z)^{-1}$ is extended to $\mathcal{H}_{+1}(C_\infty)^\perp$ by setting it equal to zero there, then $(C_n - z)^{-1}$ converges strongly to $(C_\infty - z)^{-1}$.

This yields (ii), (iii) if one notes that

$$\sup_\alpha(\psi, A_\alpha\psi) < \infty \Longleftrightarrow (\psi, \varphi) = 0. \quad \square$$

Since $\xi_\alpha(x)$ is monotone in α, and bounded by 1, it has a limit $\xi_\alpha(x)$. In fact

$$\xi_\infty(x) = \operatorname{Arg}\big(F(x + i0)\big), \qquad 0 \leq \xi_\infty(x) \leq 1 \tag{1.28}$$

since

$$\xi_\alpha(x) = \operatorname{Arg}\big(\alpha^{-1} + F(x + i0)\big).$$

By a limiting argument, $\xi_\infty(x)\, dx = \int_0^\infty (d\mu_\alpha(x))\, d\alpha$.

In addition,

$$\operatorname{Tr}\big((A - z)^{-1} - (A_\infty - z)^{-1}\big) = \lim_{\alpha\to\infty} \frac{\alpha F'(z)}{1 + \alpha F(z)} = \frac{F'(z)}{F(z)} \tag{1.29a}$$

$$= \frac{d}{dz} \ln F(z) \tag{1.29b}$$

so by mimicking the arguments in the $\alpha < \infty$ case,

$$\operatorname{Tr}\big(f(A) - f(A_\infty)\big) = -\int f'(x)\xi_\infty(x)\, dx$$

for the f's in Theorem 1.10. One place that ξ_α and ξ_∞ differ though is (compare with (1.24))

PROPOSITION 1.16 ([**25**]).

$$\int (1+|x|)^{-1}\xi_\infty(x)\,dx = \infty$$

PROOF. By monotone convergence,

$$\begin{aligned}\int (1+x)^{-1}\xi_\infty(x)\,dx &= \lim_{\alpha\to\infty}\int (1+x)^{-1}\xi_\alpha(x)\,dx\\ &= \lim_{\alpha\to\infty}\ln\bigl(1+\alpha F(-1)\bigr) = \infty. \qquad \square\end{aligned}$$

If A is bounded, it is easy to see why the integral diverges; $\xi_\infty(x) = 1$ for $x > \|A\|$ so the integral diverges logarithmically exactly as $\ln x$. This is typical of the case when $\|\varphi\| < \infty$.

THEOREM 1.17 ([**25**]). *If $\|\varphi\| < \infty$, then $\int (1+|x|)^{-1}[1-\xi_\infty(x)]\,dx < \infty$ and*

$$\ln\left[(-z)F(z)/\|\varphi\|^2\right] = \int_0^\infty \frac{[\xi_\infty(x)-1]}{x-z}\,dx. \tag{1.30}$$

PROOF. We have that

$$\frac{d}{dz}\ln F(z) = \int \frac{\xi_\infty(x)}{(x-z)^2}\,dx \tag{1.31}$$

while clearly

$$\frac{d}{dz}\ln(-z^{-1}) = \int_0^\infty \frac{1}{(x-z)^2}\,dx \tag{1.32}$$

since for $z < 0$, both sides are just $(-z)^{-1}$. (1.32) is a direct calculation but can also be regarded as (1.31) for the case $F(z) = -1/z$.

By (1.31)–(1.32):

$$\frac{d}{dz}\ln\bigl((-z)F(z)/\|\varphi\|^2\bigr) = \int_0^\infty \frac{(\xi_\infty(x)-1)}{(x-z)^2}\,dx.$$

By Theorem 1.1(iv), $\ln\bigl((-z)F(z)/\|\varphi\|^2\bigr) \to 0$ as $z \to -\infty$ along the real axis. The same monotone convergence theorem argument that we used in the proof of Theorem 1.13(i) and the fact that $1-\xi_\alpha(x) \geq 0$ implies that $(1+|x|)^{-1}(1-\xi_\infty)$ is in L^1 and that (1.30) holds. $\square$

In §4.1, we will extend this idea to some cases where $\|\varphi\| = \infty$.

How can one get a handle on a spectral measure for A_∞? At first sight, it seems a hard job since

$$\lim_{\alpha\to\infty}\int \frac{d\mu_\alpha(x)}{x-z} = \lim_{\alpha\to\infty}\frac{F(z)}{1+\alpha F(z)} = 0$$

so $d\mu_\alpha(x) \to 0$ weakly. The key is to renormalize! Define

$$d\rho_\alpha(x) = (1+\alpha^2)\,d\mu_\alpha(x).$$

One can prove the following

THEOREM 1.18 ([**25**]). (i) *There exists a measure* $d\rho_\infty$ *(which is non-zero except for the trivial case where* φ *is an eigenvector of* A*) so that*

$$\int \frac{d\rho_\infty(x)}{(1+|x|)^2} < \infty$$

and

$$\lim_{\alpha\to\infty} \int f(x)\, d\rho_\alpha(x) = \int f(x)\, d\rho_\infty(x) \tag{1.33}$$

for any $f \in C_0^\infty(\mathbb{R})$.

(ii) *There exists a vector* $\eta \in \mathcal{H}_{-2}(A_\infty)$ *so that* $d\rho_\infty$ *is the spectral measure* $d\mu^\eta_{A_\infty}$. *If* φ *is cyclic for* A, *then* η *is cyclic for* A_∞.

(iii) (1.33) *holds for* $f(x) = (x-z)^{-3}$ *but it only holds for* $f(x) = (x-z)^{-2}$ *if* $\|\varphi\| = \infty$; *explicitly,*

$$\lim_{\alpha\to\infty} \int \frac{d\rho_\alpha(x)}{(x-z)^2} = \|\varphi\|^{-2} + \int \frac{d\rho_\infty(x)}{(x-z)^2}.$$

(iv) $d\rho_\infty$ *is the boundary value of* $-F(z)^{-1}$ *in the sense that for* $f \in C_0^\infty(\mathbb{R})$

$$\int f(x)\, d\rho_\infty(x) = \lim_{\epsilon\downarrow 0} \frac{1}{\pi} \int f(x) \operatorname{Im}[-F(x+i\epsilon)^{-1}]\, dx.$$

We refer the reader to [**25**] for a complete proof but will make a series of remarks that illuminate the result and partly prove it.

REMARKS. 1. Notice that η is only stated to be in $\mathcal{H}_{-2}(A_\infty)$, not $\mathcal{H}_{-1}(A_\infty)$. Indeed, the differential equation case of the next section when $V = 0$ is one where $d\rho_\infty(x) \sim cx^{1/2}\, dx$ and $\eta \notin \mathcal{H}_{-1}(A_\infty)$. Indeed [**25**], for any $\theta < 2$, there are examples where $\int d\rho_\infty(x)/(1+|x|)^\theta = \infty$.

2. If A is bounded, $\eta = (1-P)A\varphi$. Otherwise, one has that η obeys

$$(A_\infty - z)^{-1}\eta = (1-P)F(z)^{-1}(A-z)^{-1}\varphi$$

(with the convention that $P = 0$ if $\|\varphi\| = \infty$).

3.

$$\int \frac{d\mu_\alpha(x)}{(x-z)^2} = \frac{d}{dz} \frac{F(z)}{1+\alpha F(z)} = \frac{F'(z)}{\left(1+\alpha F(z)\right)^2}$$

so

$$\lim_{\alpha\to\infty} \int \frac{d\rho_\alpha(x)}{(x-z)^2} = \frac{F'}{F^2} \tag{1.34}$$

so the $\|\varphi\|^{-2}$ term in (iii) is related to Theorem 1.1(v).

4. Notice that the right side of (1.34) is $(d/dz)(-1/F)$. This is the key to proving (iv).

There is an interesting reparametrization of the ρ's that is exactly what is used in the differential equation case. For $\theta \in (0,\pi)$, we can define $\alpha = -\cot(\theta)$ so α runs from $-\infty$ to ∞ as θ runs from 0 to π. Let

$$d\tilde\rho_\theta(x) = d\rho_{-\cot(\theta)}(x) = [\sin^2(\theta)]^{-1}\, d\mu_{-\cot(\theta)}(x).$$

Then, $d\tilde{\rho}_\theta$ has a limit as $\theta \to 0, \pi$ which is just what we called $d\rho_\infty(x)$. Moreover, we claim that

$$\int_\theta^\pi (d\tilde{\rho}_\theta(E))\, d\theta = dE \tag{1.35}$$

for

$$\int_0^\pi d\tilde{\rho}_\theta(E)\, d\theta = \int_{-\infty}^\infty (1+\alpha^2)\, d\mu_\alpha(E)\, \frac{d\alpha}{1+\alpha^2} = dE.$$

1.6. Boundary Condition Variation of ODE's. For applications to Schrödinger operators, there are two main categories of the general theory: variation of potential in the discrete (Jacobi) case, which we'll discuss in the next section, and the subject of this section, variation of boundary condition.

We begin with a lightning review of the standard Weyl m-function approach (see **[48]**) and then rephrase the set-up into the context of these lectures. We will consider a Schrödinger operator on $(0, \infty)$ (and later on $(-\infty, \infty)$) of the form

$$H_\theta u = -u''(x) + V(x)\, u(x) \tag{1.36}$$

where, for simplicity, we suppose that $V(x)$ is bounded from below but continuous. This makes H limit point at infinity so we only need a boundary condition at zero, which is where the parameter θ enters. The boundary condition at 0 is

$$u(0)\cos\theta + u'(0)\sin\theta = 0 \qquad 0 \le \theta < \pi \tag{1.37}$$

so that $\theta = 0$ is the Dirichlet boundary condition and $\theta = \pi/2$ is the Neumann boundary condition.

Explicitly, H_θ defines a self-adjoint operator with domain

$$D(H_\theta) = \left\{ u \in L^2(0,\infty) \;\middle|\; \begin{array}{l} u,\ u' \text{ are locally absolutely continuous} \\ \text{functions, } u \text{ obeys (1.37) and } H_\theta u \in L^2 \end{array} \right\}.$$

We will often be interested in solutions of the differential equation

$$-u'' + Vu = zu \tag{1.38}$$

and will consider the Wronskian $W(f, g)$ of two functions

$$W(f,g)(x) = f(x)g'(x) - f'(x)g(x).$$

As usual, if f, g solve (1.38) for the same z, then $W(f,g)(x)$ is independent of x.

Given θ, three solutions of (1.38) will concern us: ϕ which obeys the boundary condition (1.37), η defined with a "dual" boundary condition, and ψ_+ for $\operatorname{Im} z \neq 0$, the solution which is L^2 near $x = \infty$. Explicitly, ϕ, η are defined by boundary values at $x = 0$:

$$\begin{aligned} \phi_\theta(z,0) &= -\sin(\theta) & \qquad \phi'_\theta(z,0) &= \cos(\theta) \\ \eta_\theta(z,0) &= \cos\theta & \qquad \eta'_\theta(z,0) &= \sin(\theta) \end{aligned}$$

so

$$W(\eta_\theta, \phi_\theta)(x) = 1. \tag{1.39}$$

The third solution ψ_+ is defined with mixed boundary conditions; it is in $L^2\big((0,\infty)\big)$, which determines it up to a multiplicative constant, and it is normalized by

$$W(\psi_{+,\theta},\phi_\theta)=\psi_{+,\theta}(0)\cos\theta+\psi'_{+,\theta}(0)\sin\theta=1. \tag{1.40}$$

Since ϕ,η are independent solutions of (1.38), we have $\psi_+=a\eta+b\phi$. By (1.39)–(1.40), $a=1$. We therefore define the Weyl m-function, $m_\theta(z)$, by

$$\psi_{+,\theta}(z,x)=\eta_\theta(z,x)+m_\theta(z)\phi_\theta(z,x) \tag{1.41a}$$

or equivalently:

$$m_\theta(z)=W(\eta_\theta,\psi_{+,\theta}) \tag{1.41b}$$

or

$$m_\theta(z)=-\lim_{R\to\infty}\frac{\eta_\theta(z,R)}{\phi_\theta(z,R)}. \tag{1.41c}$$

The last (which follows from $\psi_+\to 0$ at infinity) is the usual definition in the limit point case because it is what Weyl used, although the first two are more useful.

It can be seen that

$$\operatorname{Im} z>0\Longrightarrow \operatorname{Im} m_\theta(z)>0. \tag{1.42}$$

We won't prove this now, since it will follow from the general theory of Borel transforms once we have set up the connection. The spectral measure $d\rho_\theta$ is defined by

$$d\rho_\theta(x)=\lim_{\epsilon\downarrow 0}\frac{1}{\pi}[\operatorname{Im} m_\theta(x+i\epsilon)]\,dx.$$

For $\theta\neq 0$, as we'll see

$$m_\theta(z)=\cot(\theta)+\int\frac{d\rho_\theta(x)}{x-z}. \tag{1.43}$$

The Green's function, $G_\theta(x,x';z)$ is the integral kernel of $(H_\theta-z)^{-1}$, that is

$$\big((H_\theta-z)^{-1}u\big)(x)=\int G_\theta(x,x';z)u(x')\,dx'.$$

Because of the normalization (1.40), one has

$$G_\theta(x,x';z)=\phi_\theta(z,x_<)\psi_{+,\theta}(z,x_>)$$

where $x_<=\min(x,x'),x_>=\max(x,x')$. In particular, by (1.41a)

$$G_\theta(0,0;z)=+\sin^2\theta\big(-\cot\theta+m_\theta(z)\big). \tag{1.44}$$

As a final general formula, we note that Wronskian gymnastics (we'll get it other ways below) shows that

$$m_\kappa(z)=\frac{-\sin(\kappa-\theta)+\cos(\kappa-\theta)m_\theta(z)}{\cos(\kappa-\theta)+\sin(\kappa-\theta)m_\theta(z)}. \tag{1.45a}$$

In particular,

$$m_0(z)=-\frac{1}{m_{\pi/2}(z)}. \tag{1.45b}$$

We want to show how to think of this in terms of rank one perturbations. Let $u \in D(H_\theta)$ be real-valued. Then, integrating by parts

$$
\begin{aligned}
(u, H_\theta u) &= \int_0^\infty (u(x))[-u'' + Vu](x)\,dx \\
&= \int_0^\infty [u'(x)^2 + V(x)u^2(x)]\,dx + u'(0)u(0) \\
&\equiv \tau(u,u) - \cot(\theta)u(0)^2
\end{aligned}
\tag{1.46}
$$

because of the boundary condition.

Let A be the $\theta = \pi/2$ Neumann boundary condition operator. A simple Sobolev estimate shows that for any $u \in \mathcal{H}_{+1}(A)$ is continuous in x and

$$|u(0)|^2 \leq c(u, (A+1)u),$$

that is, $\delta(x)$, the Dirac delta function, lies in $\mathcal{H}_{-1}(A)$. We can form

$$A_\alpha = A + \alpha\delta(x).$$

(1.46) says that $H_\theta = A_{-\cot(\theta)}$. Moreover,

$$F_\alpha(z) = (\delta, (A_\alpha - z)^{-1}\delta)$$

is exactly $G_\theta(0,0;z)$. Since G is related to m by (1.44), equation (1.45) is just the relation $F_\alpha(z) = F(z)/[1 + \alpha F(z)]$.

In addition, because of (1.44) and (1.43), we have

$$d\rho_\theta(E) = \sin^2\theta\, d\mu_{-\cot(\theta)}(E),$$

that is, the spectral measure $d\rho$ is precisely the measure we called $d\tilde{\rho}_\theta$ in the last section. The Dirichlet spectral measure, $d\tilde{\rho}_0$, is an example of a $d\rho_\infty$ with $\int d\rho_\alpha(E)/[|E|+1] = \infty$. (1.45b) is just Theorem 1.18(iv). The vector η of §1.5 is just δ'; see [**25**].

The point of this section is the spectral analysis of rank one perturbations in the next section will apply to boundary condition variation directly (rather than by analogy).

1.7. Jacobi Matrices. Jacobi matrices are operators on $\ell^2(\mathbb{Z}^\nu)$ which are discrete analogs of Schrödinger operators. We write $u \in \ell^2(\mathbb{Z}^\nu)$ as a function on $\mathbb{Z}^\nu$. Given a function, V, on $\mathbb{Z}^\nu$, we define the associated Jacobi matrix, h, by

$$(hu)(n) = \sum_{|j|=1} u(n+j) + V(n)u(n) \tag{1.47a}$$

which we write as

$$h = h_0 + V. \tag{1.47b}$$

Let δ_0 be the vector in ℓ^2 which is given by

$$\delta_0(n) = \delta_{n0}.$$

Let P be the projection on δ_0 : $(\delta_0, \cdot)\delta_0$. Fix V and let A be the Jacobi matrix associated to $V - \delta_{\cdot 0}(V(0))$. Then

$$h = A + V(0)P$$

and variations of V are precisely rank one perturbations.

As an application of this idea, let $V_\omega(n)$ be a random process where the $V_\omega(n)$ are independent, identically distributed random variables with distribution $p(v)\,dv$ where p is bounded. There is a fundamental quality called the integrated density of states, $k(\lambda)$, which is given by the following: Let $d\mu_\omega^{(0)}$ be the spectral measure for $h_0 + V_\omega$ with vector δ_0. Then

$$k(\lambda) = E_\omega\big(\mu_\omega^{(0)}(-\infty, \lambda)\big). \tag{1.48}$$

This is not the basic definition but a derived formula for k (see, e.g., [**6, 9**]). We claim

THEOREM 1.19 (WEGNER'S ESTIMATE [**49**]).

$$|k(\lambda) - k(\lambda')| \le c|\lambda - \lambda'|$$

where $c = \sup_\gamma p(\gamma)$.

PROOF. The result is equivalent to saying that dk is absolutely continuous with respect to $d\lambda$ with $dk = q(\lambda)\,d\lambda$ with $\|q\|_\infty \le c$. Fix $\big(V_\omega(n)\big)_{n\neq 0}$. Then, by Theorem 1.8

$$\int [d\mu_\omega^{(0)}(\lambda)] p(v(0))\,dv(0) \le c \int d\mu_\omega^{(0)}(\lambda)\,dv(0) = c\,d\lambda.$$

Averaging over $\big(V_\omega(n)\big)_{n\neq 0}$, we get

$$E_\omega(d\mu_\omega) \le c\,d\lambda$$

which is exactly what we want. □

2. Spectral Theory of Rank One Perturbations

Introduction. We begin our analysis with the study of the absolutely continuous spectrum, obtaining a result that also follows from the trace class theory of scattering, but without wave operators. We next discuss the fundamental work of Aronszajn [**3**] and Donoghue [**4**], which identifies spectral information for A_α in terms of $F(z)$. Next, we discuss the Simon-Wolff [**47**] work on localization and apply it to one-dimensional random Jacobi matrices. Finally, we discuss recent results of Gordon [**27**] and del Rio *et al.* [**12**] on singular spectrum.

Henceforth, we assume that φ *is a cyclic vector for* A, *that is,* $\{(A-z)^{-1}\varphi \mid z \in \mathbb{C}\}$ *is a total set for* $\mathcal{H}$. In general, if $\mathcal{H}_0$ is the closed subspace generated by these vectors, then $\mathcal{H}_0^\perp$ is an invariant space for each A_α and $A_\alpha = A$ on $\mathcal{H}_0^\perp$. Thus, the extension from the cyclic to general case is trivial. We make the assumption to be able to focus on the real issues and not have to state the theorems in convoluted forms to give the general case.

2.1. Invariance of the Absolutely Continuous Spectrum. Let $f(x)\,dx$ and $g(x)\,dx$ be two absolutely continuous (positive) measures on $\mathbb{R}$. Recall these measures are equivalent if and only if $\{x \mid f(x) \neq 0\}$ and $\{x \mid g(x) \neq 0\}$ agree up to sets of Lebesgue measure zero. Moreover, if A is multiplication by x on $L^2(\mathbb{R}, f\,dx)$ and C is multiplication by x on $L^2(\mathbb{R}, g\,dx)$, then A and C are unitarily equivalent if and only if $f\,dx$ and $g\,dx$ are equivalent. With these preliminaries

THEOREM 2.1. *For all $\alpha \neq \beta$, $(A_\alpha)_{\rm ac}$ and $(A_\beta)_{\rm ac}$, the absolutely continuous parts of A_α and A_β are unitarily equivalent.*

PROOF. Since $A_\alpha = A_\beta + (\alpha - \beta)(\varphi, \cdot)\varphi$, we can suppose for $\beta = 0$ w.l.o.g. By the above discussion and Theorem 1.6(iv), it suffices to show that $S_1 \equiv \{E \mid \operatorname{Im} F(E + i0) \neq 0\}$ and $S_2 \equiv \{E \mid \operatorname{Im} F_\alpha(E + i0) \neq 0\}$ agree up to sets of measure zero. By the Aronszajn-Krein formula $F_\alpha = \frac{F}{1+\alpha F}$, which implies that

$$\operatorname{Im} F_\alpha(z) = \frac{\operatorname{Im} F(z)}{|1 + \alpha F(z)|^2}. \tag{2.1}$$

It follows that S_1 and S_2 agree up to the sets where $F(E+i0)$ fails to exist, the set where $F(E + i0) = \infty$, and the set where $F(E + i0) = -\alpha^{-1}$. These sets all have measure zero, so S_1 and S_2 agree up to sets of measure zero. □

REMARKS. 1. The result extends to the non-cyclic case and so extends to the finite rank case. It should be possible to accommodate the trace class case. This is an area under present study.

2. Since $(A_\infty + 1)^{-1} - (A_0 + 1)^{-1}$ is rank one in the usual sense, we can apply the theorem to $(A_\infty - 1)^{-1}$ and $(A_0 + 1)^{-1}$ and so, by a spectral mapping theorem, extend the result to allow $\alpha = \infty$. One can also get $\alpha = \infty$ by using $d\rho_\infty$ and $\operatorname{Im}(-1/F) = \operatorname{Im} F/|F|^2$.

2.2. The Aronszajn-Donoghue Theory. Recall that

$$G(x) = \int d\mu(y)/\big((x-y)^2\big).$$

The following basic theorem was proven by Aronszajn [**3**] for boundary condition dependence of spectrum for Sturm-Liouville operators and then extended to (bounded) rank one perturbations by Donoghue [**14**].

THEOREM 2.2. *For $\alpha \neq 0$ ($\alpha =$ infinity allowed with $\infty^{-1} = 0$), define*

$$\begin{aligned} S_\alpha &= \{x \in \mathbb{R} \mid F(x + i0) = -\alpha^{-1}; G(x) = \infty\} \\ P_\alpha &= \{x \in \mathbb{R} \mid F(x + i0) = -\alpha^{-1}; G(x) < \infty\} \\ L &= \{x \in \mathbb{R} \mid \operatorname{Im} F(x + i0) \neq 0\}. \end{aligned}$$

Then (i) $\{S_\alpha\}_{\alpha\neq 0; |\alpha| \leq \infty}, \{P_\alpha\}_{\alpha \neq 0; |\alpha|\leq\infty}$ *and L are mutually disjoint.*

(ii) *P_α is the set of eigenvalues of A_α. In fact*

$$\begin{aligned} (d\mu_\alpha)_{\rm pp}(x) &= \sum_{x_n \in P_\alpha} \frac{1}{\alpha^2 G(x_n)} \delta(x - x_n) \qquad \alpha < \infty \\ (d\rho_\infty)_{\rm pp}(x) &= \sum_{x_n \in P_\infty} \frac{1}{G(x_n)} \delta(x - x_n) \qquad \alpha = \infty. \end{aligned}$$

(iii) *$(d\mu_\alpha)_{\rm ac}$ is supported on L, $(d\mu_\alpha)_{\rm sc}$ is supported on S_α.*

(iv) *For $\alpha \neq \beta$, $(d\mu_\alpha)_{\rm sing}$ and $(d\mu_\beta)_{\rm sing}$ are mutually singular.*

REMARK. We say that a set S "supports" a measure $d\nu$ if $\nu(\mathbb{R} \setminus S) = 0$.

PROOF. (i) is trivial and (iv) follows from (i), given (ii) and (iii). That L supports $(d\mu_\alpha)_{\rm ac}$ is proven in the last section.

By Theorem 1.6(iv), $(d\mu_\alpha)_{\rm sing}$ is supported by $\{E \mid \lim_{\epsilon\downarrow 0} |F_\alpha(E+i\epsilon)| = \infty\}$. But

$$F_\alpha(z) = \frac{F(z)}{1+\alpha F(z)}$$

says that $|F_\alpha(E+i\epsilon)| \to \infty$ if and only if $F(E+i\epsilon) \to -1/\alpha$.

(ii), (iii) thus follow by Theorem 1.6(iii) and the claim that when $F(E+i\epsilon) \to -1/\alpha$ (even if $G(E) = \infty$):

$$\lim_{\epsilon\downarrow 0} \epsilon \operatorname{Im} F_\alpha(E+i\epsilon) = 1/[\alpha^2 G(E)] \qquad (0<\alpha<\infty) \tag{2.2a}$$

$$\lim_{\epsilon\downarrow 0} \epsilon \operatorname{Im}\bigl(-F(E+i\epsilon)\bigr)^{-1} = 1/G(E) \qquad (\alpha = \infty). \tag{2.2b}$$

If $G(E) < \infty$, (2.2) follows from Theorem 1.2(v) which implies that $F(E+i\epsilon) = -1/\alpha + i\epsilon G(x) + o(\epsilon)$ and the formula for F_α. If $G(E) = \infty$ and $\alpha = \infty$, Theorem 1.2(iv) implies $\epsilon^{-1}|F(x+i\epsilon)| \to \infty$ so $|\epsilon/F(x+i\epsilon)| \to 0$ proving (2.2b) in that case. If $G(E) = \infty$ and $0<\alpha<\infty$, by Theorem 1.2(iv) again, $\epsilon^{-1}\operatorname{Im}(1+\alpha F) \to \infty$ so $\epsilon|1+\alpha F|^{-1} \to 0$. Thus, since $F \to -1/\alpha$, $\epsilon|F/1+\alpha F| \to 0$. □

This theorem has interesting consequences for one-dimensional Schrödinger operators and Jacobi matrices:

THEOREM 2.3. *Let A be a Schrödinger operator on $[0,\infty)$ with Neumann boundary conditions at $x=0$ as in §1.6. Then $G(E)<\infty$ if and only if both of the following hold*:

(i) *E is not an eigenvalue of A;*
(ii) *(1.38) (with $E=z$) has a solution which is L^2 at $+\infty$.*

PROOF. Since $G(E)$ finite implies that $F(E+i0)$ exists and is real and finite, Theorem 2.2 implies that $G(E)<\infty$ if and only if E is an eigenvalue of $A+\alpha\delta$ for some $\alpha \neq 0$ in $\mathbb{R}\cup\{\infty\}$. But this is equivalent to (i), (ii). □

THEOREM 2.4. *Let A be a Jacobi matrix on $\ell^2(\mathbb{Z})$ and $\varphi = \delta_0$ as in §1.7. Then $G(E)<\infty$ if and only if*

(i) *E is not an eigenvalue of A;*
(ii) *One of the following holds*:
   (a) *The equation*

$$u(n+1)+u(n-1)+V(n)u(n) = Eu(n) \tag{2.3}$$

   has an ℓ^2 solution on $(0,\infty)$ with $u(0)=0$;
   (b) *(2.3) has an ℓ^2 solution on $(-\infty,0)$ with $u(0)=0$;*
   (c) *(2.3) has ℓ^2 solutions $u_\pm$ on both $(0,\infty)$ and $(-\infty,0)$ with both $u_+(0) \neq 0$ and $u_-(0)\neq 0$.*

REMARKS. 1. To say u is a solution on $(0,\infty)$ means u is defined on $[0,\infty)$ but (2.3) holds for $n=1,2,\ldots$.

2. In particular, if (2.3) has ℓ^2 solutions on both $(0,\infty)$ and $(-\infty,0)$, then either E is an eigenvalue of A or $G(E)<\infty$.

PROOF. (a), (b) of (ii) hold if and only if E is an eigenvalue of A_∞. (c) holds if and only if E is an eigenvalue of some A_α with $|\alpha|<\infty$. Given these, the proof is the same as for the last two theorems. □

2.3. The Simon-Wolff Criterion. The following is useful in discussing localization for random Jacobi matrices:

THEOREM 2.5 ([**47**]). *Fix an interval $[a,b]$. Then the following are equivalent:*

(a) $G(E) < \infty$ *for a.e.* E *in* $[a,b]$ *(w.r.t. Lebesgue measure).*

(b) *For a.e.* α *(w.r.t. Lebesgue measure),* A_α *has only pure point spectrum in* $[a,b]$.

PROOF. Let $S = \{E \in [a,b] \mid G(E) = \infty\}$. Suppose (a) holds. Then $|S| = 0$ by hypothesis. By Theorem 1.8 that means $\int \mu_\alpha(S)\, d\alpha = 0$ so $\mu_\alpha(S) = 0$ for a.e. α. But if $\mu_\alpha(S) = 0$, then in the language of Theorem 2.2, $\mu_\alpha(L \cap [a,b]) = \mu_\alpha(S_\alpha \cap [a,b]) = 0$. It follows by that theorem that μ_α has only point spectrum, so (b) holds.

Conversely, if A_α has only point spectrum on $[a,b]$, then $\mu_\alpha(S) = 0$. If (b) holds, $\int \mu_\alpha(S)\, d\alpha = 0$, so by Theorem 1.8, $|S| = 0$ and (a) holds. □

As an example of how to apply this theorem, we want to prove localization for the Anderson model in one dimension following [**45**]. We'll use various facts about that model proven elsewhere.

THEOREM 2.6. *Let* $\{v_\omega(x)\}_{n\in\mathbb{Z}}$ *be bounded i.i.d.'s. with distribution* $p(\lambda)\, d\lambda$. *Let* $h_\omega = h_0 + v_\omega$ *as in §1.7. Then for a.e.* ω, h_ω *has only point spectrum.*

REMARKS. 1. The general proof following the steps below requires a lot less than bounded or independent random variables; see [**45**].

2. The proofs show that the eigenfunctions decay exponentially.

3. If $\operatorname{supp} p = [a,b]$, then $\operatorname{spec}(h_\omega) = [a-2, b+2]$ with probability one, so this is point spectrum dense in an interval.

PROOF. 1. The transfer matrix, $A_n(E,\omega)$, is the two by two matrix defined by $\binom{u(n+1)}{u(n)} = A_n(E,\omega)\binom{u(1)}{u(0)}$ where u obeys (2.3) for v_ω. General principles (the subadditive ergodic theorem; see [**45**]) imply that for each fixed E,

$$\lim_{|n|\to\infty} \frac{1}{|n|} \ln \|A_n(\omega, E)\| = \gamma(E) \tag{2.4}$$

for a.e. ω and a fixed γ. Obviously, whether the limit in (2.4) holds only depends on v_ω in a neighborhood of infinity.

2. It can be proven that $\gamma(E) > 0$ for almost all E. In the i.i.d. case under discussion, this follows from Furstenberg's theorem ([**19**]) for *all* E, but under much weaker conditions (v-non-deterministic), one gets the result for a.e. E from Kotani's theory [**32, 44**].

3. The theorem of Ruelle-Oseledec [**39, 37**] implies that when (2.4) holds, with $\gamma(E) > 0$, then there exist initial values $u_\pm(0), u_\pm(1)$ so that the corresponding solutions $u_\pm$ of (2.3) are ℓ^2 at $\pm\infty$. (In fact, the solutions decay exponentially.)

4. Theorem 2.4 then implies that if $\gamma(E) > 0$ and (2.4) holds, then either E is an eigenvalue of h_ω or $G(E,\omega) < \infty$ (with $G(E,\omega)$ associated to h_ω and δ_0).

5. By Fubini's theorem, for a.e. $\{v_\omega(n)\}_{n\neq 0}$, (2.4) holds for a.e. E with $\gamma > 0$. Since h_ω has only countably many eigenvalues, we conclude that $G(E,\omega) < \infty$ for a.e. E (for this typical set of ω). Thus, by Theorem 2.5, $h_\omega + \alpha\delta_0$ has pure point spectrum for a.e. α.

6. By the hypothesis on the absolute continuity of the distribution of $v_\omega(0)$, h_ω has point spectrum for a.e. ω. □

Essentially, the identical argument proves

THEOREM 2.7. *Let $\{V_\omega(x)\}_{-\infty<x<\infty}$ be a non-deterministic bounded ergodic process on $\mathbb{R}$. Then for a.e. ω and a.e. $\theta \in [0,\pi)$,*

$$-\frac{d^2}{dx^2} + V_\omega(x)$$

on $L^2\big((0,\infty);dx\big)$ with θ the boundary condition at 0 has only pure point spectrum.

In Section 3, we'll need the following formula of Simon-Wolff [**47**]:

THEOREM 2.8. *Suppose $\|\varphi\| = 1$. Let δ_n be an orthonormal basis for $\mathcal{H}$ with $\delta_0 = \varphi$. Then for each $E \in \mathbb{R}$, $\sum_n |(\delta_n, (A-E-i\epsilon)^{-1}\delta_0)|^2$ is monotone increasing as $\epsilon \downarrow 0$ and*

$$G(E) = \lim_{\epsilon\downarrow 0} \sum_n |(\delta_n, (A-E-i\epsilon)^{-1}\delta_0)|^2.$$

PROOF. By the Plancherel theorem, the sum is $\|(A-E-i\epsilon)^{-1}\varphi\|^2 = \int d\mu_0(x)/\big((x-E)^2+\epsilon^2\big)$. The monotonicity and identification of the limit are immediate. □

2.4. Instability of Point Spectrum. Random potentials have point spectrum embedded in the essential spectrum of h_ω. One of our results in this section is that there are locally uncountably many values of $v_\omega(0)$ which turn that point spectrum into singular continuous spectrum. These are ideas obtained by Gordon [**26, 27**], and then independently by del Rio *et al.* [**11, 12**], whose presentation we follow.

The key in many ways is Theorem 1.2(iv) which says that $\{x \mid G(x) = \infty\}$ is a dense G_δ. As an example of what this implies, we prove the following (first proven in a slightly weaker form by del Rio [**10**]):

THEOREM 2.9. *Let V be a bounded potential on $[0,\infty)$. Let S be the set of non-isolated points of $\sigma_{\rm ess}\big(-d^2/dx^2+V\big)$ (this set is independent of boundary conditions at 0). Then, there exists $T \subset S$, a dense G_δ in S so that*

(i) *$E \in T$ is not an eigenvalue of $-d^2/dx^2+V$ for any boundary condition θ at $x=0$.*

(ii) *If $E \in T$, $\big(-d^2/dx^2+V\big)u = Eu$ does not have a solution ℓ^2 at infinity and, in particular, there is no Lyapunov behavior with $\gamma(E) > 0$ for $E \in T$.*

PROOF. By Theorem 1.2(iv), the $G(E)$ associated to the Neumann operator with $\varphi = \delta(x)$ has $G(E) = \infty$ on T_0, a dense G_δ in S. Let T be T_0 minus any eigenvalues for $\theta = \pi/2$. (i), (ii) follow by the results earlier in this section. □

EXAMPLE. If V is random with $\operatorname{spec}(H) = [0,\infty)$, this implies non-Lyapunov behavior for a dense G_δ on $[0,\infty)$, even though we know there is such behavior for a.e. E. Dense G_δ sets are sometimes called Baire generic while complements of sets of measure zero are called Lebesgue generic. Both kinds of sets are locally uncountable. The theorem says the sets where one does/does not have Lyapunov behavior with $\gamma > 0$ are intertwined, locally uncountable sets!

We want to turn the $G(E) = \infty$ result into one about α values.

THEOREM 2.10 ([**12**]). *Let $d\mu$ be any measure on $\mathbb{R}$ obeying (1.4). Then the set of values $\{F(E+i0) \mid E \in \operatorname{supp}(d\mu)$ and $G(E) < \infty\}$ is the complement of a dense G_δ.*

PROOF. There is a general argument that the complement is a G_δ [**46**], so we focus on the fact that the set, S, is a countable union of nowhere dense sets. We sketch the idea behind the proof in [**12**], leaving the details for that paper.

Write $\{E \mid\in \text{supp}(d\mu) \text{ and } G(E) < \infty\}$ as a countable union of sets A_n so that

(i) $G(E)$ is close to the constant, γ_n, on A_n.

(ii) For some δ_n and all $x \in A_n$, $\int_{|x-y|<\delta_n} |x-y|^{-2}\, d\mu(y)$ is small compared to the variation of G over A_n.

(iii) For each $x \in A_n$, consider the equilateral triangle, T_x, in $\mathbb{C}$ with one vertex at x and base on the line $\text{Im}\, z = a\delta_n$ with a fixed number, a, (fixed in the proof in [**12**]). Then $\bigcup_{x\in A_n} T_x$ is connected.

By (i), (ii) $F'(E)$ is very close to the constant γ_n on each T_x so $|F(z) - F(y) - \gamma_n(z-y)| \le \gamma_n|y-z|/3$ on T_x. By a simple geometric argument using (iii), $|F(z) - F(y) - \gamma_n(z-y)| \le \frac{2}{3}\gamma_n|y-z|$ for $y, z \in \bigcup_{x\in A_n} T_x$ and so in A_n. Since A_n is nowhere dense and this inequality says that F on A_n is the restriction of a homeomorphism from $\mathbb{R}$ to $\mathbb{R}$, we see that $F[A_n]$ is nowhere dense. $\square$

Combining this with Theorem 2.2, we immediately get

THEOREM 2.11. *In the general context of rank one perturbations, suppose that* $[a,b] \subset \text{spec}(A)$ *and* $\text{spec}_{\text{ac}}(A) \cap (a,b) = \varnothing$. *Then for a dense* G_δ *of* α*'s,* A_α *has purely singular continuous spectrum on* (a,b).

We emphasize that there is a distinction between singular and point spectrum. If $\text{spec}_{\text{ac}}(A) \cap (a,b) = \varnothing$ and $(a,b) \subset \text{spec}(A)$, there are *always* many α's for which A_α has singular continuous spectrum. There may or may not be α's when there is point spectrum there. The Simon-Wolff criterion gives cases where there is point spectrum for many α's, but the criterion may or may not hold.

REMARK. This result on singular continuous spectrum for dense G_δ's is one of many recently discovered [**27, 46, 11, 29**]. For example, in [**46**], it is proven that in $C_\infty(\mathbb{R}^n)$, the continuous functions vanishing at infinity, there is a dense G_δ, S, so that if $V \in S$, then $-\Delta + V$ has purely singular continuous spectrum on $[0,\infty)$.

COROLLARY 2.12. *Let* $V_\omega(x)$ *be a non-deterministic ergodic random process for* $x \in \mathbb{R}$ *so that* $\text{spec}(-d^2/dx^2 + V) = [a,\infty)$ *for some* a. *Let* H_θ *be* $-d^2/dx^2 + V$ *on* $L^2(0,\infty)$ *with* θ *boundary condition at* $x = 0$. *Then for a dense* G_δ *of* θ*'s,* H_θ *has purely singular continuous spectrum on* $[0,\infty)$ *(while for Lebesgue almost all, it has only point spectrum there).*

COROLLARY 2.13. *Let* $\{V_\omega(n)\}_{n\in\mathbb{Z}^\nu}$ *be independent, identically distributed random variables with uniform distribution in* $[a,b]$ *and let* H_ω *be the corresponding Jacobi matrix on* $\ell^2(\mathbb{Z}^\nu)$. *Let* $|b-a|$ *be so large that* H_ω *has only point spectrum for a.e.* ω *(see Section 3). Then for a.e.* ω*, there is a dense* G_δ*,* S *in* $[a,b]$*, so that for* $\alpha \in S$

$$W(n) = \begin{cases} V_\omega(n) & n \neq 0 \\ \alpha & n = 0 \end{cases}$$

yields a Jacobi matrix with only singular continuous spectrum.

Both corollaries follow immediately from the theorem and earlier discussions of the examples.

2.5. Examples. The astute reader may have noticed that $d\mu_0$ completely determines the spectral properties of the family A_α (assuming cyclicity, of course). Moreover, given μ_0, let A be multiplication by x on $L^2(\mathbb{R}, d\mu_0)$ and let $\varphi(x) \equiv 1$.

Then $F(z)$ is just the Borel transform of $d\mu_0$. So we can describe examples by giving the measure $d\mu_0$, which we'll denote by $d\mu$.

EXAMPLE 1 (THE FRIEDRICHS MODEL). Let $d\mu = dx \restriction [0,1] + \delta_{1/2}$ be the sum of Lebesgue measure on $[0,1]$ and a point mass at $1/2$. Then $G(x) = \infty$ on $[0,1]$, and

$$\frac{1}{\pi} \operatorname{Im} F(x+i0) = \begin{cases} 1, & 0 < x < 1; x \neq 1/2 \\ 1/2, & x = 0, 1. \end{cases}$$

Thus, A_α has only a.c. spectrum on $[0,1]$ for $\alpha \neq 0$ (there is a single eigenvalue on $(-\infty, 0)$ for $\alpha < 0$ and on $(1, \infty)$ for $\alpha > 0$). The embedded eigenvalue dissolves.

EXAMPLE 2. Let $\{q_n\}_{n=1}^\infty$ be a counting of the rationals on $[0,1]$. Let

$$d\mu = dx \restriction [0,1] + \sum_{n=1}^\infty 2^{-n} \delta_{q_n}.$$

As above, all eigenvalues disappear for $\alpha \neq 0$. The interesting case is to take $d\mu = d\mu_{-1}$. The new A has no eigenvalues on $[0,1]$ but $A + (\varphi, \cdot)\varphi$ suddenly has dense point spectrum embedded in the a.c. spectrum.

EXAMPLE 3. Let $d\mu$ be conventional Cantor measure. We claim that $G(x) = \infty$ on C, the Cantor set. In fact,

$$\operatorname{Im} F(x+i\epsilon) \to \infty \tag{2.5}$$

for $x \in C$, which implies that $G(x) = \infty$. This fact about $\operatorname{Im} F$ implies that A_α has neither point nor singular spectrum on C. In fact, each A_α has one eigenvalue $e_j(\alpha)$ in each interval, I_j in $[0,1] \setminus C$. The corresponding eigenvectors, together with one, $e_0(\alpha)$, in $I_0 \equiv \mathbb{R} \setminus [0,1]$, are complete. $\operatorname{Spec}(A_\alpha) = \bigcup_{j=0}^\infty \{e_j(\alpha)\} \cup C$ since the limit points of the e's are precisely C. To prove (2.5), we claim first that

$$\operatorname{Im} F(x_0 + i\epsilon) \geq \frac{1}{2\epsilon} \mu(\{y \mid |x-y| \leq \epsilon\}) \tag{2.6}$$

for any measure μ. Moreover, if $x \in C$, then

$$\mu(\{y \mid |x-y| \leq 2(3^{-n})\}) \geq 2^{-n}$$

so (2.6) implies (2.5).

EXAMPLE 4 ([**47**]). Let $d\mu_n = (\sum_{j=1}^{2^n} \delta_{j/2^n})/2^n$ and $\mu = \sum a_n \, d\mu_n$ where $\sum a_n < \infty$. If $x \in [0,1]$, then clearly, since there is a j with $|x - (j/2^n)| \leq 2^{-n}$, we have

$$\int |x-y|^{-2} \, d\mu_n(y) \geq (2^{-n})^{-2} 2^{-n} = 2^n$$

so if $\sum 2^n a_n = \infty$, then $G(x) = \infty$ on $[0,1]$ and thus A_α has no point spectrum $[0,1]$. So A has dense point spectrum on $[0,1]$ which turns into purely singular continuous spectrum on $[0,1]$ for all $\alpha \neq 0$.

EXAMPLE 5. Let $d\nu$ be Cantor measures and let

$$d\mu(x) = \chi_{[0,1]} \sum_{n=1}^\infty n^{-2} \left[\sum_{j=1}^{2^n} d\nu(x - (j/2^n)) \right].$$

Then $\mu\{y \mid |y-x| \leq 2^{-n+1}\} \geq 3^{-n} n^{-2}$ for any $\chi \in [0,1]$ so

$$G(x) \geq (2^{-n})^{-2} 3^{-n} n^{-2}/4 = (4/3)^n n^{-2}/4$$

for any n and so $G = \infty$ on $[0,1]$. This yields an example with $\sigma(A_\alpha)$ purely singular continuous on $[0,1]$ for all α. As noted, this cannot happen for point spectrum.

EXAMPLE 6 ([**47**]). Let x_n be arbitrary points in $[0,1]$. Let a_n be a sequence with $0 \le a_n \le C_0\alpha^n$ for some $\alpha < 1$. Let

$$d\mu = \sum a_n \delta_{x_n}.$$

We claim $G(x) < \infty$ for a.e. x. For if $|x - x_n| > D^{1/2}\alpha^{n/4}$ for *all* n, then

$$G(x) < \sum_n C_0\alpha^n D^{-1}\alpha^{-n/2} = (1-\alpha^{1/2})^{-1}C_0/D.$$

Thus

$$\{x \mid G(x) > (1-\alpha^{1/2})^{-1}C_0/D\} \le \sum 2D^{1/2}\alpha^{n/4} = 2D^{1/2}(1-\alpha^{1/2})^{-1}$$

goes to zero as $D \to 0$ proving that $G(x) < \infty$ for a.e. x. In this case, by the Simon-Wolff criterion, A_α has only pure point spectrum for a.e. α, but for a dense G_δ of α, it has singular continuous spectrum, of course.

Let x_n be the counting of the dyadic rationals. Let $d\mu = \sum b_n\delta_{x_n}$. Examples 4 and 6 say that depending on the b's, A_α can have dense point or singular continuous spectrum for Lebesgue typical α. Spectral properties depend heavily on φ as well as A!

EXAMPLE 7 ([**13**]). Let $\{q_n\}_{n=1}^\infty$ be a counting of the rationals in $(0,1)$. Define $a_n = \min(3^{-n-1}, q_n, 1-q_n)$ and let $S = \bigcup_n(q_n - a_n, q_n + a_n)$ so $|S| \le 2\sum a_n \le 1/3$. Let $d\mu_0 = \chi_S\,dx$ so $\operatorname{spec}(A_0) = [0,1]$ since S is dense in $[0,1]$. Since $|[0,1]\setminus S| \ge \frac{2}{3}$ and $d\mu_{\alpha,\mathrm{ac}}([0,1]\setminus S) = 0$, there must be a positive measure set of α's with point and/or singular spectrum embedded in $\operatorname{spec}(A_\alpha)$. As in the last example, if $|x - q_n| \ge a_n + \sqrt{\epsilon}\,3^{-n/4}$ for all n, then $G(x) \le \epsilon^{-1}(1-\sqrt{1/3})^{-1}$. Thus, $\{x \in S \mid G(x) > \epsilon^{-1}(1-\sqrt{1/3})^{-1}\} = \sqrt{\epsilon}(1-\sqrt{1/3})^{-1}$ so $G(x) < \infty$ for a.e. $x \in [0,1]\setminus S$. We conclude that for a.e. α, A_α has no singular spectrum on S but it has some point spectrum there for a positive measure set of α's.

EXAMPLE 8 ([**13**]). Let $S_n = \bigcup_{j=1}^{n-1}\big(j/2^n - 1/(4n^2 2^n), j/2^n + 1/(4n^2 2^n)\big)$ and $S = \bigcup_n S_n$, so $|S_n| \le 1/(2n^2)$ and $|S| \le \pi^2/12 < 1$. Let $d\mu_0 = \chi_S\,dx$. Then, $\operatorname{spec}(A_0) = [0,1]$ but since $|[0,1]\setminus S| > 0$, there must be a positive measure set of α's with point and/or singular spectrum embedded in $\operatorname{spec}(A_\alpha)$. As in Example 4, $G(x) = \infty$ on $[0,1]$ since for any n

$$G(x) \ge (2^{-n-1})^{-2}2^{-n-1}n^{-2} = 2^n/8n^2 \to \infty.$$

Thus, for all α we have no point spectrum in $[0,1]$ but we have singular continuous spectrum embedded in the a.c. spectrum for a set of positive measure α's.

3. Localization in the Anderson Model Following Aizenman-Molchanov

In this section we'll present the approach of Aizenman-Molchanov [**2**] to the proof of localization for the large coupling Anderson model in arbitrary dimension. The proof is so simple, we won't even put it in separate sections. Results on exponential decay of Green's functions were first obtained by Fröhlich-Spencer [**18**], with later extensions and simplifications by [**17, 15**]. Even taking into account the fact that we have already developed some of the facts we need, this proof is something like a factor of ten shorter and has smaller constants.

We will prove:

THEOREM 3.1. *Let $\{V_\omega(n)\}_{n\in\mathbb{Z}^\nu}$ be independent identically distributed random variables with uniform density on $[0,1]$. Then, there is a fixed constant, K, so that if $|\lambda| > K\nu^2$, then the Jacobi matrix, $h_0 + \lambda V_\omega$, has only pure point spectrum (and it is dense in $[-2\nu, 2\nu+\lambda]$).*

REMARK. I believe that $K = 32$ is what comes out of the minimization problem below.

Suppose for a moment that we could prove that $E_\omega\big(\sum_n |G(n,0;E+i0)|^2\big) \le Ce^{-D|n|}$. Then, by Fubini's theorem, we'd get that for a.e. pair (E,ω), $\sum_n |(\delta_n,(A-E-i0)^{-1}\delta_0)|^2 < \infty$, so by Theorems 2.8 and 2.5, we'd have localization. The problem with this idea is that there is no chance that even $E_\omega(|G(0,0;E+i0)|^2)$ is finite, since after all, $G(0,0;E+i0)$ is precisely $F_\alpha(E+i0)$ for $\alpha = V_\omega(0)$ and that is $F/[1+\alpha F]$ and has a first order zero at $\alpha = -1/F$ if there is to be an eigenvalue. But then, $\int dv(0)(v(0)-a)^{-2}$ is infinite if $a \in \operatorname{supp} v(0)$. So the first key idea of [**2**] is to look at powers of $(G)^s$ with $0 < s < 1$. For simplicity, we will take $s = 1/2$. The following lemma will be useful:

LEMMA 3.2. *If $s > 1$, then for $\{x_n\}_{n=1}^N$ all positive $(N = 1,2,\dots,\infty)$:*

$$\sum_{n=1}^N {x_n}^s \le \left(\sum_{n=1}^N x_n\right)^s.$$

If $s < 1$, then

$$\sum_{n=1}^N {x_n}^s \ge \left(\sum_{n=1}^N x_n\right)^s.$$

PROOF. If $s > 1$, ${x_n}^{s-1} \le \big(\sum_{m=1}^N x_m\big)^{s-1}$

$$ {x_n}^s = x_n {x_n}^{s-1} \le x_n \left(\sum_{m=1}^N x_m\right)^{s-1}.$$

Now sum over n. If $s < 1$, ${x_n}^{s-1} \ge \big(\sum_{m=1}^N x_m\big)^{s-1}$ and the proof is identical. □

In particular,

$$\left(\sum_n |G(n,0;E+i\epsilon)|^2\right)^{1/4} \le \sum_n |G(n,0;E+i\epsilon)|^{1/2}. \tag{3.1}$$

So we need only control $\sum_n E_\omega(|G(n,0;E+i\epsilon)|^{1/2})$.

A key lemma that we will need is

LEMMA 3.3. *There is a universal constant, C_0 so that for all complex numbers, α, β:*

$$\int_0^1 |x-\alpha|^{1/2}|x-\beta|^{-1/2}\,dx \ge C_0 \int_0^1 |x-\beta|^{-1/2}\,dx. \tag{3.2}$$

REMARK. I believe $\alpha = \beta = 1/2$ is the extreme point, in which case $C_0 = 1/\sqrt{8}$.

PROOF. It is easy to see the integral on the left side of (3.2) only decreases if α is replaced by $\Re\alpha$ or if $\alpha \notin [0,1]$ and is replaced by the nearest point in $[0,1]$.

So we can suppose that $\alpha \in [0,1]$. By $x \to 1/2 - x$ symmetry, we can suppose $\alpha \in [0,1/2]$. But then

$$\int_0^1 |x-\alpha|^{1/2}|x-\beta|^{-1/2}\,dx \geq \sqrt{1/4}\int_{3/4}^1 |x-\beta|^{-1/2}\,dx$$

so we need only show that

$$\inf_{\beta\in\mathbb{C}}\left[\int_{3/4}^1 |x-\beta|^{-1/2}\,dx \bigg/ \int_0^1 |x-\beta|^{-1/2}\,dx\right] > 0.$$

Let $h(\beta)$ be the function in $[\,\cdot\,]$. It is clearly continuous on $\mathbb{C}$, non-vanishing and $\lim_{\beta\to\infty} h(\beta) = 1/4$ so the inf is strictly positive. □

One last lemma we will need concerns the free Laplacian, h_0.

LEMMA 3.4. (a) *Let $f,g \in \ell^\infty(\mathbb{Z}^\nu)$ be non-negative functions and suppose that*

$$(1-ah_0)f \leq g$$

for some $0 < a < 1/2\nu$. Then

$$f \leq (1-ah_0)^{-1}g.$$

(b) *If $0 < a < (2\nu)^{-1}$, then*

$$(1-ah_0)^{-1}(i,j) \leq (2\nu a)^{|j-i|}(1-2\nu a)^{-1}.$$

PROOF. h_0 has norm 2ν so $(1-ah_0)^{-1} = \sum_{m=0}^\infty (ah_0)^m$ if $2\nu a < 1$. This is a positivity preserving operator, so it preserves inequalities which proves (a). Moreover, by the above expression and $(ah_0)^m(i,j) = 0$ if $m < |i-j|$, we see that

$$(1-ah_0)^{-1}(i,j) = \sum_{m=|i-j|}^\infty (ah_0)^m(i,j) \leq \sum_{m=|i-j|}^\infty (2\nu a)^m$$

proving (b). □

PROOF OF THEOREM 3.1. Fix $z \in \mathbb{C}\setminus\mathbb{R}$ and let $g_\omega(n) = G_\omega(0,n;z)$ and

$$k(n) = E_\omega(|g_\omega(n)|^{1/2}).$$

We claim that (with C_0 the constant in Lemma 3.3):

$$k(n) \leq (C_0\lambda^{1/2})^{-1}\left[\left(\sum_{|j|=1} k(n+j)\right) + \delta_{n0}\right]. \tag{3.3}$$

For g_ω obeys

$$\sum_{|j|=1} g_\omega(n+j) + (\lambda v_\omega(n) - z)g_\omega(n) = \delta_{n0}.$$

Using Lemma 3.2,

$$E_\omega\left(|\lambda v_\omega(n) - z|^{1/2}|g_\omega(n)|^{1/2}\right) \leq \delta_{n0} + \sum_{|j|=1} k(n+j). \tag{3.4}$$

Fix $\{v_\omega(\ell)\}_{\ell\neq n}$. Then, by (1.14), $g_\omega(n) = C/(\lambda v_\omega(n) + d)$ with $C \neq 0$ and C, d are functions of $\{v_\omega(\ell)\}_{\ell\neq n}$. Thus averaging only over $v_\omega(n)$:

$$\begin{aligned}
\int_0^1 |\lambda v - z|^{1/2} |C/(\lambda v_\omega + d)|^{1/2}\, dv \\
&= \lambda^{1/2}(C/\lambda)^{1/2} \int_0^1 |v - z\lambda^{-1}|^{1/2} |1/v + d\lambda^{-1}|^{1/2}\, dv \\
&\geq C_0 \lambda^{1/2}(C/\lambda)^{1/2} \int_0^1 |1/v + d\lambda^{-1}|^{1/2}\, dv \\
&= C_0 \lambda^{1/2} \int_0^1 |C/\lambda v + d|^{1/2}\, dv
\end{aligned}$$

where C_0 is the constant in Lemma 3.3. Averaging over $\{v_\omega(\ell)\}_{\ell\neq n}$, we get

$$E_\omega\big(|\lambda v_\omega - z|^{1/2} |g_\omega(n)|^{1/2}\big) \geq C_0 \lambda^{1/2} k(n). \tag{3.5}$$

Obviously, (3.4/5) imply (3.3).

(3.3) can be rewritten to say

$$(1 - C_0 \lambda^{1/2} h_0) k \leq \delta_{n0}.$$

Because $z \notin \mathbb{R}$, $|g_\omega(n)| \leq (1/|\operatorname{Im} z|)^{1/2}$, so $k \in \ell^\infty$. We can therefore apply Lemma 3.4 to conclude that if $(C\lambda^{1/2})^{-1} < (2\nu)^{-1}$ (or $\lambda > 4C^{-2}\nu^2$)

$$\begin{aligned}
k(n) &\leq (C_0\lambda^{1/2})^{-1} (1 - (C\lambda^{1/2})^{-1} h_0)^{-1}(0, n) \\
&\leq (C_0\lambda^{1/2})^{-1} [(2\nu)(C_0\lambda^{1/2})^{-1}]^{|n|} (1 - 2\nu(C_0\lambda^{1/2})^{-1})^{-1}.
\end{aligned}$$

Thus, $\sum_n k(n)$ is bounded uniformly in z, so by (3.1):

$$\sup_z E_\omega\left(\left(\sum_n |G(n, 0; z)|^2\right)^{1/4}\right)$$

is bounded uniformly in z. Using Theorem 2.8, this implies that for $A = h_\omega$, $\varphi = \delta_0$, we have

$$E_\omega(G(E)^{1/4}) < \infty$$

for all $E \in \mathbb{R}$. Thus for a.e. pairs, (V, E), $G(E) < \infty$. Now apply Theorem 2.5 to get localization. $\square$

Actually, the proof shows that $E_\omega\big((e^{\delta|n|} \sum_n |G(n, 0; z)|^2)^{1/4}\big) < \infty$ for some $\delta > 0$ and this implies exponential decay of eigenfunctions (essentially by (1.14)). Indeed, one easily sees that each eigenvector φ obeys $|\varphi(n)| \leq C_{\varphi,\epsilon}\left[\nu^2(k|\lambda|)\right]^{(1-\epsilon)|n|/2}$ for each $\epsilon > 0$. For extensions of these ideas, see [**2, 1**].

4. The Xi Function

Introduction: The Trace Formula. In this section, I'll discuss a program of F. Gesztesy and mine (in part with Holden and Zhao) that deals with inverse spectral theory for one-dimensional Schrödinger and Jacobi matrices [**22**]. In the study of the inverse problem for periodic potentials, an important role is played by the trace formula originally proven by Gelfand-Levitan and exploited/refined by many authors subsequently; see [**24**] for a history.

Let me describe this formula: let $V(x+L) = V(x)$. Then $\operatorname{spec}\bigl(-d^2/dx^2 + V(x)\bigr) = [E_0, E_1] \cup [E_2, E_3], \dots$ where $E_0, E_3, E_4, \dots, E_{4n-1}, E_{4n}, \dots$ are periodic eigenvalues, and $E_1, E_2, \dots, E_{4n+1}, E_{4n+2}, \dots$ are antiperiodic eigenvalues. Let $\{\mu_j(x)\}_{j=1}^\infty$ be the eigenvalues on $L^2(x, x+L)$ with $u(x) = u(x+L) = 0$ Dirichlet boundary conditions. It can be shown that $E_{2j-1} \le \mu_j(x) \le E_{2j}$. The trace formula says that, at least when V is smooth enough (C^1 will do)

$$V(x) = E_0 + \sum_{j=1}^\infty \bigl(E_{2j} + E_{2j-1} - 2\mu_j(x)\bigr). \tag{4.1}$$

One of our goals in this section is to extend this formula to more general V's than periodic. One issue to consider immediately is a case like $V(x) = x^2 - 1$ where $V(x) \to \infty$ as $|x| \to \infty$. Then the bands, $[E_{2j}, E_{2j+1}]$ collapse to a point and we have eigenvalues $E_0 < E_1 < \cdots$ and Dirichlet eigenvalues $\{\mu_j(x)\}_{j=1}^\infty$ (with $u(x) = 0$ boundary conditions) and $E_{j-1} \le \mu_j(x) \le E_j(x)$. The formal analog of (4.1) is

$$V(x) = E_0 + \sum_{j=1}^\infty \bigl(E_{j-1} + E_j - 2\mu_j(x)\bigr). \tag{4.2}$$

As an example, take $V(x) = x^2 - 1$ and look at $\mu_j(x=0)$: $E_n = 2n$, $n = 0, 1, 2, \dots$ and $\{\mu_j(0)\}_{j=1}^\infty$ are $\{2, 2, 6, 6, 10, 10, \dots\}$. It follows that (4.2) says

$$-1 = -2 + 2 - 2 \cdots.$$

Since the Abelian sum $1 - 1 + 1 - \cdots$ is $1/2$, this is promising but it is clear that the final result will need a summability method to accommodate the formula. Indeed, our trace formula (when $V > 0$) will read: Let $\xi(x, \lambda)$ be the Krein spectral shift for infinite coupling with $A = -d^2/dx^2 + V(x)$ and $\varphi = \delta(x)$. Then

$$V(x) = \lim_{\gamma \to \infty} \int_0^\infty \frac{\gamma^2}{(\lambda + \gamma)^2} \bigl(1 - 2\xi(x, \lambda)\bigr)\, dx. \tag{4.3}$$

Note that $|1 - 2\xi(x, \lambda)| = 1$ in all discrete spectrum cases, clearly illustrating the need for a summability method.

4.1. Abstract Trace Formula. We recall some facts from §1.5,6. As usual, let $A \ge 0$, $A_\alpha = A + \alpha(\varphi, \cdot)\varphi$ with A_∞ defined as the strong resolvent limit of A_α as $\alpha \to \infty$. Define

$$F(z) = (\varphi, (A - z)^{-1}\varphi) \tag{4.4}$$

and ξ_∞ by

$$\xi_\infty(\lambda) = \frac{1}{\pi} \operatorname{Arg} F(\lambda + i0). \tag{4.5}$$

Then

$$\operatorname{Tr}\bigl(f(A) - f(A_\infty)\bigr) = -\int f'(\lambda)\xi_\infty(\lambda)\, dx \tag{4.6}$$

and

$$\int_0^\infty \frac{\xi_\infty(\lambda)\, d\lambda}{(\lambda - z)^2}\, dx = \frac{d}{dz} \ln F(z). \tag{4.7}$$

We also will consider a second problem with $A^{(0)}$, $\xi_\infty^{(0)}$ and $F^{(0)}$. The point of a comparison is that with great generality, Green's functions, $G(0,0;z)$, have the same behavior, viz $(2(-z)^{1/2})^{-1}$ at $z=-\infty$.

THEOREM 4.1 ([**24**]). *Suppose that* $\lim_{\gamma\to\infty} F(-\gamma)/F^{(0)}(-\gamma) = 1$. *Then for* $z\in\mathbb{C}\setminus[0,\infty)$

$$F(z) = F^{(0)}(z)\exp\biggl(\lim_{\gamma\to\infty}\int \frac{\gamma}{\lambda+\gamma}\,\frac{1}{\lambda-z}\,(\xi_\infty-\xi_\infty^{(0)})(x)\,dx\biggr). \tag{4.8}$$

PROOF. (4.7) for F, $F^{(0)}$ integrated says that

$$\begin{aligned}\ln\bigl(F(z)/F^{(0)}(z)\bigr) - \ln\bigl(F(-\gamma)/F^{(0)}(-\gamma)\bigr) \\ = \int\biggl(\frac{1}{\lambda-z}-\frac{1}{\lambda+\gamma}\biggr)(\xi_\infty-\xi_\infty^{(0)})(\lambda)\,d\lambda.\end{aligned} \tag{4.9}$$

Moreover, by dominated convergence, for z fixed

$$\lim_{\gamma\to\infty}\int\frac{|z|}{|\lambda-z|}\,\frac{1}{\lambda+\gamma}\,|(\xi_\infty-\xi_\infty^{(0)}(\lambda)|\,d\lambda = 0. \tag{4.10}$$

(4.9)–(4.10) and $\lim_{\gamma\to\infty}\ln\bigl(F(-\gamma)/F^{(0)}(-\gamma)\bigr)=0$ imply (4.8). $\square$

THEOREM 4.2 (THE ABSTRACT TRACE FORMULA) [**24**]. *Suppose that for some* $\epsilon>0$, $\lim_{\substack{|\gamma|\to\infty\\ |\operatorname{Arg}\gamma|<\epsilon}}\gamma\bigl(F(-\gamma)/F^{(0)}(-\gamma)-1\bigr)=a$. *Then*

$$a = \lim_{\substack{\gamma\to\infty\\ \gamma\ \text{real}}}\int\frac{\gamma^2}{(\lambda+\gamma)^2}(\xi_\infty-\xi_\infty^{(0)})(\lambda)\,d\lambda.$$

PROOF. By hypothesis

$$\lim_{\substack{|\gamma|\to\infty\\ |\operatorname{Arg}\gamma|<\epsilon}}\gamma\,\ln\biggl[\frac{F(-\gamma)}{F^{(0)}(-\gamma)}\biggr] = a.$$

By a simple contour argument

$$\lim_{\substack{\gamma\to\infty\\ \gamma\ \text{real}}}\gamma^2\left(\frac{d}{dz}\ln\biggl[\frac{F(z)}{F^{(0)}(z)}\biggr]\right)_{z=-\gamma} = a.$$

This plus (4.7) implies the result. $\square$

4.2. The Trace Formula for Schrödinger Operators. We'll postpone the proof of the following until later in this section:

LEMMA 4.3. *Suppose that* V *is bounded from below and is continuous at* x. *Let* G *be the Green's function for* $-d^2/dx^2+V(x)$ *on* $L^2(-\infty,\infty)$. *Then as* $|\gamma|\to\infty$ *with* $|\operatorname{Arg}\gamma|<\pi/2$:

$$G(x,x;-\gamma) = \frac{1}{2}(\gamma)^{-1/2}\biggl(1-\frac{1}{2}\gamma^{-1}V(x)+0(\gamma)^{-1}\biggr) \tag{4.11}$$

EXAMPLE. This will be the comparison $F^{(0)}$ so it is a critical example: Let $A^{(0)} = -d^2/dx^2$ on $L^2(\mathbb{R},dx)$. Let $\varphi=\delta_x$. Then $F^{(0)}(z)=G^{(0)}(x,x;z)=(-z)^{-1/2}/2$. Thus $F^{(0)}(x+i0)=(x)^{-1/2}i/2$ and $\xi_\infty^{(0)}(x)=1/2$, $x>0$.

We now apply the abstract theory of the last section:

THEOREM 4.4. *Suppose that V is non-negative and is continuous at x. Let $\xi(x,\lambda)$ be $\xi_\infty(x)$ for $A = -d^2/dx^2 + V; \varphi = \delta_x$. Then*

$$V(x) = \lim_{\gamma\to\infty} \int_0^\infty \frac{\gamma^2}{(\lambda+\gamma)^2}\,[1 - 2\xi(x,\lambda)]\,d\lambda \tag{4.12}$$

and

$$G(x,x;z) = \frac{1}{2}\,(-z)^{-1/2} \exp\biggl(\lim_{\gamma\to\infty} \int_0^\infty \frac{\gamma}{\gamma+\lambda}\,\frac{1}{\lambda - z}\biggl[\xi(x,\lambda) - \frac{1}{2}\biggr]dx\biggr) \tag{4.13}$$

PROOF OF LEMMA 4.3 WHEN V IS BOUNDED. W.l.o.g. suppose $x = 0$. Let $\psi(x) = (H_0+\gamma)^{-1}\delta_0(x) = \gamma^{-1/2}e^{-\gamma^{1/2}|x|}/2$, so $\|\psi_\gamma\| = 0(\gamma^{-3/4})$. By the second resolvent equation, since V is bounded

$$(\delta_0, (H_0+V+\gamma)^{-1}\delta_0) = (\delta_0, (H_0+\gamma)^{-1}\delta_0) - (\psi_\gamma, V\psi_\gamma) + R \tag{4.14}$$

where $\|R\| \le \|\psi_\gamma\|^2\gamma^{-1} = 0(\gamma^{-5/2})$. The first term in (4.14) is $\gamma^{-1/2}/2$ and the second is

$$-\frac{1}{4}\gamma^{-1}\int_{-\infty}^\infty e^{-2\gamma^{1/2}|x|}V(x)\,dx = -\frac{1}{4}\gamma^{-3/2}[V(0)+o(1)]$$

since $\int_{-\infty}^\infty e^{-2\alpha|x|}\,dx = \alpha^{-1}, \alpha > 0$. □

PROOF OF LEMMA 4.3 IN THE GENERAL CASE. We will prove with $H = H_0 + V$ that

$$e^{-tH}(x,x) = (4\pi t)^{-1/2}\bigl(1 - tV(x) + o(t)\bigr). \tag{4.15}$$

(4.11) then follows from

$$G(x,x;-\gamma) = \int_0^\infty e^{-\gamma s}e^{-sH}(x,x)\,ds$$

and

$$\int_0^\infty t^{-1/2}e^{-t}\,dt = \sqrt{\pi} \quad\text{and}\quad \int_0^\infty t^{1/2}e^{-t}\,dt = \frac{1}{2}\sqrt{\pi}.$$

To prove (4.15), let $\{\omega(s)\}_{0\le s\le 1}$ be the Brownian bridge [**43**] so the Feynman-Kac formula has the form [**43**]

$$e^{-tH}(x,x) = (4\pi t)^{-1/2}E_\omega\biggl(\exp\biggl[-t\int_0^1 V\bigl(x + 2\sqrt{2t}\,\omega(s)\bigr)\,ds\biggr]\biggr). \tag{4.16}$$

A Levy-type estimate implies that

$$E_\omega\biggl(\sup_{0\le s\le 1}\omega(s) \ge \alpha\biggr) \le C_1e^{-C_2\alpha^2}.$$

Changing V outside $(x-\delta, x+\delta)$ changes the right side of (4.16) by $0(e^{-C/t})$, so in proving (4.15), we can suppose that V is bounded. But if V is bounded, it is easy to see that the integral is $1 - tV(x) + o(t)$. □

Since we controlled $e^{-tH}(x,x)$, we can prove the slightly stronger version of (4.12), viz [**24**]:

$$V(x) = \lim_{\gamma\downarrow\infty}\int_0^\infty e^{-\gamma\lambda}\bigl(1 - 2\xi(x,\lambda)\bigr)\,d\lambda.$$

This is all with V positive. If it is only bounded below and $E_0 \leq \inf \operatorname{spec}(H)$, then

$$V(x) = E_0 + \lim_{\gamma \downarrow 0} \int_{E_0}^{\infty} e^{-\gamma\lambda}(1 - 2\xi(x,\lambda))\, d\lambda.$$

See [**23**] for higher order asymptotic formula involving moments of ξ.

4.3. Examples.

EXAMPLE 1 (PERIODIC POTENTIALS). We begin with the motivating case of periodic potentials, $V(x) = V(x+L)$. The spectrum of H is $[E_0, E_1] \cup [E_2, E_3] \cup \cdots$. General principles imply that $\xi(x,\lambda)$ obeys

$$\xi(x,\lambda) = \begin{cases} 1/2, & E_{2n} < \lambda < E_{2n+1} \\ 1, & E_{2n-1} < \lambda < \mu_n(x) \\ 0, & \mu_n(x) < \lambda < E_{2n} \text{ or } \lambda < E_0. \end{cases}$$

If V is smooth enough (e.g., $V \in C^1$), $\sum |E_{2n} - E_{2n-1}| < \infty$ so $\int |2 - 2\xi(x,\lambda)|\, d\lambda < \infty$ and the summability method is not needed. Clearly,

$$\int_{E_{2n-1}}^{E_{2n}} (1 - 2\xi)\, d\lambda = -(\mu_n - E_{2n-1}) + (E_{2n} - \mu_n) = E_{2n} + E_{2n-1} - 2\mu_n$$

$$\int_0^{E_0} (1 - 2\xi)\, d\lambda = E_0$$

and we get

$$V(x) = E_0 + \sum_{n=1}^{\infty} (E_{2n} + E_{2n-1} - 2\mu_n(x)).$$

EXAMPLE 2 (HARMONIC OSCILLATOR). $V(x) = x^2 - 1$; the eigenvalues are $0, 2, 4 \ldots$ and

$$\xi(0,\lambda) = \begin{cases} 1, & 0 < \lambda < 2;\ 4 < \lambda < 6, \ldots \\ 0, & 2 < \lambda < 4;\ 6 < \lambda < 8, \ldots \end{cases}$$

$$\begin{aligned} \int_0^{\infty} e^{-\alpha\lambda}(-2\xi(0,\lambda))\, d\lambda &= \alpha^{-1}(1 - e^{-2\alpha}) \sum_{n=0}^{\infty} (-1)^{n+1} e^{-2n\alpha} \\ &= \alpha^{-1}(1 - e^{-2\alpha})(-1)/[1 + e^{-2\alpha}] \end{aligned}$$

converges to -1 as $\alpha \downarrow 0$. Thus the summability method yields $V(0)$. In general,

THEOREM 4.5. *Let $V(x) \to \infty$ as $|x| \to \infty$. Let $E_0 \leq \mu_1(x) \leq E_1 \leq \cdots$ be the eigenvalues of $-d^2/dx^2 + V$ on all of $L^2(\mathbb{R}, dx)$ and with a Dirichlet boundary condition at x. Then*

$$V(x) = E_0 + \lim_{\gamma \downarrow 0} \sum_{n=1}^{\infty} [2e^{-\gamma\mu_n(x)} - e^{-\gamma E_n} - e^{-\gamma E_{n-1}}]\gamma^{-1}.$$

EXAMPLE 3 (SHORT RANGE POTENTIALS). Suppose $|V(x)| \leq C(1+|x|)^{-1-\epsilon}$. Let $R_\ell(\lambda)$ be the usual reflection coefficient and $f_\ell(x,\lambda)$ the Jost function at $-\infty$. Then [**21**]

$$\xi(x,\lambda) = \frac{1}{2} + \operatorname{Arg}\left[1 + R_\ell(\lambda) f_\ell(x,\lambda)^2 / |f_\ell(x,\lambda)|^2\right], \qquad \lambda > 0.$$

We see that if $\int |R_\ell(\lambda)|\,d\lambda < \infty$, then $\int |1-2\xi(x,\lambda)|\,dx < \infty$ and one can take the limit inside the integral. See [**21**] for other short range examples.

As a preliminary to the next example, we need a formula about how Dirichlet eigenvalues $\mu(x)$ move as x changes.

THEOREM 4.6. *Let (a,b) be an interval in $\mathbb{R}\setminus \operatorname{spec}(H)$ (i.e., a spectral gap of H) and suppose $\mu(x)$ is an eigenvalue of H_x, the operator with a Dirichlet boundary condition at x with $\mu(x)\in(a,b)$. Let $g(x,\lambda)\equiv G(x,x;\lambda)$.*

(i) *$\partial g/\partial x(x,\lambda)\big|_{\lambda=\mu(x)} = \pm 1$ with the value equal $+1$ (resp. -1) if the eigenvalue lies on $(-\infty,0]$ (resp. $[0,\infty)$).*

(ii) *With the same sign as in* (i)

$$\mu'(x) = \pm\left(\frac{\partial g}{\partial \mu}(x,\mu(x))\right)^{-1}. \tag{4.17}$$

PROOF. (i) By general principles with $\varphi_\pm$ L^2 at $\pm\infty$:

$$g(x,\lambda)\equiv \frac{\varphi_-(x)\varphi_+(x)}{\varphi_+(x)\varphi_-'(x)-\varphi_-(x)\varphi_+'(x)}$$

so

$$\frac{\partial g}{\partial x}(x,\lambda) = \frac{\varphi_+(x)\varphi_-'(x)+\varphi_-(x)\varphi_+'(x)}{\varphi_+(x)\varphi_-'(x)-\varphi_-(x)\varphi_+'(x)}.$$

At $\mu(x)$, either $\varphi_+(x)=0$, in which case $\partial g/\partial x=-1$, or $\varphi_-(x)=0$, in which case $\partial g/\partial x = 1$ as claimed.

(ii) Immediate from (i), given the implicit function theorem that says that $\mu' = -[\partial g/\partial x]/[\partial g/\partial \mu]$. $\square$

EXAMPLE 4 (THE SOLITON). A *reflectionless* potential is one with $\xi(x,\lambda) = 1/2$ for *all* x and a.e. λ in $\operatorname{spec}_{\text{ac}}(H)$. Let us look for reflectionless potentials with a single eigenvalue at $E=-1$. Let $\mu(x)$ be the Dirichlet eigenvalue in $(-1,0]$. Then

$$\xi(x,\lambda)=\begin{cases}0, & \lambda<-1 \text{ or } \mu(x)<\lambda<0\\ 1, & -1<\lambda<\mu(x)\\ \frac{1}{2}, & \lambda>0.\end{cases}$$

The formulas for g and V then equal

$$V(x) = 2(-1-\mu(x))$$
$$g(x,\lambda) = \frac{1}{2}(-\lambda)^{-1/2}\,\frac{\mu(x)-\lambda}{-1-\lambda}.$$

(4.17) reads

$$\mu'(x) = \pm 2(-\mu)^{1/2}(1+\mu). \tag{4.18}$$

Let $c(x) = (-\mu)^{1/2}$. Then (4.18) becomes

$$c' = \pm(1-c^2)$$

whose solution is $c(x) = \pm\tanh(x)$ so we get

$$V(x) = -2(1-\tanh^2(x)) = \frac{-2}{\cosh^2(x)},$$

the familiar soliton! This argument does *not* show this potential is reflectionless, only that it is the only candidate. However, a comparison with the formula for $\xi(x,\lambda)$ in Example 3 indeed verifies that $R_\ell(\lambda)=0, \lambda>0$ in the present case.

4.4. The Trace Formula for Jacobi Matrices.

THEOREM 4.7. *Let $H=H_0+V$ be a Jacobi matrix on $\mathbb{Z}^\nu$ with V bounded. Let $\xi(n,\lambda)$ be ξ_∞ for $A=H_0+V$ and $\varphi=\delta_n$. Let $E_\pm = {}^{\sup}_{\inf}\operatorname{spec}(H)$. Then*

$$V(n)=\frac{1}{2}(E_+ + E_-)+\int_{E_-}^{E_+}\left(\xi(n,\lambda)-\frac{1}{2}\right)d\lambda. \tag{4.19}$$

PROOF. (4.19) is unchanged if we add a constant, c, to V (since then V, E_+, and E_- all increase by c), so w.l.o.g. we can suppose that $H\geq 0$. Since $\xi=0$ (resp. 1) for $\lambda<E_-$ (resp. $\lambda>E_+$), we need only prove (4.19) for some $E_+>\sup\operatorname{spec}(H)$ and $E_-<\inf\operatorname{spec}(H)$.

Next we note that (4.19) can be rewritten

$$\begin{aligned}V(n)&=E_-+\int_{E_-}^{E_+}(\xi(n,\lambda)-1)\,d\lambda=E_-+\int_{E_-}^{\infty}(\xi(n,\lambda)-1)\,d\lambda\\&=E_++\int_{-\infty}^{E_+}\xi(n,x)\,dx\end{aligned}$$

so it suffices to prove

$$V(n)=\int_0^\infty(\xi(n,x)-1)\,dx$$

for $H\geq 0$.

But noting that

$$(\delta_n,(H+\gamma)^{-1}\delta_n)=\gamma^{-1}-\gamma^{-2}V(n)+O(\gamma^{-3}),$$

this follows from (1.30) (Theorem 1.17). □

4.5. A Regularity Theorem for the A.C. Spectrum. In this section we'll prove the following:

THEOREM 4.8 ([**24**]). *Let $V_m(x)$ and $V_\infty(x)$ be functions on $\mathbb{R}$ (resp. $\mathbb{Z}$) so that*

(i) $\inf_{\substack{x\in\mathbb{R}\\ m\geq 1}} V_m(x)>-\infty$ *and for each* $R>0$, $\sup_{\substack{|x|\in R\\ m\geq 1}}|V_m(x)|<\infty$ *(resp.* $\sup_{\substack{x\in\mathbb{Z}\\ m\geq 1}}|V_m(x)|<\infty$*).*

(ii) $V_m\to V_\infty$ *uniformly on compacts (resp. pointwise).*

(iii) *Each V_m is periodic (but the period can be m-dependent).*

Let H_m, H_∞ be the Hamiltonians $-d^2/dx^2+V_m$ (resp. the Jacobi matrices on $\ell^2(\mathbb{Z})$, h_0+V_m). Then (with $|\cdot|=$ Lebesgue measure):

$$|\sigma_{\rm ac}(H_\infty)|\geq\overline{\lim}\,|\sigma_{\rm ac}(H_m)|.$$

The most interesting application of this result is to prove a slightly strengthened version of a striking result of Last [**34**]:

COROLLARY 4.9. *Let* $V(n) = \lambda\cos(\pi\alpha n + \theta)$ *and let* h *be the corresponding Jacobi matrix. Then* $|\sigma_{\rm ac}(h)| \geq 4 - 2|\lambda|$.

PROOF. If $\alpha = p/q$ is rational, this is proven in [**7**]. If α is arbitrary, let $p_m/q_m \to \alpha$ and let $V_m(n) = \lambda\cos(\pi_m\alpha n + \theta)$. Then V, V_m obey the hypotheses of Theorem 4.8. So, by the theorem, we can take limits. □

REMARK. If α is a Liouville number (irrational but well approximated by rationals), it is known for $\lambda > 2$, that h has only singular continuous spectrum and before Last [**34**], some believed that was true also for $|\lambda| < 2$. Last showed that if $|\lambda| < 2$, $|\sigma_{\rm ac}(h)| > 0$.

To prove Theorem 4.8, we begin with

THEOREM 4.10. *Under the hypotheses* (i), (ii) *of Theorem* 4.8,

$$\xi_\infty^{(m)}(x,\lambda)\,dx \to \xi_\infty^{(\infty)}(x,\lambda)\,dx$$

weakly as measures where $\xi_\infty^{(m)}$ *is the spectral shift for* $-d^2/dx^2 + V_m$ *with* $\varphi = \delta(x)$.

PROOF. It suffices to prove that for all $z \in \mathbb{C}\setminus\mathbb{R}$

$$\int \frac{\xi_\infty^{(m)}(x,\lambda)\,d\lambda}{(\lambda - z)^2} \longrightarrow \int \frac{\xi_\infty^{(\infty)}(x,\lambda)\,d\lambda}{(\lambda - z)^2}$$

by a standard density argument. By (1.29) and the Cauchy formula, it suffices to prove that $F_m(z) \to F_\infty(z)$ where $F_m(z) = G_{V_m}(0,0;z)$. Writing

$$\big[(H_n - z)^{-1} - (H_\infty - z)^{-1}\big](H_\infty - z)\psi = (H_n - z)^{-1}(V_\infty - V_n)\psi$$

for $\psi \in C_0^\infty(\mathbb{R})$, we see that

$$\text{s-lim}(H_n - z)^{-1} = (H_\infty - z)^{-1}. \tag{4.20}$$

Let $H_0 = -d^2/dx^2$. Then $(H_\infty + 1)^{+1/2}(H_n - z)^{-1}(H_0 + 1)^{1/2}$ is uniformly bounded, so (4.20) implies that

$$\text{s-lim}(H_\infty + 1)^{1/2}(H_n - z)^{-1}(H_0 + 1)^{1/2} = (H_0 + 1)^{1/2}(H_\infty - z)^{-1}(H_0 + 1)^{1/2}.$$

Since $(H_0 + 1)^{-1/2}\delta_0$ is in L^2, we can take matrix elements in this vector and conclude that $F_m(z) \to F_\infty(z)$ is as required. □

PROOF OF THEOREM 4.8. Since V_m is periodic, $\xi_m(x,\lambda) = 0$, 1 or 1/2 for a.e. λ. Fix $[a,b] \subset \mathbb{R}$ bounded and let $A_m = \{\lambda \mid \xi_m(x,\lambda) = 0\} \cap [a,b]$ for $m = 1$, 2, ... , ∞. Then

$$\begin{aligned} 0 = \int_{A_\infty} \xi_\infty(x,\lambda)\,d\lambda &= \lim \int_{A_\infty} \xi_m(x,\lambda)\,d\lambda \\ &\geq \lim \frac{1}{2}\,|A_\infty \setminus A_m| \geq \frac{1}{2}\,(|A_\infty| - \underline{\lim}\,|A_m|). \end{aligned}$$

We conclude that

$$|A_\infty| \leq \underline{\lim}\,|A_m|.$$

A similar argument with $\xi_m = 1$ shows that

$$|\{\lambda \mid 0 < \xi_\infty(x,\lambda) < 1\} \cap [a,b]| \geq \overline{\lim}\left|\left\{\lambda \mid \xi_m(x,\lambda) = \frac{1}{2}\right\} \cap [a,b]\right|.$$

But $\xi_\infty(x,\lambda) = \operatorname{Arg}(F_\infty(\lambda+i0)) \ / \ \pi$ lies in $(0,1)$ if and only if $\operatorname{Im} F_\infty > 0$ (at points with $F_\infty(\lambda+i0)$ finite). So

$$|\sigma_{\rm ac}(H_m)\cap[a,b]| \geq |\{\lambda \mid 0<\xi_m(x,\lambda)<1\}\cap[a,b]|$$

with equality if m is finite. $\square$

4.6. Inverse Problems. These ideas play an important role in understanding inverse spectral theory. We'll give an example here; see [**24**] (and a paper in preparation) for further examples. We'll rely on the Gelfand-Levitan theory [**20, 35**] which allows one to recover $V(x)$ from any $m_\alpha(z)$ and, in particular, from $G_{\theta=\pi/2}(0,0;z)$. In 1952 Borg [**8**] and Marčenko [**36**] proved that:

THEOREM 4.11. *Let $V(x)\to\infty$ to infinity. Let $\{E_n^{(\alpha)}\}_{n=1}^\infty$ be the eigenvalues for $-d^2/dx^2+V(x)$ with θ boundary conditions at $x=0$ where $\alpha=-\cot(\theta)$. Then V can be recovered from $\{E_n^{(\alpha)}\}_{n=1}^\infty$ for any two distinct values of α.*

Actually, Marčenko [**36**] proved in addition that the two values of α can be recovered as well from the two sets $\{E_n^{(\alpha)}\}_{n=1}^\infty$.

We'll prove a generalization of this.

THEOREM 4.12. *Suppose that V is bounded below. Let $\xi_{\alpha,\beta}(\lambda)$ $(\alpha>\beta)$ be the spectral shift for going from A_β to A_α. Then α, β, and V can be recovered from $\xi_{\alpha,\beta}(\lambda)$.*

REMARK. This implies Theorem 4.11. For in that case $E_n^{(\beta)}\leq E_n^{(\alpha)}\leq E_{n+1}^{(\beta)}$ and

$$\xi_{\alpha,\beta}(\lambda)=\begin{cases}1, & E_n^{(\beta)}<\lambda<E_n^{(\alpha)}\\ 0, & E_n^{(\alpha)}<\lambda<E_{n+1}^{(\beta)}.\end{cases}$$

PROOF. Let $F(z)$ be the Neumann-Green's function, $G_{\theta=\pi/2}(0,0;z)$, from which $V(x)$ can be recovered by Gelfand-Levitan. We have that $\xi_{\alpha,\beta}(\lambda)=\xi_\alpha(\lambda)-\xi_\beta(\lambda)$ and

$$\int\frac{\xi_\alpha(\lambda)}{(\lambda-z)}\,d\lambda=\ln\big(1+\alpha F(z)\big)\qquad \alpha<\infty \tag{4.21}$$

$$\int\frac{\xi_\infty(\lambda)}{(\lambda-z)^2}\,d\lambda=\frac{F'(z)}{F(z)}\qquad \alpha=\infty \tag{4.22}$$

and

$$F(z)=(-z)^{-1/2}(1+O(|z|)^{-1}) \tag{4.23}$$

by (1.16), (1.19), (1.29), and (4.11).

Since $\int\xi_\alpha(\lambda)/(|\lambda|+1)=\infty$ if and only if $\alpha=\infty$, we see that $\int\xi_{\alpha,\beta}(\lambda)/(|\lambda|+1)\,d\lambda=\infty$ if and only if α is infinite, and we can read that fact off of knowing $\xi_{\alpha,\beta}(\lambda)$. Consider first the case with α, β finite. By (4.23)

$$\begin{aligned}\ln\big(1+\alpha F(z)\big)&=\alpha F(z)-\frac12\,\alpha^2F(z)^2+O(F(z)^3)\\&=\alpha(-z)^{-1/2}-\frac12\,\alpha^2(-z)^{-1}+O(z^{-3/2}).\end{aligned}$$

Thus, by (4.21)

$$\int\frac{\xi_{\alpha,\beta}(\lambda)}{(\lambda-z)}\,d\lambda\sim(\alpha-\beta)(-z)^{-1/2}-\frac12\,(\alpha^2-\beta^2)(-z)^{-1}+O(z^{-3/2})$$

so we can find $\alpha - \beta$ and $\alpha^2 - \beta^2$ and so α and β from $\xi_{\alpha,\beta}$. Moreover, since

$$\int \frac{\xi_{\alpha,\beta}(\lambda)}{\lambda - z}\, d\lambda = \ln\left(\frac{1+\alpha F(z)}{1+\beta F(z)}\right)$$

$\xi_{\alpha,\beta}$ determines $F(z)$ and so V by Gelfand-Levitan.

Now consider the case $\alpha = \infty$. Then by (4.23) and a contour integral:

$$\begin{aligned}\frac{F'(z)}{F(z)} &= -\frac{1}{2}\,(-z)^{-1} + O(z^{-2})\\ \frac{\beta F'(z)}{1+\beta F(z)} &= -\frac{\beta}{2}\,(-z)^{-3/2} + O(z^{-2}).\end{aligned}$$

Thus by (4.22) and the discussion of (4.21):

$$\int \frac{[\xi_\infty(\lambda) - \xi_\beta(\lambda)]}{(x-z)^2}\, d\lambda = \frac{\beta F'(z)}{1+\beta F(z)} - \frac{F'(z)}{F(z)} = \frac{1}{2}\,(-z)^{-1} - \frac{\beta}{2}\,(-z)^{-3/2} + O(z^{-2})$$

so we can read β off of $\xi_{\infty,\beta}(\lambda)$. By the above

$$\begin{aligned}\int \frac{\xi_{\infty,\beta}(\lambda)}{(\lambda - z)^2} &= \frac{d}{dz}\ln(\beta + F(z)^{-1})\\ &= -\frac{1}{2}\,z^{-1} + \frac{d}{dz}\ln\big((-z)^{1/2}[F(z)^{-1} + \beta]\big)\\ &\equiv -\frac{1}{2}\,z^{-1} + \frac{d}{dz}\,[H_\beta(z)].\end{aligned}$$

By the above $H_\beta(-\gamma) \to 0$ as $\gamma \to \infty$. Thus

$$H_\beta(z) = \lim_{\gamma\to\infty} \int \frac{(\xi_{\infty,\beta} - 1/2)(\lambda)}{(\lambda - z)}\left(\frac{\gamma}{\gamma+\lambda}\right) d\lambda$$

and we can recover H_β and so F from $\xi_{\infty,\beta}(\lambda)$. Gelfand-Levitan completes the proof. □

References

1. M. Aizenman, *Localization at weak disorder: Some elementary bounds*, Rev. Math. Phys. **6** (1994) (preprint).
2. M. Aizenman and S. Molchanov, *Localization at large disorder and at extreme energies: An elementary derivation*, Comm. Math. Phys. **245** (1993), 245–277.
3. N. Aronszajn, *On a problem of Weyl in the theory of Sturm-Liouville equations*, Amer. J. Math. **79** (1957), 597–610.
4. N. Aronszajn and W.F. Donoghue, *On exponential representations of analytic functions in the upper half-plane with positive imaginary part*, J. Analyse Math. **5** (1957), 321–388.
5. J. Avron, R. Seiler, and B. Simon, *The index of a pair of projections*, J. Funct. Anal. **120** (1994), 220–237.
6. J. Avron and B. Simon, *Almost periodic Schrödinger operators* II. *The integrated density of states*, Duke Math. J. **50** (1983), 369–391.
7. J. Avron, B. Simon, and P. van Mouche, *On the measure of the spectrum for the almost Mathieu equation*, Comm. Math. Phys. **132** (1990), 103–118.
8. G. Borg, *Uniqueness theorems in the spectral theory of* $y'' + (\lambda - q(x))y = 0$, Proc. of the 11th Scandinavian Congress of Mathematicians, Johan Grundt Tanums Forlag, Oslo, 1952, pp. 276–287.
9. H. Cycon, R. Froese, W. Kirsch, and B. Simon, *Schrödinger operators with appplication to quantum mechanics and global geometry* (1987), Springer-Verlag, Berlin.

10. R. del Rio, *A forbidden set for embedded eigenvalues*, Proc. Amer. Math. Soc. **121** (1994), 77–82.
11. R. del Rio, S. Jitomirskaya, N. Makarov, and B. Simon, *Singular continuous spectrum is generic*, Bull. Amer. Math. Soc. (N.S.) **31** (1994), 208–212.
12. R. del Rio, N. Makarov, and B. Simon, *Operators with singular continuous spectrum* II. *Rank one operators*, Comm. Math. Phys. (to appear).
13. R. del Rio, B. Simon, and G. Stolz, *Stability of spectral types for Sturm-Liouville operators*, Math. Research Lett. **1** (1994), 437–450.
14. W. Donoghue, *On the perturbation of the spectra*, Comm. Pure Appl. Math. **18** (1965), 559–579.
15. H. von Dreifus and A. Klein, *A new proof of localization in the Anderson tight binding model*, Comm. Math. Phys. **124** (1989), 285-299.
16. E. Effros, *Why the circle is connected*, Math. Intelligencer **11** (1989), 27–35.
17. J. Fröhlich, F. Martinelli, E. Scoppola, and T. Spencer, *Constructive proof of localization in the Anderson tight binding model*, Comm. Math. Phys. **101** (1985), 21–46.
18. J. Fröhlich and T. Spencer, *Absence of diffusion in the Anderson tight binding model for large disorder or low energy*, Comm. Math. Phys. **88** (1983), 151–184.
19. H. Furstenberg, *Non-commuting random products*, Trans. Amer. Math. Soc. **108** (1963), 377–428.
20. I. M. Gelfand and B. M. Levitan, *On a simple identity for eigenvalues of a second order differential operator*, Dokl. Akad. Nauk SSSR **88** (1953), 593–596 (Russian); English transl. I. M. Gelfand, *Collected papers* I (S. G. Gindikin, V. W. Guillemin, A. A. Kirillov, B. Konstant, and S. Sternberg, eds.), Springer, Berlin, 1987, pp. 457–461.
21. F. Gesztesy, H. Holden, and B. Simon, *Absolute summability of the trace relation for certain Schrödinger operators*, Comm. Math. Phys (to appear).
22. F. Gesztesy, H. Holden, B. Simon, and Z. Zhao, *Trace formulae and inverse spectral theory for Schrödinger operators*, Bull. Amer. Math. Soc. (N.S.) **29** (1993), 250–255.
23. ———, *Higher order trace relations for Schrö-dinger operators* (preprint).
24. F. Gesztesy and B. Simon, *The xi function* (preprint).
25. ———, *Rank one pertubations at infinite coupling*, J. Funct. Anal. (to appear).
26. A. Gordon, *On exceptional value of the boundary phase for the Schrödinger equation of a half-line*, Russian Math. Surveys **47** (1992), 260–261.
27. ———, *Pure point spectrum under 1-parameter perturbations and instability of Anderson localization*, Comm. Math. Phys. **164** (1994), 489–505.
28. V.A. Javrjan, *A certain inverse problem for Sturm-Liouville operators*, Izv. Akad. Nauk Armyan. SSR Ser. Mat. **6** (1971), 246–251.
29. S. Jitomirskaya and B. Simon, *Operators with singular continuous spectrum*III. *Almost periodic Schrödinger operators*, Ann. of Math. (2) (to be submitted).
30. T. Kato, *Perturbation theory for linear operators*, 2nd ed., Springer,, Berlin, 1980.
31. Y. Katznelson, *An introduction to harmonic analysis*, Dover, New York, 1976.
32. S. Kotani, *Ljapunov indices determine absolutely continuous spectra of stationary random one-dimensional Schrödinger operators*, Stochastic Analysis (K. Ito, ed.), North Holland, Amsterdam, 1984, pp. 225–248.
33. M.G. Krein, *Perturbation determinants and a formula for the traces of unitary and self-adjoint operators*, Soviet Math. Dokl. **3** (1962), 707–710.
34. Y. Last, *A relation between a.c. spectrum of ergodic Jacobi matrices and the spectra of periodic approximants*, Comm. Math. Phys. **151** (1993), 183–192.
35. B. Levitan and M. Gasymov, *Determination of a differential equation by two of its spectra*, Russian Math. Surveys **19** (1964), 1–63.
36. V. A. Marčenko, *Some questions in the theory of one-dimensional linear differential operators of the second order* I, Trudy Moskov. Mat. Obshch. **1** (1952), 327–420 (Russian); Engl. transl., Amer. Math. Soc. Transl. Ser. 2 **101** (1973), 1–104.
37. V. Oseledec, *A multiplicative ergodic theorem. Lyapunov charateristic numbers for dynamical systems*, Trans. Moscow Math. Soc. **19** (1968), 197–231.
38. M. Reed and B. Simon, *Methods of modern mathematical physics* II. *Fourier analysis, self-adjointness*, Acadmic Press, London, 1975.
39. D. Ruelle, *Ergodic theory of differentiable dynamical systems*, Inst. Hautes Études Sci. Publ. Math. **50** (1979), 275–306.

40. S. Saks, *Theory of the integral* (1937), Hafner, New York.
41. B. Simon, *A canonical decomposition for quadratic forms with applications to monotone convergence theorems*, J. Funct. Anal. **28** (1978), 377–385.
42. ———, *Trace Ideals and Their Applications*, Cambridge University Press, New York, 1979.
43. ———, *Functional integration and quantum physics* (1979), Academic Press, New York.
44. ———, *Kotani theory for one-dimensional stochastic Jacobi matrices*, Comm. Math. Phys. **89** (1983), 227–234.
45. ———, *Localization in general one dimensional random systems I. Jacobi matrices*, Comm. Math. Phys. **102** (1985), 327–336.
46. ———, *Operators with singular continuous spectrum*, I. *General operators*, Ann. of Math. (2) (to appear).
47. B. Simon and T. Wolff, *Singular continuous spectrum under rank one perturbations and localization for random Hamiltonians*, Comm. Pure Appl. Math. **39** (1986), 75–90.
48. E. C. Titchmarsh, *Eigenfunction expansions*, 2nd ed., Oxford University Press, Oxford, 1962.
49. F. Wegner, *Bounds on the density of states in disordered systems*, Z. Phys. B **44** (1981), 9–15.

DIVISION OF PHYSICS, MATHEMATICS, AND ASTRONOMY, CALIFORNIA INSTITUTE OF TECHNOLOGY, 253-37 PASADENA, CA 91125

E-mail address: 76004.1664@compuserve.com

Centre de Recherches Mathématiques
CRM Proceedings and Lecture Notes
Volume 8, 1995

Inverse Potential Scattering: A Geometrical Approach

Volker Enss and Ricardo Weder

ABSTRACT. In multidimensional short-range potential scattering the high energy limit of the scattering operator determines the potential uniquely. For given long-range potential the high energy behavior of the Dollard modified scattering operator determines the potential uniquely.

We use geometrical, time-dependent methods for the proof and obtain a reconstruction formula as well.

1. Results

In short-range potential scattering the wave operators and the scattering operator S are naturally uniquely defined and have the expected properties. For space dimension $\nu \geq 2$ we show in this note by a simple time-dependent geometrical proof that the high velocity limit of the scattering operator determines the potential uniquely. In the long-range case one has to know (the tail part of) the long-range potential before one can even define modified wave operators and a corresponding scattering operator. Thus, given a long-range potential we show that the short-range part (and thus all) of the potential is uniquely determined by the high velocity behavior of the scattering operator. The result is known in the short-range case [**9**], but our approach seems to be simpler and it easily generalizes to include long-range potentials. Moreover, it extends to N-body systems interacting by pair potentials [**4**].

On the Hilbert space $\mathcal{H} = L^2(\mathbb{R}^\nu), \nu \geq 2$ the self-adjoint free Hamiltonian is $H_0 = -(1/2m)\Delta$. For simplicity of presentation we give here the proof for bounded continuous potentials, the general case including singular potentials is treated in [**4**]. The class of short-range potentials consists here of multiplication operators V^s from the set

$$\mathcal{V}_{SR} := \Big\{ V \in C(\mathbb{R}^\nu) \;\Big|\; \sup_{|\mathbf{x}| \geq R} |V(\mathbf{x})| \in L^1\big([0,\infty), dR\big) \Big\}. \tag{1.1}$$

The interacting Hamiltonian $H = H_0 + V^s$ is self-adjoint on the same domain as H_0. It is well known that the wave operators

$$\Omega_\pm(H_0, H_0 + V^s) := \operatorname*{s-lim}_{t \to \pm\infty} e^{i(H_0+V^s)t} e^{-iH_0 t} \tag{1.2}$$

1991 *Mathematics Subject Classification.* Primary 35P25, 81U40, 81U05; Secondary 35Q40, 47A40.

The second author is Fellow, Sistema Nacional de Investigadores.

Research partially supported by Deutscher Akademischer Austauschdienst.

This is the final form of the paper.

exist and are complete and that the scattering operator

$$S(H_0, H_0 + V^s) := \big[\Omega_+(H_0, H_0 + V^s)\big]^* \Omega_-(H_0, H_0 + V^s) \tag{1.3}$$

is unitary on $\mathcal{H}$ etc., see e.g. [**8**] as a general reference for mathematical scattering theory. We will consider now the scattering operator as a mapping from the potentials into bounded (even unitary) operators and call it the "scattering map".

THEOREM 1.1 (SHORT-RANGE). *The scattering map*

$$\begin{aligned} S\colon \mathcal{V}_{SR} &\longrightarrow L(\mathcal{H}),\\ V^s &\longrightarrow S(H_0, H_0 + V^s) \end{aligned}$$

is injective.

Actually, the high energy behavior of S is enough to determine V^s, see Theorem 1.4 below for the explicit formula.

This result has a long history beginning with Faddeev [**5**] and Berezanskii [**1**]. Under essentially the same assumptions as above it was shown by Saito [**9**], see [**4**] for further references.

The differentiable long-range potential is assumed to satisfy

$$\lim_{|\mathbf{x}|\to\infty} V^\ell(\mathbf{x}) = 0, \quad \big|(\nabla V^\ell)(\mathbf{x})\big| \le \text{const}\big(1 + |\mathbf{x}|\big)^{-\gamma}, \quad \gamma > \frac{3}{2}. \tag{1.4}$$

The splitting $V = V^\ell + V^s$ into long- and short-range parts is not unique. By [**7**, Lemma 3.3] one can choose it *without loss of generality* such that V^ℓ is smooth and in addition

$$\big|(\Delta V^\ell)(\mathbf{x})\big| \le \text{const}\big(1 + |\mathbf{x}|\big)^{-(2+\varepsilon)}, \quad \varepsilon > 0 \tag{1.5}$$

is satisfied. We will use this convention in the sequel. Under these assumptions the ordinary wave operators between $H^\ell := H_0 + V^\ell$ and $H = H^\ell + V^s$ exist on $P^{\text{cont}}(H^\ell)\mathcal{H}$, the continuous spectral subspace of H^ℓ, and are complete.

THEOREM 1.2 (LONG-RANGE). *The scattering map*

$$\begin{aligned} S^\ell\colon \mathcal{V}_{SR} &\longrightarrow L(\mathcal{H}),\\ V^s &\longrightarrow S(H^\ell, H^\ell + V^s) \end{aligned}$$

is injective.

This result is simple to state but it has the disadvantage that we cannot control the evolution generated by H^ℓ effectively and it is of little use in physics. We will only sketch its proof. However, an explicitly known time evolution like the Dollard evolution requires the prior knowledge of the long-range part of the potential. Without the latter it is impossible to even define modified wave operators and the scattering operator. Therefore, from now on a long-range part V^ℓ of the potential which satisfies (1.4) and (1.5) is given. The question how V^ℓ can be obtained from information about scattering of wave packets will be addressed in a later paper.

As a modified Dollard time evolution we use

$$U^D(t,\tau) := e^{-iH_0(t-\tau)} \exp\left\{-i\int_\tau^t ds\, V^\ell(s\,\mathbf{p}/m)\right\} \tag{1.6}$$

which is generated by $H^D(t) := H_0 + V^\ell(t\mathbf{p}/m)$. The corresponding modified Dollard wave operators are

$$\Omega^D_\pm(H_0, H_0 + V^\ell + V^s) := \underset{t\to\pm\infty}{\text{s–lim}}\, e^{i(H_0+V^\ell+V^s)t} U^D(t,0). \tag{1.7}$$

They are well known to exist and to be complete for any $V^s \in \mathcal{V}_{SR}$. The unitary modified Dollard scattering operator is

$$\begin{aligned} S^D(H_0, H_0 + V^\ell + V^s) \\ = \left[\Omega^D_+(H_0, H_0 + V^\ell + V^s)\right]^* \Omega^D_-(H_0, H_0 + V^\ell + V^s). \end{aligned} \tag{1.8}$$

The non-uniqueness of the splitting $V = V^\ell + V^s$ and of the definition (1.6) for U^D does not cause problems. The wave operators establish a connection between scattering states (given by their initial condition at $t = 0$) and asymptotic configurations. The latter can be represented by vectors in a Hilbert space in many different but equivalent ways. The scattering operator is a mapping from incoming to outgoing asymptotic configurations. Its concrete form clearly depends on the chosen representation. (1.6)–(1.8) are a convenient and common choice.

THEOREM 1.3 (LONG-RANGE). *The Dollard modified scattering map*

$$\begin{aligned} S^D \colon \mathcal{V}_{SR} &\longrightarrow L(\mathcal{H}) \\ V^s &\longrightarrow S^D(H_0, H_0 + V^\ell + V^s) \end{aligned}$$

is injective.

Again the high velocity behavior is enough to reconstruct the potential. To make this precise we need some notation. Let Ψ_0 be a normalized asymptotic configuration with compact momentum support in the open ball of radius $m\widehat{R}$ around the origin

$$\widehat{\psi}_0 \in C_0^\infty\big(B_{m\widehat{R}}(0)\big), \quad \|\Psi_0\| = 1. \tag{1.9}$$

For later use we mention that there is a function $g \in C_0^\infty\big(B_{m\widehat{R}}(0)\big)$ such that $g(\mathbf{p})\Psi_0 = \Psi_0$ (i.e. $g = 1$ on the momentum support of Ψ_0). The boosted configuration ($\mathbf{x}$ the position operator)

$$\Psi_\mathbf{v} := e^{im\mathbf{v}\cdot\mathbf{x}}\Psi_0 \iff \widehat{\psi}_\mathbf{v} \in C_0^\infty\big(B_{m\widehat{R}}(m\,\mathbf{v})\big) \tag{1.10}$$

has velocity support of radius $\widehat{R}$ around $\mathbf{v}$. We will use the high velocity limit in an arbitrary fixed direction $\hat{\mathbf{v}} = \mathbf{v}/|\mathbf{v}|$, $v := |\mathbf{v}| \to \infty$.

THEOREM 1.4 (RECONSTRUCTION FORMULA). *For $\Psi_\mathbf{v}, \Phi_\mathbf{v}$ as defined in* (1.9), (1.10) *the following high velocity limit exists for any $\widehat{R}, \widehat{\mathbf{v}}$*

$$\begin{aligned} &\lim_{v\to\infty} iv\big(\Phi_\mathbf{v}, (S^D - \mathbf{1})\,\Psi_\mathbf{v}\big) \\ &\quad = \left(\Phi_0, \int_{-\infty}^{\infty} d\tau\, \left[V^s(\mathbf{x} + \widehat{\mathbf{v}}\tau) + V^\ell(\mathbf{x} + \widehat{\mathbf{v}}\tau) - V^\ell(\widehat{\mathbf{v}}\tau)\right] \Psi_0\right). \end{aligned} \tag{1.11}$$

This contains short-range scattering as a special case with $V^\ell = 0$ and $S^D = S$. With the notation $\mathbf{x}^\perp := \mathbf{x} - (\mathbf{x}\cdot\widehat{\mathbf{v}})\widehat{\mathbf{v}} \equiv \mathbf{x} - x^\parallel\widehat{\mathbf{v}}$ the integrals

$$W^s(\mathbf{x}^\perp;\widehat{\mathbf{v}}) := \int d\tau\, V^s(\mathbf{x}+\widehat{\mathbf{v}}\tau),$$
$$W^\ell(\mathbf{x}^\perp;\widehat{\mathbf{v}}) := \int d\tau\, \left[V^\ell(\mathbf{x}+\widehat{\mathbf{v}}\tau) - V^\ell(\widehat{\mathbf{v}}\tau)\right]$$

exist and are continuous by (1.1) and (1.4), respectively, and W^ℓ is given (possibly zero).

Varying $\widehat{R}$ in (1.9) the set of Φ_0, Ψ_0 is rich enough to determine for any $\widehat{\mathbf{v}}$ the continuous function $W^s(\cdot;\widehat{\mathbf{v}})$ uniquely from the r.h.s. of (1.11):

$$\int dx\, \overline{\phi_0(\mathbf{x})}\psi_0(\mathbf{x})W^s(\mathbf{x}^\perp;\widehat{\mathbf{v}}). \tag{1.12}$$

In $\nu = 2$ dimensions $W^s(\mathbf{x}^\perp;\widehat{\mathbf{v}})$ is the Radon transform of $V^s(\mathbf{x})$, the latter is square integrable. It is known that it determines V^s uniquely [**6**, p. 115f]. In $\nu \geq 3$ dimensions it is the X-ray transform. One can fix arbitrarily $(x^3, \cdots, x^\nu)$ and apply the same to the resulting two-dimensional function. Thus, varying $\widehat{\mathbf{v}}$ in a plane is actually sufficient to reconstruct $V^s(\mathbf{x})$ from $W^s(\mathbf{x}^\perp;\widehat{\mathbf{v}})$. Alternatively, one can see this intuitively from looking at (1.12) in momentum space:

$$\int dp^\parallel\, dp^\perp\, d\tilde{p}^\perp \overline{\hat{\phi}_0(p^\parallel, \mathbf{p}^\perp)}\widehat{V}^s(0, \mathbf{p}^\perp - \tilde{\mathbf{p}}^\perp)\widehat{\psi}_0(p^\parallel, \tilde{\mathbf{p}}^\perp). \tag{1.13}$$

To obtain $\widehat{V}^s(\mathbf{q})$ choose $\widehat{\mathbf{v}}$ orthogonal to $\mathbf{q}$ such that $\mathbf{q} = \mathbf{q}^\perp$. Then with a normalized one-dimensional χ let $\widehat{\psi}_0(p^\parallel, \tilde{\mathbf{p}}^\perp) = \chi(p^\parallel)\delta_n(\tilde{\mathbf{p}}^\perp)$, $\hat{\phi}_0(p^\parallel, \mathbf{p}^\perp) = \chi(p^\parallel)\delta_n(\mathbf{p}^\perp - \mathbf{q}^\perp)$ for a "δ-sequence" of square integrable functions δ_n of $\nu - 1$-dimensional argument. Clearly, $V^s(\mathbf{x})$ or $\widehat{V}^s(\mathbf{q})$ need not be determined pointwise but it is enough to know their action on a (form) core for H_0. This fixes $H = H_0 + V^s$ uniquely. See [**4**] for a complete proof.

Thus Theorems 1.1 and 1.3 follow from the more detailed Theorem 1.4. One can prove Theorem 1.2 directly but here it is easier to use the chain rule for wave operators.

Remark. For applications of our method to multiparticle systems we mention that completeness of the wave operators and unitarity of the scattering operator have not been used in the proofs. All we need is existence of (modified) wave operators on states whose Fourier transform has compact support away from the origin in momentum space. Actually, this follows as a byproduct of our estimates.

In the next section we give the simple and intuitive proof of Theorem 1.4 in the short-range case, the slightly more technical modifications to include long-range potentials follow in Section 3.

2. Proof in the Short-Range Case

We use the shorthands $\Omega_\pm = \Omega_\pm(H_0, H_0 + V^s)$ and $S = S(H_0, H_0 + V^s)$. The wave operators are well known to be isometric: $\Omega_\pm^*\Omega_\pm = \mathbf{1}$ and to satisfy the intertwining relations $\Omega_\pm H_0 = H\Omega_\pm$. Moreover, the following identity holds

$$\Omega_+ - \Omega_- = i\int_{-\infty}^{\infty} dt\, e^{iHt}V^s\, e^{-iH_0 t}, \tag{2.1}$$

see e.g. [**8**]. Thus

$$
\begin{aligned}
(2.2)\qquad i(S-\mathbf{1}) &= i(\Omega_+^*\,\Omega_- - \Omega_-^*\Omega_-)\\
&= i(\Omega_+ - \Omega_-)^*\Omega_-\\
&= \int dt\, e^{iH_0t}V^s e^{-iHt}\Omega_-\\
&= \int dt\, e^{iH_0t}V^s\Omega_- e^{-iH_0t}\\
&= \int dt\, e^{iH_0t}V^s e^{-iH_0t} + \int dt\, e^{iH_0t}V^s(\Omega_- - \mathbf{1})e^{-iH_0t}.
\end{aligned}
$$

The first term on the r.h.s. of (2.2)—the first Born approximation is the leading term which decays as v^{-1} if the velocity $\mathbf{v}$ goes to infinity. It will be used to recover the potential. The second summand—representing "multiple scattering"—is of order $o(v^{-1}) = O(v^{-2})$ and can therefore be neglected asymptotically.

Intuitively it is clear that each instance of scattering by a short-range potential yields a factor v^{-1}: the strength of the interaction as measured by $(S - \mathbf{1})$ or $(\Omega_\pm - \mathbf{1})$ is proportional to the time which the particle spends in the interaction region where the potential is strong. This time is inverse proportional to the speed v.

Under translations in momentum or configuration space, generated by $\mathbf{x}$ or $\mathbf{p}$, respectively, we get

$$
(2.3)\qquad e^{-im\mathbf{v}\cdot\mathbf{x}} f(\mathbf{p}) e^{im\mathbf{v}\cdot\mathbf{x}} = f(\mathbf{p} + m\mathbf{v}),
$$

in particular,

$$
(2.4)\qquad e^{-im\mathbf{v}\cdot\mathbf{x}} e^{-iH_0t} e^{im\mathbf{v}\cdot\mathbf{x}} = e^{-i\mathbf{p}\cdot\mathbf{v}t} e^{-iH_0t} e^{-imv^2t/2},
$$

$$
(2.5)\qquad e^{i\mathbf{p}\cdot\mathbf{v}t} f(\mathbf{x}) e^{-i\mathbf{p}\cdot\mathbf{v}t} = f(\mathbf{x} + \mathbf{v}\,t).
$$

With $\Phi_\mathbf{v}, \Psi_\mathbf{v}$ as defined in (1.9), (1.10)

$$
\begin{aligned}
(2.6)\qquad & v\int dt\, (\Phi_\mathbf{v}, e^{iH_0t}V^s(\mathbf{x})e^{-iH_0t}\Psi_\mathbf{v})\\
&= \int d(v\,t)\big(\Phi_\mathbf{0}, e^{iH_0t}V^s(\mathbf{x} + \mathbf{v}\,t)e^{-iH_0t}\Psi_\mathbf{0}\big)\\
&= \int d\tau\, \big(\Phi_\mathbf{0}, e^{iH_0\tau/v}V^s(\mathbf{x} + \widehat{\mathbf{v}}\tau)e^{-iH_0\tau/v}\Psi_\mathbf{0}\big)
\end{aligned}
$$

with the substitution $\tau := vt$ and $\widehat{\mathbf{v}} = \mathbf{v}/v$ the arbitrary fixed unit vector in the direction of $\mathbf{v}$. For any $\tau \in \mathbb{R}$ we have $\lim_{v\to\infty} \exp\{-iH_0\tau/v\}\Psi_\mathbf{0} = \Psi_\mathbf{0}$. Thus the pointwise limit of the integral in (2.6) yields the desired limit (1.11) in Theorem 1.4 (without V^ℓ). To justify it we need a bound for all large enough v

$$
(2.7)\qquad \left|\big(\Phi_\mathbf{0}, e^{iH_0\tau/v}V^s(\mathbf{x} + \widehat{\mathbf{v}}\tau)e^{-iH_0\tau/v}\Psi_\mathbf{0}\big)\right| \le h(\tau)
$$

where h is integrable. The left hand side is dominated by

$$
(2.8)\qquad \left\|V^s(\mathbf{x} + \widehat{\mathbf{v}}\tau)F\left(|\mathbf{x}| \le \frac{|\tau|}{2}\right)\right\| + \|V^s\| \left\|F\left(|\mathbf{x}| \ge \frac{|\tau|}{2}\right) e^{-iH_0\tau/v}\Psi_\mathbf{0}\right\|.
$$

The first summand is bounded by $h_1(\tau) := \sup_{|\mathbf{y}|\geq\tau/2} |V(\mathbf{y})|$ which is independent of v and integrable by (1.1). If $v > 2\widehat{R}$ ($\widehat{R}$ the radius of the velocity support of $\Psi_\mathbf{0}$) the second term describes free propagation into the "classically forbidden region" which is rapidly decaying: With g as introduced following (1.9), i.e. $g(\mathbf{p})\Psi_\mathbf{0} = \Psi_\mathbf{0}$, $g \in C_0^\infty(B_{m\widehat{R}}(0))$, we get for $v \geq 4\widehat{R}$

$$\begin{aligned}(2.9) \quad & \left\| F\left(|x| \geq \frac{|\tau|}{2}\right) e^{-iH_0\tau/v}\Psi_\mathbf{0} \right\| \\ & \leq \left\| F\left(|x| \geq \frac{|\tau|}{8}\right)\Psi_\mathbf{0} \right\| \\ & \quad + \left\| F\left(|x| \geq \frac{|\tau|}{8} + \widehat{R}\frac{|\tau|}{v} + \frac{|\tau|}{8}\right) e^{-iH_0\tau/v} g(p) F\left(|x| \leq \frac{|\tau|}{8}\right)\right\| \\ & = h_2(\tau) + h_3(\tau).\end{aligned}$$

Both terms decay rapidly in $(1+|\tau|)$ uniformly for large v: For the first it is obvious because $\psi_\mathbf{0} \in \mathcal{S}(\mathbb{R}^\nu)$; for the second it is a standard non-stationary phase estimate, see e.g. [**2**, Proposition 2.10] where $r = |\tau|/8$ is the distance from the classically allowed region. We have with $h = h_1 + h_2 + h_3$ an integrable bound for all $v \geq 4\widehat{R}$. By the dominated convergence theorem we may take pointwise limits and obtain

$$(2.10) \quad \lim_{v\to\infty} v \int dt\, (\Phi_\mathbf{v}, e^{iH_0t}V^s e^{-iH_0t}\Psi_\mathbf{v}) = \left(\Phi_\mathbf{0}, \int d\tau\, V^s(\mathbf{x} + \widehat{\mathbf{v}}\tau)\Psi_\mathbf{0}\right).$$

Remark. The free time evolution of a wave packet can be considered as consisting of two parts: a purely kinematical part of translation with the average velocity and an additional evolutionary part which changes the shape of the packet (spreading etc.). Locally in the interaction region the translational part is dominant in the high velocity limit yielding the simple formula. The evolutionary part which is absent for improper states with sharp energy can be neglected asymptotically likewise as in the stationary approach. This remains true for interacting time evolutions as well.

To finish the proof we have to show that the second "multiple scattering" term in (2.2) is small compared to the leading first one.

$$\begin{aligned}(2.11) \quad & \left| \int_{-\infty}^{\infty} dt\, \left(\Phi_\mathbf{v}, e^{iH_0t}V^s(\Omega_- - \mathbf{1})e^{-iH_0t}\Psi_\mathbf{v}\right)\right| \\ & = \left| \int_{-\infty}^{\infty} dt\, \left(V^s e^{-iH_0t}\Phi_\mathbf{v}, \int_{-\infty}^{0} dT e^{iHT} V^s e^{-iH_0(T+t)}\Psi_\mathbf{v}\right)\right| \\ & \leq \int_{-\infty}^{\infty} dt \|V^s e^{-iH_0t}\Phi_\mathbf{v}\| \int_{-\infty}^{\infty} dT \|V^s e^{-iH_0T}\Psi_\mathbf{v}\|.\end{aligned}$$

With the same calculations (2.3)–(2.5) as above

$$\begin{aligned}(2.12) \quad & v\int dt\, \|V^s(\mathbf{x})e^{-iH_0t}\Phi_\mathbf{v}\| = v\int dt\, \|V^s(\mathbf{x} + \mathbf{v}t)e^{-iH_0t}\Phi_\mathbf{0}\| \\ & \leq \int d(vt) \left\{ \left\| V^s(\mathbf{x} + \mathbf{v}t) F\left(|\mathbf{x} + \mathbf{v}t| > \frac{|tv|}{2}\right)\right\| + \left\| F\left(|x| > \frac{|tv|}{2}\right) e^{-iH_0t}\Phi_\mathbf{0}\right\|\right\}.\end{aligned}$$

The first integral is finite uniformly in v by (1.1) and the same applies to the second by (2.9). This shows that (2.11) is $O(v^{-2})$ at infinity and thus negligible compared to the leading term (2.10) in the high velocity limit. This finishes the proof of Theorem 1.4 in the case $V^\ell = 0$. It implies also Theorem 1.1. □

Remark. In the situation of Theorem 1.2 one has $[\Omega_\pm^\ell]^*\Omega_\pm^\ell = P^{\text{cont}}(H^\ell)$, $H^\ell = H_0 + V^\ell$, and one obtains analogously

$$\lim_{v\to\infty} iv\Big(\Phi_{\mathbf{v}}, \big(S^\ell - P^{\text{cont}}(H^\ell)\big)\Psi_{\mathbf{v}}\Big) = \Big(\Phi_0, \int d\tau\, V^s(\mathbf{x}+\hat{\mathbf{v}}\tau)\Psi_0\Big). \tag{2.13}$$

Asymptotically, $\Psi_{\mathbf{v}}$ is orthogonal to all bound states of H^ℓ. It is easy to verify that for fixed τ

$$\begin{aligned}\lim_{v\to\infty} & e^{imv\tau/2} e^{i\mathbf{p}\cdot\hat{\mathbf{v}}\tau} e^{-im\mathbf{v}\cdot\mathbf{x}} e^{-i(H^\ell)\tau/v} e^{im\mathbf{v}\cdot\mathbf{x}}\Psi_0 \\ &= \lim_{v\to\infty} e^{i\mathbf{p}\cdot\hat{\mathbf{v}}\tau} e^{-i\mathbf{p}\cdot\hat{\mathbf{v}}\tau + i(H^\ell)\tau/v}\Psi_0 = \Psi_0.\end{aligned}$$

However, the uniform bound corresponding to (2.7) is harder in this case. See the end of the next section for a different approach which relies on results which we prove anyway for Theorems 1.3. and 1.4.

3. Proof in the Long-Range Case

With the shorthands $\Omega_\pm^D$ and S^D we obtain in this case instead of (2.2)

$$\begin{aligned} i(S^D - \mathbf{1}) &= i(\Omega_+^D - \Omega_-^D)^*\Omega_-^D \\ &= \int dt\, U^D(t,0)^* \Big\{V^s(\mathbf{x}) + V^\ell(\mathbf{x}) - V^\ell\Big(\mathbf{p}\frac{t}{m}\Big)\Big\} e^{-iHt}\Omega_-^D \\ &= \int dt\, U^D(t,0)^* \Big\{V^s(\mathbf{x}) + V^\ell(\mathbf{x}) - V^\ell\Big(\mathbf{p}\frac{t}{m}\Big)\Big\} U^D(t,0) \\ &\quad + \int dt\, U^D(t,0)^* \Big\{V^s(\mathbf{x}) + V^\ell(\mathbf{x}) - V^\ell\Big(\mathbf{p}\frac{t}{m}\Big)\Big\} \\ &\qquad \times \big(e^{-iHt}\Omega_-^D - U^D(t,0)\big). \end{aligned} \tag{3.1}$$

Again the first summand will give the leading contribution while the second is asymptotically negligible. Analogous to (2.6) we obtain

$$\begin{aligned} & v\int dt\Big(\Phi_{\mathbf{v}}, U^D(t,0)^* \Big\{V^s(\mathbf{x}) + V^\ell(\mathbf{x}) - V^\ell\Big(\mathbf{p}\frac{t}{m}\Big)\Big\} U^D(t,0)\Psi_{\mathbf{v}}\Big) \\ &= \int d(vt)\Big(\Phi_0, \widetilde{U}(\mathbf{v};t,0)^* e^{iH_0t}\Big\{V^s(\mathbf{v}t+\mathbf{x}) + V^\ell(\mathbf{v}t+\mathbf{x}) \\ &\qquad - V^\ell\Big(\mathbf{v}t + \mathbf{p}\frac{t}{m}\Big)\Big\} e^{-iH_0t}\widetilde{U}(\mathbf{v};t,0)\,\Psi_0\Big) \\ &= \int d\tau\,\Big(\Phi_0, \widetilde{U}\Big(\mathbf{v};\frac{\tau}{v},0\Big)^* e^{iH_0\tau/v}\Big\{V^s(\hat{\mathbf{v}}\tau+\mathbf{x}) + V^\ell(\hat{\mathbf{v}}\tau+\mathbf{x}) \\ &\qquad - V^\ell\Big(\hat{\mathbf{v}}\tau + \mathbf{p}\frac{\tau}{mv}\Big)\Big\} e^{-iH_0\tau/v}\widetilde{U}\Big(\mathbf{v};\frac{\tau}{v},0\Big)\Psi_0\Big). \end{aligned} \tag{3.2}$$

Here the translated correction term in the Dollard evolution is

$$\widetilde{U} \equiv \widetilde{U}(\mathbf{v};t,0) = \exp\left\{-i\int_0^t ds V^\ell\left(\mathbf{v}s + \mathbf{p}\frac{s}{m}\right)\right\} \tag{3.3}$$

and its rescaled version $\widetilde{U}(\mathbf{v};\tau/v,0)$ tends for fixed τ strongly to $\mathbf{1}$ in the limit $v \to \infty$, similarly as $\exp\{\pm iH_0\tau/v\}$ do. Moreover, $V^\ell(\hat{\mathbf{v}}\tau + \mathbf{p}\tau/mv)$ converges strongly to $V^\ell(\hat{\mathbf{v}}\tau)$.

Thus, taking a pointwise limit of the integrand on the r.h.s. of (3.2) we obtain as $v \to \infty$ the limit

$$\left(\Phi_0, \int d\tau \, \{V^s(\hat{\mathbf{v}}\tau + \mathbf{x}) + V^\ell(\hat{\mathbf{v}}\tau + \mathbf{x}) - V^\ell(\hat{\mathbf{v}}\tau)\} \, \Psi_0\right) \tag{3.4}$$

as claimed in Theorem 1.4. To justify this we show several estimates. We include the details for the convenience of the reader to make the article self-contained although some of them are standard in long-range scattering theory [**3, 8**].

PROPOSITION 3.1. *For $\Phi_{\mathbf{v}}$ as given in* (1.9), (1.10) *there is a function $h \in L^1(\mathbb{R})$ such that for $v \geq 4\widehat{R}$*

$$\left\|\left\{V^s(\mathbf{x}) + V^\ell(\mathbf{x}) - V^\ell\left(\mathbf{p}\frac{t}{m}\right)\right\} U^D(t,0)\Phi_{\mathbf{v}}\right\| \leq h(vt). \tag{3.5}$$

PROOF. It is convenient to split the long-range potential V^ℓ into an inner part and an outer part with easily controllable decay properties. Let $\chi \in C^\infty$ satisfy $\chi(\lambda) = 1$ for $\lambda \geq 1/2$ and $\chi(\lambda) = 0$ for $\lambda \leq 1/4$ and define

$$V^\ell_{vt}(\mathbf{y}) := V^\ell(\mathbf{y})\,\chi\left(\left|\frac{\mathbf{y}}{vt}\right|\right). \tag{3.6}$$

Then $\operatorname{supp}(V^\ell - V^\ell_{vt}) \subset B_{|vt|/2}(0)$, $\|V^\ell - V^\ell_{vt}\| \leq \|V^\ell\|$, and the decay of V^ℓ for large arguments translates into decay w.r.t. $|vt|$ independent of the argument: from (1.4) and (1.5) follows

$$\|\nabla V^\ell_{vt}\| = \sup_{\mathbf{y}} |(\nabla V^\ell_{vt})(\mathbf{y})| \leq \text{const}\,|vt|^{-\gamma}, \quad \gamma > \frac{3}{2}, \tag{3.7}$$

$$\|\Delta V^\ell_{vt}\| \leq \text{const}\,|vt|^{-(2+\varepsilon)}, \quad \varepsilon > 0. \tag{3.8}$$

The norm in (3.5) equals

$$\begin{aligned} &\left\|\left\{V^s(\mathbf{x}) + V^\ell(\mathbf{x}) - V^\ell\left(t\frac{\mathbf{p}}{m}\right)\right\} e^{-iH_0t} \exp\left\{-i\int_0^t ds V^\ell\left(s\frac{\mathbf{p}}{m}\right)\right\} e^{im\mathbf{v}\cdot\mathbf{x}}\Phi_0\right\| \\ &\leq \left\|V^s(t\mathbf{v} + \mathbf{x})e^{-iH_0t}\widetilde{U}(\mathbf{v};t,0)\Phi_0\right\| \\ &\qquad + \left\|(V^\ell - V^\ell_{vt})(t\mathbf{v} + \mathbf{x})e^{-iH_0t}\widetilde{U}(\mathbf{v};t,0)\Phi_0\right\| \\ &\qquad + \left\|\left\{V^\ell_{vt}(t\mathbf{v} + \mathbf{x}) - V^\ell_{vt}\left(t\mathbf{v} + t\frac{\mathbf{p}}{m}\right)\right\} e^{-iH_0t}\widetilde{U}(\mathbf{v};t,0)\Phi_0\right\|. \end{aligned} \tag{3.9}$$

We have used (2.3)–(2.5), the shorthand (3.3), and the fact that $V^\ell(t\mathbf{v} + t\mathbf{p}/m) = V^\ell_{vt}(t\mathbf{v} + t\mathbf{p}/m)$ on the momentum support of Φ_0 for $v > 2\widehat{R}$.

Note that $|t\mathbf{v}+\mathbf{x}| < |tv|/2$ implies $|x| > |tv|/2$, thus the summand with V^s is estimated by

$$\text{(3.10)}\quad \left\| V^s(t\mathbf{v}+\mathbf{x})F\left(|t\mathbf{v}+\mathbf{x}| > \frac{|tv|}{2}\right)\right\| \\ + \|V^s\| \left\| F\left(|x| > \frac{|tv|}{2}\right) e^{-iH_0t}\widetilde{U}\Phi_{\mathbf{0}}\right\|.$$

The first term decays integrably in vt by (1.1), the integrable decay of the second will be shown later in Lemma 3.2. The second line in (3.9) is bounded as well by

$$\text{(3.11)}\quad \|V^\ell\| \left\| F\left(|x| > \frac{|tv|}{2}\right) e^{-iH_0t}\widetilde{U}\Phi_{\mathbf{0}}\right\|$$

due to the support of $(V^\ell - V^\ell_{vt})$. For the tail part of the long-range potential in (3.9) we use $\exp(iH_0t)f(\mathbf{x})\exp(-iH_0t) = f(\mathbf{x}+t\mathbf{p}/m)$ and obtain

$$\begin{aligned}\text{(3.12)}\quad &\left\| \left\{ V^\ell_{vt}\left(t\mathbf{v}+t\frac{\mathbf{p}}{m}+\mathbf{x}\right) - V^\ell_{vt}\left(t\mathbf{v}+t\frac{\mathbf{p}}{m}\right)\right\}\widetilde{U}\Phi_{\mathbf{0}}\right\| \\ &\le \int_0^1 ds \left\| \frac{d}{ds}V^\ell_{vt}\left(t\mathbf{v}+t\frac{\mathbf{p}}{m}+s\mathbf{x}\right)\widetilde{U}\Phi_{\mathbf{0}}\right\| \\ &= \int_0^1 ds \left\| (\nabla V^\ell_{vt})\left(t\mathbf{v}+t\frac{\mathbf{p}}{m}+s\mathbf{x}\right)\cdot \mathbf{x}\widetilde{U}\Phi_{\mathbf{0}} - \left[\frac{it}{(2m)}\right](\Delta V^\ell_{vt})(\cdots)\widetilde{U}\Phi_{\mathbf{0}}\right\| \\ &\le \text{const}\left[|vt|^{-\gamma}\|\mathbf{x}\widetilde{U}\Phi_{\mathbf{0}}\| + |t||vt|^{-(2+\varepsilon)}\right].\end{aligned}$$

The gradient-term is the same as for commuting arguments of the potential. The correction with the Laplacian of the potential (at the same argument) is obtained e.g. using the Fourier representation and the Baker-Campbell-Hausdorff formula: $\exp(A+B) = \exp(A)\exp(B)\exp([A,B]/2)$ if $[A,B]$ commutes with A and B. The formula is used to isolate the terms depending on s in the exponent.

$$\text{(3.13)}\quad |t||vt|^{-(2+\varepsilon)} = v^{-1}|vt|^{-(1+\varepsilon)}$$

is integrable in vt uniformly for large v. We show in Lemma 3.2 that

$$\text{(3.14)}\quad \|\mathbf{x}\widetilde{U}\Phi_{\mathbf{0}}\| \le \text{const}\, v^{-2}|vt|^{2-\gamma} + \|\mathbf{x}\Psi_{\mathbf{0}}\|.$$

Thus we get as a bound for (3.12) for large $|vt|$

$$\text{(3.15)}\quad \text{const}\left[v^{-2}|vt|^{2-2\gamma} + |vt|^{-\gamma}\|\mathbf{x}\Psi_{\mathbf{0}}\| + v^{-1}|vt|^{-(1+\varepsilon)}\right]$$

and uniform boundedness for all v, t. Since $\gamma > 3/2$ we have shown integrability in vt of this contribution uniformly for $v \ge 2\widehat{R}$. □

LEMMA 3.2. *For $\Phi_{\mathbf{0}}$ as given in (1.9), $\widetilde{U}(\mathbf{v};t,0)$ as defined in (3.3) and $v > 4\widehat{R}$*

$$\text{(3.16)}\quad \|\mathbf{x}\widetilde{U}(\mathbf{v};t,0)\Phi_{\mathbf{0}}\| \le \|\mathbf{x}\Phi_{\mathbf{0}}\| + \text{const}\, v^{-2}|vt|^{2-\gamma},$$

$$\begin{aligned}\text{(3.17)}\quad &\left\| F\left(|\mathbf{x}| > \frac{|tv|}{2}\right) e^{-iH_0t}\widetilde{U}(\mathbf{v};t,0)\Phi_{\mathbf{0}}\right\| \\ &\le \text{const}\left[1 + v^{-2}|vt|^{2-\gamma} + v^{-3}|vt|^{1-\varepsilon} + v^{-4}|vt|^{4-2\gamma}\right](1+|vt|^2)^{-1}.\end{aligned}$$

Both constants depend on V^ℓ, the one in (3.17) *depends in addition on g, $\|\mathbf{x}\,\Phi_0\|$, and $\|\mathbf{x}^2\,\Phi_0\|$. In particular* (3.17) *is integrable in vt uniformly for large v.*

Remark. With a longer proof one shows decay faster than any inverse power of vt in (3.17), see [**2**, Corollary 2.5].

PROOF. We can use that V^ℓ may be replaced by V^ℓ_{vs} in the exponent of $\widetilde{U}$ on states with momentum support in $B_{m\widehat{R}}(0)$, i.e. on the range of $g(\mathbf{p})$ as defined following (1.9), if $v > 2\widehat{R}$.

$$\text{(3.18)}\qquad \left[\mathbf{x},\widetilde{U}(\mathbf{v};t,0)\right]g(\mathbf{p}) = \int_0^t ds\frac{s}{m}(\nabla V^\ell_{vs})\left(\mathbf{v}s+\mathbf{p}\frac{s}{m}\right)\widetilde{U}(\mathbf{v};t,0)g(\mathbf{p}),$$

$$\text{(3.19)}\qquad \left\|\left[\mathbf{x},\widetilde{U}(\mathbf{v};t,0)\right]g(\mathbf{p})\right\| \le \text{const}\, v^{-\gamma}\left|\int_0^t ds\, s^{1-\gamma}\right| \le \text{const}\, v^{-2}|vt|^{2-\gamma},$$

$$\text{(3.20)}\ \left\|\left[\mathbf{x},[\mathbf{x},\widetilde{U}(\mathbf{v};t,0)]\right]g(\mathbf{p})\right\| \le \left|\int\cdots\right|^2 + \left|\int_0^t ds\left(\frac{s}{m}\right)^2(\Delta V^\ell_{vs})\left(\mathbf{v}s+\mathbf{p}\frac{s}{m}\right)\right|$$
$$\le \text{const}\left[v^{-4}|vt|^{4-2\gamma}+v^{-3}|vt|^{1-\varepsilon}\right].$$

The constants depend only on V^ℓ. We get (3.16) from

$$\|\mathbf{x}\widetilde{U}(\mathbf{v};t,0)\Phi_0\| \le \|\mathbf{x}\Phi_0\| + \left\|[\mathbf{x},\widetilde{U}(\mathbf{v};t,0)]g(\mathbf{p})\right\|.$$

For (3.17) observe that

$$\text{(3.21)}\qquad \left\|F\left(|\mathbf{x}|>\frac{|tv|}{2}\right)e^{-iH_0t}\widetilde{U}\Phi_0\right\|$$
$$\le \left\|F\left(|\mathbf{x}|>\frac{|tv|}{2}\right)e^{-iH_0t}g(\mathbf{p})F\left(|\mathbf{x}|<\frac{|tv|}{8}\right)\right\| + \left\|F\left(|\mathbf{x}|>\frac{|tv|}{8}\right)\widetilde{U}\Phi_0\right\|.$$

Again, the first summand has rapid decay in $|vt|$, e.g.

$$\text{(3.22)}\qquad \left\|F\left(|\mathbf{x}|>\frac{|tv|}{2}\right)e^{-iH_0t}g(p)F\left(|\mathbf{x}|<\frac{|tv|}{8}\right)\right\| \le \text{const}(1+|vt|^2)^{-1}$$

for all $v \ge 4\widehat{R}$ by non-stationary phase estimates for propagation into classically forbidden regions, cf. (2.9).

$$\text{(3.23)}\qquad \left(\frac{|tv|}{8}\right)^2\left\|F\left(|\mathbf{x}|>\frac{|tv|}{8}\right)\widetilde{U}\Phi_0\right\|$$
$$\le \left\|F\left(|\mathbf{x}|>\frac{|tv|}{8}\right)\mathbf{x}^2\widetilde{U}\Phi_0\right\|$$
$$\le \left\|\left[\mathbf{x},[\mathbf{x},\widetilde{U}]\right]g(\mathbf{p})\right\| + 2\left\|[\mathbf{x},\widetilde{U}]g(\mathbf{p})\right\|\,\|\mathbf{x}\Phi_0\| + \|\mathbf{x}^2\Phi_0\|$$
$$\le \text{const}\left[v^{-4}|vt|^{4-2\gamma}+v^{-3}|vt|^{1-\varepsilon}+v^{-2}|vt|^{2-\gamma}+1\right]$$

for a constant which depends on g, V^ℓ, $\|\mathbf{x}\,\Phi_0\|$, and $\|\mathbf{x}^2\,\Phi_0\|$. □

PROPOSITION 3.3. *For $\Psi_{\mathbf{v}}$ as given in (1.9), (1.10) there is a bound $M < \infty$ such that*

$$\sup_{t\in\mathbb{R}} v\left\| \left(e^{-iHt}\Omega_\pm^D - U^D(t,0)\right)\Psi_{\mathbf{v}}\right\| \le M. \tag{3.24}$$

PROOF.

$$\begin{aligned}\left(e^{-iHt}\Omega_\pm^D - U^D(t,0)\right) &= \lim_{t'\to\pm\infty} e^{-iHt}e^{iHt'}U^D(t',0) - U^D(t,0)\\ &= \lim_{T\to\pm\infty} e^{iHT}U^D(T+t,0) - U^D(t,0)\\ &= \int_0^{\pm\infty} ds e^{iHs} i\left\{H - H^D(s+t)\right\}U^D(s+t,0).\end{aligned}$$

This yields the uniform bound

$$\begin{aligned}&\sup_{t\in\mathbb{R}} v\|\left(e^{-iHt}\Omega_\pm^D - U^D(t,0)\right)\Psi_{\mathbf{v}}\|\\ &\qquad\le v\int_{-\infty}^{\infty} ds \left\|\left\{V^s(\mathbf{x}) + V^\ell(\mathbf{x}) - V^\ell\left(\mathbf{p}\frac{s}{m}\right)\right\}U^D(s,0)\Psi_{\mathbf{v}}\right\|\\ &\qquad\le \int_{-\infty}^{\infty} d(vs)h(vs) = M < \infty\end{aligned}$$

by Proposition 3.1. □

Now we are ready to finish the proof of Theorem 1.4. The integrable bound for (3.2) uniform for large v is a direct consequence of Proposition 3.1. By the dominated convergence theorem pointwise limits as $v \to \infty$ yield the correct result (3.4). To show that the second term in (3.1) is $O(v^{-2})$—and thus small compared to the first one—we estimate it with $vt = \tau$ and Propositions 3.1, 3.3

$$\begin{aligned}v^2 \int dt &\left\|\left\{V^s(\mathbf{x}) + V^\ell(\mathbf{x}) - V^\ell\left(\mathbf{p}\frac{t}{m}\right)\right\}U^D(t,0)\Phi_{\mathbf{v}}\right\|\\ &\times \sup_{t\in\mathbb{R}}\|\left(e^{-iHt}\Omega_\pm^D - U^D(t,0)\right)\Psi_{\mathbf{v}}\| \le \int d\tau\, h(\tau)M = M^2.\end{aligned} \tag{3.25}$$

This finishes the proof of Theorem 1.4 and as a consequence of Theorem 1.3. □

It remains to show Theorem 1.2. For shorter writing let $H^\ell := H_0 + V^\ell$, $H := H_0 + V^\ell + V^s$. Since all wave operators exist and are complete we may use the chain rule (see e.g. [**8**]) and obtain

$$\Omega_\pm(H^\ell, H) = \Omega_\pm^D(H_0, H)\left[\Omega_\pm^D(H_0, H^\ell)\right]^*.$$

This yields for the scattering operator

$$\begin{aligned}S(H^\ell, H) &= \Omega_+^D(H_0, H^\ell)\left[\Omega_+^D(H_0, H)\right]^*\Omega_-^D(H_0, H)\left[\Omega_-^D(H_0, H^\ell)\right]^*\\ &= \Omega_+^D(H_0, H^\ell)S^D(H_0, H)\left[\Omega_-^D(H_0, H^\ell)\right]^*.\end{aligned}$$

Since the ranges of $\left[\Omega_\pm^D(H_0, H^\ell)\right]^*$ are all of $\mathcal{H}$ we conclude that the knowledge of $S(H^\ell, H)$ determines $S^D(H_0, H)$ uniquely. By Theorem 1.3 this fixes V^s as well. □

References

1. Yu. M. Berezanskii, *The uniqueness theorem in the inverse problem of spectral analysis for the Schrödinger equation*, Trudy Moscov. Mat. Obshch. **7** (1958), 3–62; English transl. in Amer. Math. Soc. Transl., Series 2, **35** (1964), 167–235.
2. V. Enss, *Propagation properties of quantum scattering states*, J. Funct. Anal. **52** (1983), 219–251.
3. V. Enss, *Asymptotic completeness for quantum mechanical potential scattering* II. *Singular and long-range potentials*, Ann. Physics **119** (1979), 117-132; *Long-range scattering of two- and three-body quantum systems*, Section 2, Journées "Équations aux dérivées partielles" Saint Jean de Monts 1989, Publications École Polytechnique, Palaiseau, 1989, pp. 1–31.
4. V. Enss and R. Weder, *The geometrical approach to multidimensional inverse scattering*, preprint no. 27, IIMAS-UNAM (1994), 24 pages.
5. L. D. Faddeev, *Uniqueness of the solution of the inverse scattering problem*, Vestnik Leningrad Univ. **11** (1956), no. 7, 126–130.
6. S. Helgason, *Groups and geometric analysis*, Academic Press, Orlando, 1984.
7. L. Hörmander, *The existence of wave operators in scattering theory*, Math. Z. **146** (1976), 69–91.
8. M. Reed and B. Simon, *Methods of modern mathematical physics* III. *Scattering theory*, Academic Press, New York, 1979.
9. Y. Saito, *An asymptotic behavior of the S-matrix in the inverse scattering problem*. J. Math. Phys. **25** (1984), 3105–3111.

Institut für Reine und Angewandte Mathematik, Rheinisch-Westfälische Technische Hochschule Aachen, D-52056 Aachen, Germany

E-mail address: enss@rwth-aachen.de

Instituto de Investigaciones en Matemáticas Applicadas y en Sistemas, Universidad Nacional Autónoma de México, Apartado Postal 20-726, México, D.F., 01000

E-mail address: weder@redvax1.dgsca.unam.mx

Centre de Recherches Mathématiques
CRM Proceedings and Lecture Notes
Volume 8, 1995

New Results on the Long-Range Scattering

Christian Gérard

1. Introduction

We would like to describe in this lecture some new results obtained jointly with Jan Derezinski about the quantum scattering for evolutions generated by time-dependent Hamiltonians of the form

$$H(t) = \frac{1}{2}D^2 + V(t,x), \tag{1.1}$$

for long-range potentials V. We will mainly be interested in the properties of the modified wave operators (among which unitarity is of course the most important) and of other observables which appear naturally in scattering theory.

This class of problems is probably not as well motivated physically as the one associated with time-independent Hamiltonians

$$H = \frac{1}{2}D^2 + V(x). \tag{1.2}$$

Nevertheless, from the mathematical point of view, time-dependent systems constitute a very natural class to study the scattering theory. Most of the results about the scattering for systems that belong to this class can be formulated and developed in a particularly clean and simple way. Another reason for their study is that in practice, if one is studying a time-independent system like (1.2) it is very convenient to reduce the problem to a time-dependent one. For example the simplest scheme of proof of asymptotic completeness for long-range time-independent potentials consists in two steps. In the first step one proves a rough propagation estimate which pinpoints the difference between bound states and scattering states. The Mourre estimate [**M**] or the RAGE theorem are two examples of tools used in this first step (see for example [**E1, E2, SS, Gr**]). In the second step, one has to prove some sharper propagation estimates for scattering states which are inspired by classical mechanics. This second step can be carried over in a rather simple way by reducing oneself to a time-dependent problem.

Let us now give the content of this lecture. In Section 2, we will consider the scattering theory for time-dependent forces in classical mechanics. This serves as a guideline for what to expect in the quantum case. In Section 3 we study time-dependent Hamiltonians under minimal smoothness assumptions on the potential. We prove here existence and completeness of the modified wave operators. In Section 4, we study for smooth long-range potentials the properties of the wave

---

1991 *Mathematics Subject Classification.* Primary: 81U05; Secondary: 35S05, 35S30.

This results will appear in a book published by Springer Verlag in the collection Texts and Monographs in Physics.

operators and of the asymptotic momentum and asymptotic position observables. In Section 5, we give an alternative definition of the wave operators. We prove also that the propagator and the wave operators can be written as Fourier integral operators. Finally in Section 6, we give a rather standard counterexample to asymptotic completeness for long-range potentials.

2. Classical Time-Dependent Forces

In this section we would like to study the scattering theory for the solutions of Newton's equations with a time-dependent force. Let us first introduce some notations. We consider the equations of motion in the presence of a time-dependent force $F(t,x)$

$$\begin{cases} \partial_t x(t) = \xi(t), \\ \partial_t \xi(t) = F\big(t, x(t)\big). \end{cases} \tag{2.1}$$

Note that the force F does not need to be conservative, i.e. to be the gradient of a potential. Similarly we will see that in the quantum case only the conditions on the gradient of the potential are important.

Sufficient conditions for the global existence of the solutions to (2.1) are for example

$$\int_{T_1}^{T_2} |F(t,0)|\, dt < \infty, \tag{2.2}$$

$$\int_{T_1}^{T_2} \|\nabla_x F(t,\cdot)\|_\infty \, dt < \infty, \tag{2.3}$$

for all $T_1 \le T_2 < \infty$. The solution of (2.1) with initial condition $x(s) = y$, $\xi(s) = \eta$, will be denoted by

$$\big(x(t,y,\eta), \xi(t,y,\eta)\big).$$

We will interchangeably use the word 'trajectory' for the phase-space trajectory $\big(x(t), \xi(t)\big)$ and for the configuration space trajectory $x(t)$.

We will start our exposition of the scattering theory for (2.1) with a construction of the so-called *asymptotic momentum*.

THEOREM 2.1. *Assume that*

$$\int_0^{+\infty} \|F(t,\cdot)\|_\infty \, dt < \infty. \tag{2.4}$$

Then for every (y,η) *there exists the uniform limit*:

$$\lim_{t\to+\infty} \xi(t,y,\eta) =: \xi^+(y,\eta). \tag{2.5}$$

Moreover

$$\lim_{t\to+\infty} \frac{x(t,y,\eta)}{t} = \xi^+(y,\eta).$$

Theorem 2.1 gives a partial classification of all trajectories. Namely, they are classified by the limit (2.5). One would also like to classify all trajectories corresponding to a given asymptotic momentum. Obviously the asymptotic momentum alone is not enough to determine a trajectory, since it is an $n-$dimensional quantity and a trajectory depends on $2n$ initial conditions.

In the *short-range* case (see for example [**Si**]), it is possible to classify them in a natural way by elements of the configuration space X. Indeed one can compare a trajectory with asymptotic momentum ξ with the free trajectory with the same asymptotic momentum

$$x_0(t) = t\xi.$$

In fact if we assume that

$$\int_0^{+\infty} \langle t\rangle \|F(t,\cdot)\|_\infty \, dt < \infty,$$

then the limit

$$\lim_{t\to+\infty} x(t,0,y,\eta) - t\xi(t,0,y,\eta) =: x_{\rm s}^+(y,\eta) \tag{2.6}$$

exists and is equal to

$$\lim_{t\to+\infty} x(t,0,y,\eta) - t\xi^+(y,\eta).$$

The mapping

$$\begin{aligned} \mathcal{F}_{\rm s}^+ : X\times X' &\to X\times X' \\ (y,\eta) &\mapsto \left(\xi^+(y,\eta), x_{\rm s}^+(y,\eta)\right) \end{aligned}$$

is called the inverse wave transformation. Under the additional assumption

$$\int_0^{+\infty} \langle t\rangle \|\nabla_x F(t,\cdot)\|_\infty \, dt < \infty,$$

the mapping $\mathcal{F}_{\rm s}^+$ exists and is equal to

$$\lim_{t\to+\infty} \phi(0,t)\phi_0(t),$$

where the free flow $\phi_0(t)$ is defined by

$$\phi_0(t)(x,\xi) = (x+t\xi,\xi).$$

In the long-range case, the limit (2.6) typically does not exist. The trajectories with a given asymptotic momentum cannot be compared with free trajectories. However they can be classified in a natural way by elements of the affine space X. This classification is the subject of the next theorem, which was first proven by I. Herbst [**He**] under more restrictive conditions on the forces.

THEOREM 2.2. *Assume that the force $F(t,x)$ satisfies*

$$\int_0^{\infty} \langle t\rangle^{|\alpha|} \|\partial_x^\alpha F(t,\cdot)\|_\infty \, dt < \infty, \quad |\alpha| = 0,1. \tag{2.7}$$

i) *Let* $(y_1(s), \eta_1(s))$ *and* $(y_2(s), \eta_2(s))$ *be two trajectories such that*

$$\lim_{s\to+\infty} \eta_1(s) = \lim_{s\to+\infty} \eta_2(s).$$

Then there exists

$$\lim_{s\to+\infty} \big(y_1(s) - y_2(s)\big).$$

ii) *Let* $(y_1(s), \eta_1(s))$ *be a trajectory and* $x \in X$. *Then there exists a unique trajectory* $(y_2(s), \eta_2(s))$ *such that*

$$\begin{cases} \lim_{s\to+\infty} \eta_1(s) = \lim_{s\to+\infty} \eta_2(s), \\ \lim_{s\to+\infty} \big(y_1(s) - y_2(s)\big) = x. \end{cases}$$

Theorem 2.2 means that the space of trajectories with a given asymptotic momentum has the structure of an *affine space*. To turn it into a vector space, we have to fix an 'origin' in this space. Unlike in the short-range case, where we could use the free trajectory $x_0(t,\xi) = t\xi$ to fix the origin in a natural way, in the long-range case, we have to make an arbitrary choice. Namely, we need to choose for each $\xi \in X'$, a trajectory $Y(t,\xi)$, with the asymptotic momentum ξ.

To make this more precise, let us introduce some notation. For $t_1 \le t_2 \le \infty$ and $(x,\xi) \in X \times X'$, we denote by

$$\tilde{y}(s, t_1, x, \xi)$$

the trajectory for the force F such that

$$\tilde{y}(t_1) = x, \quad \partial_t \tilde{y}(t_2) = \xi.$$

Such a trajectory exists under the hypothesis (2.7) if t_1 is large enough. We will denote for T large enough

$$Y(t,\xi) := \tilde{y}(t, T, \infty, 0, \xi). \tag{2.8}$$

So $Y(t,\xi)$ is the unique trajectory with asymptotic momentum ξ starting at the origin at time T. Note that the choice of $t_1 = T$ and of $x = 0$ is completely arbitrary.

We can now use $Y(t,\xi)$ as an origin to classify the trajectories having asymptotic momentum equal to ξ. Using Theorem 2.2, we can define

DEFINITION 2.3. We define the *asymptotic position* by

$$x_Y^+(y,\eta) := \lim_{t\to+\infty} x(t,y,\eta) - Y\big(t, \xi^+(y,\eta)\big).$$

The mapping

$$\begin{aligned} (\mathcal{F}_Y^+)^{-1} \colon T^\star X &\to T^\star X, \\ (y,\eta) &\mapsto \big(x_Y^+(y,\eta), \xi^+(y,\eta)\big), \end{aligned}$$

is called the *inverse long-range wave transformation*.

It turns out that there is an alternative definition of $(\mathcal{F}_Y^+)^{-1}$ which may be more familiar to the reader. Let us denote

$$\widetilde{Y}(t,\xi) := \tilde{y}(t,T,t,0,\xi).$$

By standard Hamilton-Jacobi theory, one can see that if $F(t,x) = -\nabla_x V(t,x)$, and if $S(t,\xi)$ is the solution of the Hamilton-Jacobi equation

$$\begin{cases} \partial_t S(t,\xi) = \xi^2/2 + V\big(t, \partial_\xi S(t,\xi)\big), \\ S(T,\xi) = 0, \end{cases} \tag{2.9}$$

then one has

$$\widetilde{Y}(t,\xi) = \partial_\xi S(t,\xi). \tag{2.10}$$

Then one has the following theorem

THEOREM 2.4. *Under the hypotheses* (2.7), *one has*

$$x_Y^+(y,\eta) = \lim_{t\to+\infty} x(t,y\,\eta) - \widetilde{Y}\big(t,\xi(t,y,\eta)\big).$$

This theorem allows to recover the wave transformation with a formula which is familiar from the quantum case. Indeed if we denote by $\phi_{\widetilde{Y}}(t)$ the following *modified free flow*

DEFINITION 2.5. The *modified free flow* is

$$\phi_{\widetilde{Y}}(t)(x,\xi) = (x + \widetilde{Y}(t,\xi),\xi).$$

then one has

$$\mathcal{F}_Y^+(x,\xi) = \lim_{t\to+\infty} \phi(0,t)\phi_{\widetilde{Y}}(t)(x,\xi).$$

Note that the modified free flow conserves the momentum. In the conservative case, due to (2.10), it is moreover symplectic.

3. Quantum Time-Dependent Hamiltonians

In this section, we consider time-dependent Hamiltonians

$$H(t) = \frac{1}{2}D^2 + V(t,x).$$

As explained in the introduction, this allows to simplify the exposition of many points of scattering theory. However it also leads to a technical difficulty: when does a time-dependent Hamiltonian generate a unitary dynamics, and what does it mean?

It turns out that the following notion of a solution of the time-dependent Schrödinger equation is sufficient to make sense of the propagation estimates that we will need below and at the same time allows for a rather large class of examples.

We fix an unbounded self-adjoint operator B with a dense domain in a Hilbert space $\mathcal{H}$. We will say that the time-dependent Hamiltonian

$$t \mapsto H(t)$$

generates the dynamics

$$(t,s) \mapsto U(t,s)$$

if the following conditions are satisfied:

(a) $U(t,s)$ is unitary for all t, s;
(b) $U(s,s) = 1$;
(c) $U(u,s)U(s,t) = U(u,t)$;
(d) $U(t,s)(B+i)^{-1}$ is strongly differentiable in t;
(e) $\partial_t U(t,s)(B+i)^{-1} \in L^1_{\mathrm{loc}}(\mathbb{R}, B(\mathcal{H}))$;
(f) $H(t)$ is a self-adjoint operator on $\mathcal{H}$ for almost all t with $\mathcal{D}(B) \subset \mathcal{D}(H(t))$;
(g) $H(t)(B+i)^{-1} \in L^1_{\mathrm{loc}}(\mathbb{R}, B(\mathcal{H}))$;
(h) $\partial_t U(t,s)(B+i)^{-1} = H(t)U(t,s)(B+i)^{-1}$.

As in the classical case, we star our exposition of the scattering theory with the construction of the *asymptotic momentum.*

For a function $V(t,x)$, we will denote by $\partial_{x_i} V(t,x)$ the (possibly unbounded) operator $[D_i, iV(t,x)]$.

THEOREM 3.1. *Suppose that*

$$V(t,x) = V_0(t,x) + V_1(t,x)$$

such that

$$\int_0^{+\infty} \|V_0(t,\cdot)\|_\infty \, dt < \infty,$$

$$\int_0^{+\infty} \|\langle D\rangle^{-1} \nabla_x V_1(t,x) \langle D\rangle^{-2}\| \, dt < +\infty, \tag{3.1}$$

$$\lim_{R\to+\infty} R^{-1} \int_0^{+\infty} \left\| \left(1 + \frac{D^2}{R^2}\right)^{-1} \nabla_x V_1(t,x) \left(1 + \frac{D^2}{R^2}\right)^{-1} \right\| dt = 0.$$

Then for any $g \in C_\infty(X)$ there exists

$$\lim_{t\to\infty} U(0,t) g(D) U(t,0). \tag{3.2}$$

There exists a vector of commuting self-adjoint operators P^+ with a dense domain such that (3.2) *equals $g(P^+)$.*

Moreover for $f \in C_\infty(X)$ the strong limit

$$\operatorname*{s-lim}_{t\to\infty} U(0,t) f\left(\frac{x}{t}\right) U(t,0) \tag{3.3}$$

exists and equals $f(P^+)$.

Let us point out that the last condition in (3.1) is needed only to ensure that D^+ has a dense domain. A sufficient condition is for example

$$\int_0^{+\infty} \|\langle D\rangle^{-\sigma} \nabla_x V_1(t,x) \langle D\rangle^{-\sigma}\| \, dt < +\infty,$$

for some $\sigma < 1/2$.

The asymptotic momentum observable gives a classification of all states according to their asymptotic behaviour under the evolution $U(t,0)$. Clearly to understand the evolution $U(t,0)$ we need first of all to study the spectral properties of D^+.

From this point of view, one can reformulate the basic problem of scattering theory, existence and unitarity of wave operators, as follows: when is the asymptotic momentum unitarily equivalent to the momentum ?

One can view this property as a definition of asymptotic completeness in the long-range case which is independent of a specific choice of a modified free evolution. It turns out that to have this property one needs also an estimate on the first derivative of the force $F(t,x)$. In fact we will give in Section 6 an example of a time-dependent potential $V(t,x)$ such that 0 belongs to the point spectrum of D^+.

The most commonly used construction of unitary operators implementing this unitary equivalence, which we will give in this section, resembles that of usual wave operators of the short-range case. Therefore, these unitary operators are usually called *modified wave operators.*

The main result of this section is the following theorem:

THEOREM 3.2. *Assume that* $V(t,x) = V_0(t,x) + V_1(t,x) + V_2(t,x)$ *where*

$$\begin{aligned} &\int_0^\infty \|V_0(t,\cdot)\|_\infty \, dt < \infty, \\ &\int_0^\infty \langle t\rangle \|\nabla_x V_1(t,\cdot)\|_\infty \, dt < \infty, \end{aligned} \tag{3.4}$$

and

$$\int_0^\infty \langle t\rangle^{|\alpha|-1} \|\partial_x^\alpha V_2(t,\cdot)\|_\infty \, dt < \infty, \quad |\alpha| = 1, 2. \tag{3.5}$$

Then there exist a real function $S(t,\xi)$ *such that the limits*

$$\operatorname*{s-lim}_{t\to\infty} U(0,t) e^{-iS(t,D)} \tag{3.6}$$

and

$$\operatorname*{s-lim}_{t\to\infty} e^{iS(t,D)} U(t,0) \tag{3.7}$$

exist.

If we denote (3.6) *by* Ω^+, *then* (3.7) *equals* Ω^{+*}. *Moreover,*

$$P^+ = \Omega^+ D \Omega^{+*}. \tag{3.8}$$

Under the above conditions, the limit

$$\lim_{t\to+\infty} U(0,t)\big(x - \nabla_x S(t,D)\big) U(t,0) =: x_1^+ \tag{3.9}$$

exists in strong resolvent sense. The observable x_1^+ *is a commuting family of self-adjoint operators and is called the* asymptotic position. *Moreover one has*

$$x_1^+ = \Omega^+ x \Omega^{+*}. \tag{3.10}$$

Let us first point out that conditions (3.5) are exactly the same as the ones needed to have existence and completeness of the wave transformations in the classical case. So we have a very close analogy between the classical and the quantum case. Moreover as the counterexample constructed in Section 6 shows, these conditions are very close to be optimal for the completeness of the wave operators.

If one applies a rather standard propagation estimate as for example in [**Gr**], one can deduce from Theorem 3.2 the following result for time-independent potentials which improves a result of L. Hörmander [**Hö**]:

THEOREM 3.3. *Let*

$$H = \frac{1}{2}D^2 + V(x),$$

such that $V(x) = V_{\mathrm{s}}(x) + V_{\mathrm{l}}(x)$ *where*

$$\int_0^\infty \|\langle D\rangle^{-2} V_{\mathrm{s}} F(|x| \geq R)\| \, dR < \infty,$$

$$\int_0^\infty \langle R\rangle^{|\alpha|-1} \sup_{|x|\geq R} \|\partial_x^\alpha V_{\mathrm{l}}(x)\| \, dR < \infty, \quad |\alpha| = 1, 2.$$

Then there exist a real function $S(t,\xi)$ *such that the limits*

$$\operatorname*{s-lim}_{t\to\infty} e^{itH} e^{-iS(t,D)} \tag{3.11}$$

and

$$\operatorname*{s-lim}_{t\to\infty} e^{iS(t,D)} e^{-itH} E_{\mathrm{c}}(H) \tag{3.12}$$

exist. In other words the modified wave operators exist and are complete.

In [**Hö**] the existence and completeness of modified wave operators was proven using a stationary method. The assumptions on the long-range part of the potential were stronger, corresponding to replace a condition $f(R) \in L^1(dR)$ by $f(R) \in O(R^{-1-\epsilon})$ for $\epsilon > 0$.

Let us finally notice that using a regularization as in [**Hö**], a potential $V(t,x)$ can satisfy the conditions in Theorem 3.2 without being twice differentiable. For example if $V(t,x)$ satisfies:

$$\int_0^\infty \langle t\rangle^{1/2} \|\nabla_x V(t,\cdot)\|_\infty \, dt < \infty, \tag{3.13}$$

then is satisfies the hypotheses of Theorem 3.2. Note that the above condition implies also the existence and completeness of *Dollard wave operators.*

One may want to know what is the relationship between the function $S(t,\xi)$ and the potentials appearing in the statement of Theorem 3.2. It turns out that if we strenghten the assumptions of Theorem 3.2, then $S(t,\xi)$ can be taken equal to a solution of the Hamilton-Jacobi equation with the potential V_2:

THEOREM 3.4. *Assume that the potential* V_2 *satisfy one of the following two conditions*

$$\int_0^\infty \langle t\rangle^{|\alpha|-1} \|\partial_x^\alpha V_2(t,\cdot)\|_\infty \, dt < \infty, \quad |\alpha| = 1, 2, 3, \tag{3.14}$$

$$\begin{aligned} &\int_0^\infty \langle t\rangle^{1/2} \|\partial_x^\alpha V_2(t,\cdot)\|_\infty \, dt < \infty, \quad |\alpha| = 1, \\ &\int_0^\infty \langle t\rangle \|\partial_x^\alpha V_2(t,\cdot)\|_\infty \, dt < \infty, \quad |\alpha| = 2, \end{aligned} \tag{3.15}$$

Then the function $S(t,\xi)$ in Theorem 3.2 can be taken equal to the solution of the Hamilton-Jacobi equation

$$\begin{cases} \partial_t S(t,\xi) = \xi^2/2 + V\big(t, \partial_\xi S(t,\xi)\big), \\ S(T,\xi) = 0, \end{cases}$$

which exists for large enough T.

To end this section, we would like to formulate a conjecture about the *asymptotic position observable.* In Theorem 3.2, the asymptotic position is obtained as the strong resolvent limit of

$$x_t - \nabla_\xi S(t, D_t),$$

where

$$x_t := U(0,t)xU(t,0), \quad D_t := U(0,t)DU(t,0).$$

This definition corresponds in the classical case to the formula for the asymptotic position given in Theorem 2.4. In view of Definition 2.3, we conjecture that for u in a dense subspace of $L^2(X)$

$$x_1^+ u = \lim_{t\to+\infty} x_t u - Y(t, D^+)u. \tag{3.16}$$

Here the function $Y(t,\xi)$ is defined in (2.8), and the observable $Y(t,D^+)$ is defined through the functional calculus for a commuting family of selfadjoint operators. This conjecture is probably difficult to prove under the minimal assumptions of Theorem 3.2. However even for the smooth potentials considered in Section 4, a proof of (3.16) would be of much interest.

4. Smooth Long-Range Potentials

In this Section we would like to study the smoothness properties of the various objects that we constructed in the preceeding Section, like the wave operators, the asymptotic momentum and the asymptotic position observables. Before stating the results, we would like to introduce some notations.

Let $S(dx^2 + d\xi^2)$ denote the set of functions on $X \times X'$ such that

$$|\partial_x^\alpha \partial_\xi^\beta a(x,\xi)| \le C_{\alpha,\beta}, \quad \alpha,\beta \in \mathbb{N}^n.$$

Let $\Psi(dx^2 + d\xi^2)$ denote the * algebra of operators A such that

$$ad_x^\alpha ad_D^\beta A \in B(\mathcal{H}), \quad \alpha,\beta \in \mathbb{N}^n.$$

It is an important result of R. Beals [**Be**] that

$$A \in \Psi(dx^2 + d\xi^2) \iff A = a^w(x,D), \text{ for some } a \in S(dx^2 + d\xi^2),$$

where $a^w(x,D)$ is the Weyl quantization

$$a^w(x,D)u(x) := (2\pi)^{-n} \iint e^{i\langle x-y,\xi\rangle} a\Big(\frac{x+y}{2}, \xi\Big) u(y)\, dy\, d\xi.$$

So the algebra $\Psi(dx^2+d\xi^2)$ is a non-commutative analog of $S(dx^2+d\xi^2)$. Of course $\Psi(dx^2+d\xi^2)$ is nothing else than the class $\mathrm{Op}(S^0_{0,0})$ of pseudodifferential operators. This class is probably the most natural class of pseudodifferential operators to consider. Note that there exists no asymptotic calculus in this class, since the associated Planck constant is equal to 1.

In all this section and in the next one, we will assume that the potential V satisfies the following *smooth long-range condition*

$$\int_0^{+\infty} \langle t\rangle^{|\alpha|-1}\|\partial_x^\alpha V(t,\cdot)\|_\infty\, dt < \infty, \quad |\alpha| \geq 1. \tag{4.1}$$

Note that (4.1) is only a condition on the force $-\nabla_x V$. The main result of this section is the following theorem.

THEOREM 4.1. *Assume* (4.1). *Then*

$$D^+ - D \in \Psi(dx^2 + dx^2), \tag{4.2}$$

there exist $A, B \in \Psi(dx^2 + d\xi^2)$ *such that*

$$x_1^+ - x = Ax + B, \tag{4.3}$$

$$\|\langle x\rangle^{-m}\langle x_1^+\rangle^m\| \leq C, \quad m \in \mathbb{R}, \tag{4.4}$$

$$\|\langle x\rangle^m ad_D^\alpha(\Omega_1^+)\langle x\rangle^{-m}\| \leq C, \quad m \in \mathbb{R}, \alpha \in \mathbb{N}^n. \tag{4.5}$$

As a consequence of Theorem 4.1, we get the following boundedness result of the wave operator between weighted Sobolev spaces

$$\|\langle x\rangle^m\langle D\rangle^k\Omega_1^+\langle D\rangle^{-k}\langle x\rangle^{-m}\| \leq C, \quad m, k \in \mathbb{R}.$$

Note also that the identity (4.3) is the quantum reminder of the identity

$$x_Y^+(0,\eta) = 0,$$

for the classical problem (check it!).

Finally it is instructive to compare these results with the ones one can obtain in the short-range case. If we assume instead of (4.1) the following *smooth short-range condition*

$$\int_0^{+\infty} \langle t\rangle^{|\alpha|}\|\partial_x^\alpha V(t,\cdot)\|_\infty\, dt < \infty, \quad |\alpha| \geq 1,$$

then one can prove that for

$$\theta(t) = \int_0^t V(s,0)\, ds,$$

the (renormalized) short-range wave operator

$$\operatorname*{s-lim}_{t\to+\infty} U(0,t)e^{-itH_0 - i\theta(t)}$$

for $H_0 = \frac{1}{2}D^2$ is a unitary pseudodifferential operator in the class $\Psi(dx^2 + d\xi^2)$. The asymptotic momentum satisfies again (4.2). The asymptotic position defined as

$$x_s^+ := \operatorname*{s-lim}_{t\to+\infty} U(0,t)\,(x - tD)\,U(t,0)$$

satisfies

$$x_s^+ - x \in \Psi(dx^2 + d\xi^2).$$

So in the short-range case the wave operators are unitary pseudodifferential operators of order 0. We will give in Section 5 a corresponding result about the nature of the long-range wave operators.

As indicated above there is no asymptotic calculus in the algebra $\Psi(dx^2+d\xi^2)$. To get pseudodifferential operators in a class with an asymptotic calculus, we have to let the various observables depend on a large initial time s. Accordingly we introduce the following notations. We put

$$\begin{aligned} x(t,s) &:= U(s,t)xU(t,s),\\ D(t,s) &:= U(s,t)DU(t,s),\\ x_1(t,s) &:= x(t,s) - \nabla_\xi S\big(t, D(t,s)\big). \end{aligned}$$

We also need to define new classes of symbols. For a positive real function f, we denote by $\Psi\big(f(t), g_0(t)\big)$ the algebra of operators $A = a^w(t,x,D)$ with

$$|\partial_x^\alpha \partial_\xi^\beta a(t,x,\xi)| \le C_{\alpha,\beta} f(t)\langle t\rangle^{-|\alpha|}, \quad \alpha,\beta \in \mathbb{N}^n.$$

Note that the Planck constant associated to the algebra $\Psi\big(f(t), g_0(t)\big)$ is equal to $\langle t\rangle^{-1}$. Similarly we denote by $L^1\big(f(t)\,dt, \Psi\big(g_0(dt)\big)\big)$ the set of operators $A = a^w(t,x,D)$ with

$$\int_0^\infty f(t)^{-1}\langle t\rangle^{|\alpha|}\|\partial_x^\alpha\partial_\xi^\beta a(t,\cdot)\|_\infty\,dt < \infty, \quad \alpha,\beta\in\mathbb{N}^n.$$

Then we have the following result:

THEOREM 4.2. *Assume* (4.1). *Then*

$$\begin{aligned} D(t,s) - D &\in \Psi\big(o(s^0), g_0(s)\big),\\ x(t,s) - x - (t-s)D &\in \Psi\big(|t-s|o(s^0), g_0(s)\big), \end{aligned}$$

there exist $A(t,s)$, $B(t,s) \in \Psi\big(o(s^0), g_0(s)\big)$ *such that*

$$\begin{aligned} x_1(t,s) - x_1(s,s) &= A(t,s)x_1(s,s) + B(t,s),\\ \|\langle x_1(t,s)\rangle^{-m}\langle x_1(s,s)\rangle^{m}\| &\le C, \quad m\in\mathbb{R}, \end{aligned}$$

Theorems 4.1 and 4.2 are rather direct consequences of a version of *Egorov's theorem* for solutions of Heisenberg equation which is of independent interest and which we would like to describe here. Let $M(t) \in L^1\big(dt, \Psi\big(g_0(t)\big)\big)$. Denote

$$M(t,s) := U(s,t)M(t)U(t,s),$$

which solves the Heisenberg equation

$$\begin{cases} \partial_s M(t,s) - [H(s), iM(t,s)] = 0,\\ M(t,t) = M(t), \end{cases} \tag{4.6}$$

Then one can prove the following fact

$$M(t,s) \in L^1\big(dt, \Psi\big(g_0(s)\big)\big),$$

uniformly for $s \le t$. This fact is proved using Beals criterion and the equation (4.6). Note that one could give a more complicated proof similar to the original proof of Egorov's theorem by writing the propagator $U(t,s)$ as a Fourier integral operator as we will see in the next section.

5. Wave Operators and Fourier Integral Operators

In this section we will give an alternative definition of the wave operator which is similar to the one considered in the time-independent case by Isozaki and Kitada [**IK**]. We will then write the propagator and the wave operators as Fourier integral operators.

The alternative definition of the wave operators is based on a solution of the *eikonal equation.* This function is obtained as the limit of the difference of two solutions of the Hamilton-Jacobi equation. More precisely, one can prove the following facts. For $T \leq s \leq t < \infty$ and T large enough, there exist a unique function $S(s,t,x,\xi)$ such that

$$\begin{cases} \partial_t S(s,t,x,\xi) = \xi^2/2 + V\big(t, \nabla_\xi S(s,t,x,\xi)\big), \\ S(t,t,x,\xi) = \langle x, \xi\rangle, \end{cases} \tag{5.1}$$

$$\begin{cases} -\partial_s S(s,t,x,\xi) = \big(\nabla_x S(s,t,x,\xi)\big)^2/2 + V(s,x), \\ S(t,t,x,\xi) = \langle x, \xi\rangle. \end{cases} \tag{5.2}$$

The function $S(t,\xi)$ used to define the long-range wave operator is equal to $S(T,t,0,\xi)$. Then we have the following proposition:

PROPOSITION 5.1. *The following limit exists*:

$$\lim_{t\to\infty} \big(S(s,t,x,\xi) - S(t,\xi)\big) =: \Phi(s,x,\xi).$$

It satisfies the following properties:

i) $-\partial_s \Phi(s,x,\xi) = \big(\nabla_x \Phi(s,x,\xi)\big)^2/2 + V(s,x)$,

ii) $(\nabla_\xi \Phi(s,x,\xi), \xi) = (\mathcal{F}_Y^+)^{-1}\big(\phi(0,s)\big(x, \nabla_x\Phi(s,x,\xi)\big)\big)$,

iii) *uniformly for* $T \leq s$, $(x,\xi) \in X \times X'$ *we have*

$$\left|\partial_x^\alpha \partial_\xi^\beta \left(\Phi(s,x,\xi) - \langle x,\xi\rangle + \frac{1}{2}s\xi^2\right)\right| \leq o(s^{1-|\alpha|}), \quad |\alpha| \geq 1.$$

So $\Phi(s,x,\xi)$ is the generating function for the canonical transformation $\phi(s,0)\circ \mathcal{F}_Y^+$. We define the following Fourier integral operator

$$J(s)u(x) := (2\pi)^{-n} \iint e^{i\Phi(s,x,\xi) - i\langle y,\xi\rangle} u(y)\,dy\,d\xi. \tag{5.3}$$

Then

THEOREM 5.2. *Assume* (4.1). *Then the following results hold*:

i) *the operator* $J(s)$ *is uniformly bounded on* $L^2(X)$ *for* $s \geq T$.

ii) *the norm limit*

$$\Omega_J^+ := \lim_{s\to+\infty} U(0,s)J(s)$$

exist.

iii) Ω_J^+ *is equal to* Ω^+ *defined in Theorem* 3.2.

Note that the limit in the above theorem is a norm limit so the convergence in ii) is much better than usual. Note also that this theorem is the exact quantum analog of Proposition 5.1 ii).

We will next give exact expressions of the propagator and of the wave operators as Fourier integral operators. We have the following result:

THEOREM 5.3. *Assume* (4.1). *Then there exist a time T large enough and a symbol $a(t,x,\xi)$ with*

i) $a(+\infty,x,\xi) := \lim_{t\to+\infty} a(t,x,\xi)$ *exists.*

ii) $|\partial_x^\alpha \partial_\xi^\beta a| \le C_{\alpha,\beta}, \quad \forall T \le s \le t \le +\infty,$

such that

$$U(t,T) = (2\pi)^{-n} \iint e^{i\langle x,\xi\rangle - iS(T,t,y,\xi)} a(t,y,\xi)u(y)\,dy\,d\xi, \tag{5.4}$$

$$U(T,0)\Omega_1^+ = (2\pi)^{-n} \iint e^{i\Phi(T,x,\xi) - i\langle y,\xi\rangle} a(+\infty,x,\xi)u(y)\,dy\,d\xi. \tag{5.5}$$

So we have written the propagator $U(t,T)$ and the time-translated wave operator $U(T,0)\Omega_1^+$ as Fourier integral operators associated respectively with the canonical transformations $\phi(t,T)$ and $\phi(T,0)\circ\mathcal{F}_Y^+$. Again we have a perfect analogy between the classical and quantum problem. Note that the identities in Theorem 5.3 are exact and not asymptotic. Similar asymptotic expressions for the propagator were given earlier by Kitada [**K**].

One may ask if one can put $T=0$ in the above theorem. In general one cannot extend the solution of the Hamilton-Jacobi equation (5.1) up to arbitrary time s because of the possible presence of caustics. However one can still write $U(t,0)$ uniformly for $t\ge 0$ as a Fourier integral operator associated with the canonical relation $\phi(t,0)$ if one uses more integration variables in (5.4). This follows by writing first $U(t,0)$ as a Fourier integral operator for $0\le t\ll 1$ and using then the group property to write $U(T,0)$ for some large but finite T.

6. A Counterexample to Asymptotic Completeness

In this section, we will give a counterexample to asymptotic completeness which is inspired by a construction of Yajima [**Ya**]. Another class of counterexamples has been given by Yafaev [**Y**].

We start with the observation that the unitary evolution $U_1(t)$ generated by the time-dependent Hamiltonian

$$H_1(t) := \frac{1}{2}D^2 + \frac{Cx^2}{t^2}$$

can be constructed explicitely. Indeed if we put

$$T_t u(x) := t^{-n/4} u\Big(\frac{x}{t^{1/2}}\Big),$$
$$U_0 u(x) := e^{-ix^2/4} u(x),$$

it is an easy computation to check that

$$U_1(t,1)u = T_t U_0 e^{-i\ln t H_1} u,$$

for

$$H_1 = \frac{1}{2}D^2 + \Big(C - \frac{1}{8}\Big)x^2.$$

Let us take $C > 1/8$ and pick an eigenfunction ϕ_0 of H_1 for the eigenvalue λ_0. We put

$$\begin{aligned} u_t(x) :=& T_t U_0 e^{-i\lambda_0 \ln t} \phi_0 \\ =& t^{-n/4} e^{-i\lambda_0 \ln t + ix^2/4t} \phi_0\left(\frac{x}{t^{1/2}}\right), \end{aligned}$$

which solves

$$\begin{cases} i\partial_t u_t = H_1(t) u_t, \\ u_1 = \phi_0. \end{cases} \tag{6.1}$$

Let us now consider the time-dependent potential

$$V(t,x) := \frac{Cx^2}{t^2} \chi\left(\frac{|x|}{t^{1/2}\delta(t)}\right),$$

for some cutoff function $\chi \in C_0^\infty(\mathbb{R})$ equal to 1 near the origin and

$$\delta(t) = (\alpha \ln t)^{1/2}, \quad \alpha > 0. \tag{6.2}$$

We denote by $U(t,s)$ the unitary evolution generated by the Hamiltonian

$$H(t) = \frac{1}{2} D^2 + V(t,x).$$

Note that one has

$$\begin{aligned} &\|\nabla_x V(t,\cdot)\|_\infty \sim t^{-3/2} (\ln t)^{1/2}, \\ &\|\partial_x^2 V(t,\cdot)\|_\infty \sim t^{-2}, \end{aligned}$$

so V almost (but not quite) satisfies (3.5) and (3.13). Using the exponential decay of ϕ_0, we see that

$$\left\| \left(V(t,x) - \frac{Cx^2}{t^2} \right) u_t \right\| \leq \frac{C_\epsilon}{t} e^{-(1-\epsilon)\delta^2(t)}, \quad \forall \epsilon > 0.$$

From (6.2) we have

$$\left\| \left(V(t,x) - \frac{Cx^2}{t^2} \right) u_t \right\| \in L^1(dt). \tag{6.3}$$

From (6.1), (6.3) and the Cook argument we deduce immediatly that

$$\lim_{t \to +\infty} U(0,t) u_t =: \phi_\infty \tag{6.4}$$

exists. Using (6.4) and the explicit form of u_t, we also see that if D^+ is the asymptotic momentum for the evolution $U(t,0)$, one has

$$E_{\{0\}}(D^+)\phi_\infty = \phi_\infty,$$

so that $0 \in \sigma_{\rm pp}(D^+)$. Since we can choose any eigenfunction ϕ_0, we see that 0 is actually an eigenvalue of infinite multiplicity. Obviously D^+ cannot be unitarily equivalent to D and asymptotic completeness is violated. However it is easy to see that the restriction of D^+ to $(\operatorname{Ker} D^+)^\perp$ is unitarily equivalent to D.

References

[Be] R. Beals, *Weighted distribution spaces and pseudodifferential operators*, J. Analyse Math. **39** (1981), 130–187.

[E1] V. Enss, *Asymptotic completeness for quantum-mechanical potential scattering* I *Short range potentials*, Comm. Math. Phys. **61** (1978), 285–291.

[E2] ——, *Long range scattering of two- and three-body quantum systems*, Proceedings of the Conference 'Équations aux dérivées partielles' (Saint Jean de Monts), École Polytechnique, 1989.

[Gr] G. M. Graf, *Asymptotic completeness for N-body short range quantum systems: A new proof*, Comm. Math. Phys. **132** (1990), 73–101.

[He] I. Herbst, *Classical scattering with long range forces*, Comm. Math. Phys. **35** (1974), 193–214.

[Hö] L. Hörmander, *The analysis of linear partial differential operators*, Vol. IV, Chapter 30, Springer-Verlag, Berlin-Heidelberg-New York, 1985.

[IK] H. Isozaki and H. Kitada, *Modified wave operators with time independent modifiers*, J. Fac. Sci. Univ. Tokyo Sect. IA Math. **32** (1985), 77–104.

[K] H. Kitada, *Scattering theory for time-dependent long-range potentials*, J. Fac. Sci. Univ. Tokyo Sect. IA Math. **29** (1982) ,353–369.

[M] E. Mourre, *Absence of singular continuous spectrum for certain selfadjoint operators*, Comm. Math. Phys. **78** (1981), 519–567.

[SS] I. M. Sigal and A. Soffer, *The N-particle scattering problem: Asymptotic completeness for short range quantum systems*, Ann. of Math. **125** (1987), 35–108.

[Si] B. Simon, *Wave operators for classical particle scattering*, Comm. Math. Phys. **23** (1971), 37–48.

[Y] D. Yafaev, *On the violation of unitarity in time-dependent scattering*, Soviet Math. Dokl. **19** (1978), no. 6, 1517–1521.

[Ya] K. Yajima, *The surfboard Schrödinger equation*, Comm. Math. Phys. **96** (1984), 349–360.

Centre de Mathématiques, École Polytechnique, 91128 Palaiseau Cedex, France

E-mail address: gerard@orphee.polytechnique.fr

Centre de Recherches Mathématiques
CRM Proceedings and Lecture Notes
Volume 8, 1995

Spectral Properties of the Kac Operator in Large Dimension

Bernard Helffer

ABSTRACT. The aim of this survey is to present recent results obtained by J. Sjöstrand or myself for the Schrödinger or the Kac Operator attached to a potential $V^{(m)}$ defined on $\mathbb{R}^m$ appearing as the sum of the harmonic potential $\left(\sum_j x_j^2\right)/2$ and of some interaction, for example of the form: $I^{(m)}(x) = \beta \sum_j W(x_j, x_{j+1})$ where W is a potential on $\mathbb{R}^2$ with bounded gradient and β can be considered as the inverse of the temperature. The study of the behavior as $m \to \infty$ of the first eigenvalue or the study of statistical quantities (like the correlation pair functions) attached to the measure $(u_1^{(m)})^2 dx$ on $\mathbb{R}^m$, where $u_1^{(m)}$ denotes the first eigenfunction is immediately related to the study of Laplace integrals where techniques based on the maximum principle give strong results as demonstrated by J. Sjöstrand (and earlier by Singer, Wong, Yau, and Yau). In particular we shall present new results concerning the splitting in the convex case and give a new insight on the classical FKG inequalities or the Brascamp-Lieb inequalities.

1. Introduction

In different recent publications we have given a new insight in old problems coming from statistical mechanics. Most of the time we were presenting results on Laplace integrals or on spectral properties attached to a Schrödinger operator. We want to come back in this survey to one of the original motivations which was to understand some course of M. Kac. The aim of this paper will be to review what is already known. In some sense this is the continuation of our first paper with M. Brunaud [**3**] but can actually be understood independently. The semiclassical point of view will not play here an important role but could be useful in order to give more precise results. Our main object of study is the Kac operator associated to a potential V by the formula:

$$K_V = \exp\left(-\frac{1}{2}V\right) \cdot \exp \Delta \cdot \exp\left(-\frac{1}{2}V\right) \tag{1.1}$$

where $V = V^{(m)}$ is a family of potentials defined on $\mathbb{R}^m$ by:

$$V = \frac{1}{2}\sum_{j=1}^{m} x_j^2 + \sum_j W(x_j, x_{j+1}), \tag{1.2}$$

W is a C^∞ function defined on $\mathbb{R}^2$ s.t.:

$$|D_x^\alpha W| \leq C_\alpha, \quad \forall \alpha \text{ s.t. } |\alpha| \geq 1, \tag{1.3}$$

1991 *Mathematics Subject Classification.* Primary: 81; Secondary: 35.
This is the final form of the paper.

and

$$\Delta = \Delta^{(m)} = \sum_j (\partial_{x_j})^2 \tag{1.4}$$

is the usual Laplacian on $\mathbb{R}^m$.

Sometimes the interaction W will depend on a parameter β (corresponding to the inverse of the temperature). For example we can replace the above W by βW and we observe that in this case the potential V becomes convex for β small enough.

In parallel to this study we shall sometimes recall corresponding results obtained earlier in the case of the Schrödinger operator.

Let us now briefly recall how this operator appears in a specific context in statistical mechanics as a transfert matrix and we shall try to extract from this example due to Kac [**24**] some general questions concerning the Kac operator.

KAC MODELS ON $\mathbb{Z}^2$. The model we want to consider is called Model A in Section 7 of [**24**]. We consider the following Hamiltonian:

$$E_{\Lambda(n,m)}(\sigma) = - \sum_{(P,Q)\in\Lambda(n,m)\times\Lambda(n,m)} v_{P,Q}\sigma_P\sigma_Q \tag{1.5}$$

with:

$$\Lambda(n,m) = [1,\dots,n]\times(\mathbb{Z}/m\mathbb{Z}), \quad \sigma_P \in \{-1,+1\}, \quad J\in\mathbb{R}_+^\star, \quad \gamma\in\mathbb{R}_+^\star$$

with:

$$v_{P,P} = 0$$

and

$$v_{P,Q} = J\gamma\exp(-\gamma|k-k'|)\left(\delta_{\ell',\ell} + \frac{1}{2}(\delta_{\ell',\ell+1}+\delta_{\ell',\ell-1})\right)$$

if $P=(k,\ell)\neq Q=(k',\ell')$.

In this lattice, two spins interact if they are on the same column or nearest neighbors, but with exponential decay with respect to the distance between the lines. We are interested in the asymptotic behavior of this model when the boxes $\Lambda(n,m)$ tend to ∞ in a suitable sense. More precisely, let us first consider the partition function $Z_{\Lambda(n,m)} = Z_{n,m}$ which is defined by

$$Z_{n,m} = \sum_{\sigma\in\{-1,+1\}^{\Lambda(n,m)}} \exp\bigl(-\beta E_{\Lambda(n,m)}(\sigma)\bigr).$$

We get a natural probability $P^{(n,m)}$ on $\{-1,+1\}^{\Lambda(n,m)}$ such that

$$P^{(n,m)}(\{\sigma\}) = \frac{1}{Z_{n,m}}\exp\bigl(-\beta E_{\Lambda(n,m)}(\sigma)\bigr).$$

We then consider the energy per site $-\beta\Psi$ in the thermodynamical limit

$$\beta\Psi := \lim_{m\to\infty}\left(\frac{1}{m}\cdot\left[\lim_{n\to\infty}\frac{1}{n}\ln Z_{n,m}\right]\right). \tag{1.6}$$

M. Kac remarks that this free energy per site can be computed by the formula:

$$-\beta\Psi = \ln 2 - \frac{\beta\gamma}{2} + \lim_{m\to\infty}\left(\frac{\ln\mu_1^{(m)}}{m}\right) \tag{1.7}$$

where $\mu_1^{(m)}$ is the largest eigenvalue of a compact integral operator K defined by:

$$K = K_{Q^{(m)}} := \exp\Big(-\frac{Q^{(m)}}{2}\Big)\exp(\gamma\Delta^{(m)})\Big(-\frac{Q^{(m)}}{2}\Big) \tag{1.8}$$

with

$$Q^{(m)}(y) = \left(\frac{1}{2}\tanh\left(\frac{\gamma}{2}\right)\right)\sum_{k=1}^{m} y_k^2 - \sum_{k=1}^{m}\ln\cosh\Big(\sqrt{\frac{\nu\gamma}{2}}(y_k + y_{k+1})\Big)$$

and

$$\nu = J\beta.$$

More generally, we shall denote by $\mu_j^{(m)}$ the decreasing sequence of the eigenvalues of K.

As γ tends to 0, the operator is in a certain sense well approximated by the operator $\exp(\gamma\Delta^{(m)} - Q^{(m)})$ (Confer [**3**] or [**15**] for rigorous results in this direction) and it is consequently natural, after a scaling,

$$x_k = \gamma^{1/2} y_k,$$

to study the limit as $m \to \infty$ of $\lambda_1(m; h, \nu)/m$ where $\lambda_1(m; h, \nu)$ is the smallest eigenvalue of the Schrödinger operator:

$$S^{(m)}(x, hD_x; \nu) = -h^2\Delta^{(m)} + V^{(m)}(x; \nu) \tag{1.9}$$

with:

$$V^{(m)}(x, \nu) = \frac{1}{4}\sum_{k=1}^{m} x_k^2 - \sum_{k=1}^{m}\ln\cosh\Big(\sqrt{\frac{\nu}{2}}(x_k + x_{k+1})\Big), \tag{1.10}$$

where $x_{m+1} = x_1$.

h is essentially equivalent to γ as γ tends to 0 and we arrive to a more traditional semi-classical problem for the Schrödinger operator.

In this context, two natural correlations appear in statistical mechanics. We follow here the course by M. Kac [**24**]. We can first consider the correlation between two spins in a column separated by a distance r:

$$\mathrm{Cor}_{\mathrm{row}}(r) = \lim_{m\to\infty} \mathrm{Cor}_{\mathrm{row}}^{(m)}(r) \tag{1.11}$$

where

$$\mathrm{Cor}_{\mathrm{row}}^{(m)}(r) = \lim_{n\to\infty} \langle \sigma_{k,\ell} \cdot \sigma_{(k+r,\ell)} \rangle_{n,m} \tag{1.12}$$

and $\langle\cdot\rangle_{n,m}$ corresponds to the mean value for the probability measure $P^{(n,m)}$. The computations of $\mathrm{Cor}_{\mathrm{row}}^{(m)}(r)$ (cf. formula (7.32) in [**24**]) gives:

$$\mathrm{Cor}_{\mathrm{row}}^{(m)}(r) = \sum_{j=2}^{\infty}\left(\frac{\mu_j^{(m)}}{\mu_1^{(m)}}\right)^r \Bigg[\int_{\mathbb{R}^m} u_1^{(m)}(x)u_j^{(m)}(x) \times \tanh\left(\sqrt{\frac{\nu\gamma}{2}}(x_1 + x_2)\right) dx\Bigg]^2. \tag{1.13}$$

Its asymptotic behavior as r tends to ∞ is immediately related to the behavior as $m \to \infty$ of $\mu_2^{(m)}/\mu_1^{(m)}$.

Note that if one thinks to the correspondence Kac-Schrödinger this quotient corresponds to $\exp\big(-(\lambda_1 - \lambda_2)\big)$ where λ_1 and λ_2 correspond to the two first eigenvalues of the Schrödinger operator.

Let us now consider the correlation of two spins in a row separated by a distance t:

$$\operatorname{Cor}_{\mathrm{col}}(t) = \lim_{m \to \infty} \operatorname{Cor}^{(m)}_{\mathrm{col}}(t) \tag{1.14}$$

where

$$\operatorname{Cor}^{(m)}_{\mathrm{col}}(t) = \lim_{n \to \infty} \langle \sigma_{k,\ell} \cdot \sigma_{(k,\ell+t)} \rangle_{n,m}. \tag{1.15}$$

The computation of $\operatorname{Cor}^{(m)}_{\mathrm{col}}(t)$ (cf formula (7.36) in [**24**]) gives:

$$\begin{aligned}\operatorname{Cor}^{(m)}_{\mathrm{col}}(t) = \int_{\mathbb{R}^m} u^{(m)}_1(x)^2 \cdot \tanh\Big(\sqrt{\frac{\nu\gamma}{2}}(x_1 + x_2)\Big) \\ \cdot \tanh\Big(\sqrt{\frac{\nu\gamma}{2}}(x_{1+t} + x_{2+t})\Big)\, dx.\end{aligned} \tag{1.16}$$

This appears as the mean value of the function:

$$x \to \tanh\Big(\sqrt{\frac{\nu\gamma}{2}}(x_1 + x_2)\Big) \cdot \tanh\Big(\sqrt{\frac{\nu\gamma}{2}}(x_{1+t} + x_{2+t})\Big)$$

with respect to the probability measure $(u^{(m)}_1)(x)^2\, dx$. This will be studied in Subsection 6.3.

From the study of this example we select three natural problems.

PROBLEM 1 (THE THERMODYNAMIC LIMIT). We want to study the existence of the limit (and the speed of convergence) of:

$$\lambda = -\lim_{m \to \infty} (\ln \mu^{(m)}_1)/m. \tag{1.17}$$

PROBLEM 2 (THE CORRELATION FUNCTIONS). If g is a function of k variables, we shall study the convergence of the mean value of g as m tends to ∞:

$$\langle g \rangle_\infty = \lim_{m \to \infty} \langle g \rangle_m \tag{1.18}$$

where

$$\langle g \rangle_m = \frac{\int g\, d\mu^{(m)}}{\int d\mu^{(m)}}. \tag{1.19}$$

As a particular but important case, we shall analyze the case when

$$g_{ij}(x) = x_i \cdot x_j, \tag{1.20}$$

and will be interested to the behavior as $|i - j| \to \infty$ of the correlation:

$$\operatorname{Cor}(i,j) = \langle g_{ij} \rangle_\infty - \langle x_i \rangle_\infty \langle x_j \rangle_\infty. \tag{1.21}$$

We shall prove in the convex case (in connection with results of Sokal [**40**] mentioned in the book of Ellis [**6**]) an exponential decay with respect to $|i - j|$. The problem appearing in the study of the asymptotic behavior of (1.14)–(1.16) is of this type.

PROBLEM 3 (THE SPLITTING). As can be seen in formula (1.13) it is interesting to study the limit as m tends to ∞ of the quotient $\mu_2^{(m)}/\mu_1^{(m)}$. In the convex case the problem has been recently solved in [**17**] (the corresponding problem for Schrödinger was solved by J. Sjöstrand [**35**]). The non convex case is still open for this Kac model (see [**26**]) but there are interesting conjectures.

All these problems are related to the properties of the measure $u_1^{(m)}(x)^2\,dx$ whose study is unfortunately far to be complete. We shall present here new results in this direction: logconcavity of the first eigenfunction, ferromagnetic properties, but we have not completely done the connection with all the assumptions of [**21**]. We remark here that the results concerning the Schrödinger equation are much more complete (see [**36**], [**38**] and [**39**]) but sometimes a direct approach of the Kac operator or in some case of the initial problem leading to the introduction of this Kac operator appears to be simpler.

This survey is organized as follows. Section 2 is devoted to the existence of a solution for the first "basic" equation:

$$g - \langle g\rangle = (\nabla\Phi\nabla - \Delta)u,$$

where Φ is defined above and g is a suitable C^∞ function with bounded gradient and to the introduction of precise estimates using the maximum principle in the spirit of [**33**] and [**36**]. Most of the material comes from [**21**] or from the more recent preprint of Sjöstrand [**39**] where one can find more precized estimates.

In Section 3 we present a new point of view on the Brascamp-Lieb inequalities and give some applications.

Section 4 is devoted to the FKG inequalities and presents some new aspects coming from [**21**] and from [**22**].

Section 5 will give the proof of the analogue of the Brascamp-Lieb results for Schrödinger in the case of the Kac operator. The precized estimate of the logconcavity is inspired by the proofs of [**33**] and of [**36**] in the case of Schrödinger.

Section 6 considers first the problem of the existence of the thermodynamic limit. It is of course wellknown in the case of the motivating example to get the existence of such a limit as a consequence of general results for spin systems but we give two different proofs whose general character goes in another direction. We shall also give in the same way as for the second proof precise estimates on the correlations attached to the measure associated to the square of the first eigenfunction by the formula: $d\mu^{(m)} = u_1^{(m)}(x)^2\,dx^{(m)}$.

Finally we present in Section 7 the recent result obtained in [**17**] on the splitting for the Kac operator which extends naturally the previous result of Sjöstrand concerning the Schrödinger operator.

2. PDE Techniques and the Maximum Principle

The main goal of this section will be to prove the following theorem:

THEOREM 2.1. *Let Φ in $C^\infty(\mathbb{R}^m;\mathbb{R})$ satisfy:*

$$\operatorname{Hess}\Phi(x) \geq \sigma, \quad \forall x \in \mathbb{R}^m \tag{2.1}$$

$$|\partial^\alpha\Phi(x)| \leq C, \quad \forall\alpha \text{ s.t. } |\alpha| \geq 2. \tag{2.2}$$

Let g in $C^1(\mathbb{R}^m;\mathbb{R})$ s.t. $x \to \nabla g(x)$ is bounded. Then the equation:

$$g - \langle g \rangle = \nabla_x \Phi \nabla u - \Delta u \text{ for } x \in \mathbb{R}^m, \tag{2.3}$$

admits a unique temperate solution u such that:

$$u \in L^2\big(\mathbb{R}^m, \exp(-\Phi \cdot dx)\big) \tag{2.4}$$

and

$$\langle u \rangle = 0. \tag{2.5}$$

Moreover this solution is with bounded gradient, and we have the estimate:

$$\sup_x \|\nabla u(x)\| \leq \frac{1}{\sigma} \sup_x \|\nabla g(x)\|. \tag{2.6}$$

We emphasize that as for example in (2.6) all the constants appearing in our estimates are controlled very explicitly, once we have an explicit control of the convexity property (like for example in (2.1)) and once we have chosen the norms on $\mathbb{R}^m$. Here the norm appearing in (2.6) is the ℓ^2-norm but an important point is to understand that we will have to play later with other norms as we shall see at the end of this section.

SKETCH OF THE PROOF. As observed in [**21**], we consider the operator A:

$$u \to \nabla_x \Phi \nabla u - \Delta u \tag{2.7}$$

which appears to be essentially selfadjoint (starting from C_0^∞) on the weighted space: $L^2(\mathbb{R}^m, \exp(-\Phi) \cdot dx)$. Using the conjugation: $f \to f \exp -\Phi/2$, the problem of existence of solutions is reduced to the study of a "pseudo" harmonic oscillator:

$$h \to -\Delta h + \frac{1}{4}|\nabla \Phi|^2 h - \Delta \Phi h. \tag{2.8}$$

Under the weaker condition that for some $C > 0$:

$$\operatorname{Hess} \Phi(x) \geq \sigma, \quad \forall x \in \mathbb{R}^m \setminus B(O, C), \tag{2.9}$$

and condition (2.2)), we observe using standard techniques that this new operator $\tilde{A}$ admits a unique selfadjoint extension which defines a positive operator whose domain is characterized as:

$$B^2(\mathbb{R}^m) = \{u \in L^2(\mathbb{R}^m) \mid x^\beta \partial_x^\alpha u \in L^2(\mathbb{R}^m)\, \forall \beta \text{ s.t. } |\alpha| + |\beta| \leq 2\}$$

and consequently with compact resolvent. The smallest eigenvalue is 0 and the first eigenfunction is the function $x \to \exp\big(-\Phi(x)\big)/2$. The first part of the statement can consequently seen as a consequence of the Fredhom alternative.

For the other statements, we come back to A and need a regularization argument in order to work with the maximum principle.

We shall first work with a function Φ of the form $x^2/2 + \Psi$ with Ψ compactly supported, and we need in this case a more precise information for the solution constructed in the first part of the proof. As is proved in [**36**] (see also [**21**] and [**39**]), we observe that if in equation (2.3) g belongs to C_0^∞ then the solution u has the property that $\nabla u(x)$ tends to 0 as $|x|$ tends to ∞ and that the same property is true for the higher derivatives of u. One way to prove this property, which will be useful also in order to apply the maximum principle, is to give the following remark:

Remark 2.2. If u is a solution of (2.3), then $v =: \nabla u$ is a solution of

$$w = (-\Delta + \nabla\Phi \cdot \nabla)v + (\text{Hess}\,\Phi)v \tag{2.10}$$

with

$$w = \nabla g. \tag{2.11}$$

The equation (2.2) was called the "basic" equation and studied more directly in [**21**]. We shall call it here the second basic equation.

PROOF OF (2.6), THE MAXIMUM PRINCIPLE. In order to prove the last part of the theorem we shall apply the maximum principle to the solution of the second basic equation. Let x_0 one point where $\|v(x)\|$ is maximal. We then take the scalar product with $v(x_0)$ in (2.10) and applying the maximum principle at the function $x \to \langle v(x) \mid v(x_0)\rangle$ and using the assumption (2.1) we obtain (2.6). Note that the argument is valid because $\|v(x)\|$ tends to 0 at ∞. □

But a regularization argument is needed in order to treat the general case which is explained in [**36**], [**21**] or more recently [**39**]. Let us sketch briefly the argument.

THE REGULARIZATION ARGUMENT. The proof given before is valid under the additional assumption that $\Phi(x) = (x^2/2) + \Psi(x)$ with Φ strictly convex and Ψ with compact support. In order to treat the general case we introduce the family:

$$\Phi_t = \left(\frac{x^2}{2}\right) + \chi_t(x)\Psi(x)$$

where $\chi_t(x) = \chi(t \ln|x|)$ and χ is with compact support.

We observe that Φ_t is a family which is strictly convex if Φ is. We have indeed:

$$\nabla\Phi_t = x + \chi_t(x)\nabla\Psi + \mathcal{O}(t|x|).$$

Similarly:

$$\text{Hess}\,\Phi_t = I + \chi_t\,\text{Hess}\,\Psi + O(t) = (1-\chi_t)I + \chi_t\,\text{Hess}\,\Phi + \mathcal{O}(t)$$

and the uniform strict convexity is then easy to verify. Similarly we have to replace g by $g_t = \chi_t g$. We then let t tend to 0, in order to get the result. □

THE GENERAL MAXIMUM PRINCIPLE OF SJÖSTRAND. As already mentioned, more sophisticated versions of the maximum principle will be useful in different questions. Here we follow the presentation given by [**39**]. Let B is an $n-$dimensional $(n < \infty)$ real normed space and $B^\star$ be the dual space. Let $A\colon B \to B$ be a linear map and let $\delta \geq 0$. We say that A satisfies mp(δ) with respect to B (where "mp" is in reference to "maximum principle"), if we have

$$\text{if } t \in B, s \in B^\star \text{ and } \langle t, s\rangle = \|t\|_B\|s\|_{B^\star}, \text{ then } \langle At, s\rangle \geq \delta\|t\|_B\|s\|_{B^\star}. \tag{2.12}$$

Examples.

Example 2.3. We take $n = m$, $B = B^\star = \ell^2$ and $A = \text{Hess}\,V$ where V is a C^∞ strictly convex function satisfying:

$$\text{Hess}\,V(x) \geq \sigma > 0.$$

Then A satisfies mp(σ) with respect to B.

Example 2.4. We take $n = mp$, $B = \bigoplus_{j=1}^{p} E_j$ with $E_j = \ell^\infty_\rho(\mathbb{Z}/m\mathbb{Z}) = \{y \in \mathbb{R}^m \mid \rho(j)y_j \in \ell^\infty\}$, where ρ is a positive weight on $\mathbb{Z}/m\mathbb{Z}$, and the norm on B is defined by:

$$\|x\|_B = \sqrt{\sum_j \|x_j\|^2_{E_j}}.$$

Our matrix A is now $\operatorname{Hess}\Phi^{(p)}$ where:

$$\Phi^{(p)}(x_1, \ldots, x_p) = \sum_{j=1}^{p} \Big(V(x_j) + \frac{1}{4}(x_{j+1} - x_j)^2 \Big), \tag{2.13}$$

with $x_j \in \mathbb{R}^m$ (and where we take the convention that $x_{p+1} = x_1$).

Then, if

$$\operatorname{Hess} V(y) = D + C(y), \tag{2.14}$$

where D is diagonal and satisfies:

$$D \geq \sigma > 0 \tag{2.15}$$

and

$$\|C(y)\|_{\mathcal{L}(\ell^\infty_\rho(\mathbb{Z}/m\mathbb{Z}))} \leq \epsilon < \sigma, \tag{2.16}$$

for all ρ on $(\mathbb{Z}/m\mathbb{Z})$ satisfying

$$\exp(-\kappa) \leq \frac{\rho(i)}{\rho(i+1)} \leq \exp\kappa, \tag{2.17}$$

A satisfies $\operatorname{mp}(\sigma - \epsilon)$ with respect to all the associated B.

Let us verify this property for this example. We observe first that $B^\star = \bigoplus_{j=1}^{p} E_j^\star$ with $E_j^\star = \ell^1_{1/\rho}(\mathbb{Z}/m\mathbb{Z})$ and the norm on $B^\star$ can be defined by:

$$\|x\|_{B^\star} = \sqrt{\sum_j \|x_j\|^2_{E_j^\star}}.$$

We observe now that if $\langle t \mid s\rangle = \|t\| \cdot \|s\|$ is satisfied for some t in B and some s in $B^\star$ then we have necessarily:

$$\langle t_j \mid s_j\rangle = \|t_j\| \cdot \|s_j\|$$

and

$$\|t_j\| = \lambda\|s_j\|.$$

For the second property we have used the property of $\ell^2(\mathbb{Z}/p\mathbb{Z})$. The end of the proof is then easy using these two properties.

Example 2.5 (*cf.* [**17**]). We take $n = mp$, $B = \bigoplus_{j=1}^{p} E_j$ with $E_j = \ell^2(\mathbb{Z}/m\mathbb{Z})$ and the norm on B is defined by:

$$\|x\|_B = \sup_j(\rho(j)\|x_j\|_{E_j}).$$

Our matrix A is again $\operatorname{Hess}\Phi^{(p)}$. Then if $\operatorname{Hess} V(x) \geq \sigma$, $\forall x$, and if ρ is a weight on $\mathbb{Z}/p\mathbb{Z}$ such that (2.17) is satisfied if κ satisfies:

$$\cosh\kappa < 1 + \sigma, \tag{2.18}$$

A satisfies mp(δ) with respect to B with:

$$\delta = -\cosh\kappa + 1 + \sigma. \tag{2.19}$$

Let us verify this property for this example. We observe first that $B^\star = \oplus_{j=1}^p E_j$ where the norm on $B^\star$ can be defined by:

$$\|x\|_{B^\star} = \sum_j \frac{1}{\rho(j)} \|x_j\|_{E_j}.$$

We observe now that if (2.12) is satisfied for some t in B and some s in $B^\star$ then we have necessarily:

$$t_j = \lambda_j s_j, \text{ for } \lambda_j > 0.$$

For this property we have used the Hilbertian property of E_j. The end of the proof is then easy using this property.

Application of the mp-property (cf. [39]). We consider the situation we analyzed above for the second basic equation and we assume that Hess $\Phi(x)$ satisfies mp(σ) for all $x \in \mathbb{R}^n$ for some fixed $\sigma > 0$. We also assume that $v(x) = \nabla u(x)$ tends to 0 as $|x| \to \infty$. Let $x_0 \in \mathbb{R}^n$ be a point where $\|v(x)\|_B$ is maximal:

$$M_v = \sup_x \|v(x)\|_B = \|v(x_0)\|_B.$$

Let s in $B^\star$ a unit vector such that:

$$\langle v(x) \mid s\rangle = M_v.$$

In particular, we have the property:

$$\langle v(x) \mid s\rangle = \|v(x)\|_B \cdot \|s\|_{B^\star}.$$

We now take the duality product with s in the second basic equation. At the point x_0, we observe that:

$$\langle -\Delta v(x_0) \mid s\rangle \geq 0,$$

and

$$\langle \nabla\Phi \cdot \nabla v(x_0) \mid s\rangle = 0.$$

Using the mp(σ)-property for Hess $\Phi(x_0)$, we get finally:

$$\sigma M_v \leq M_w,$$

or more explicitely the estimate:

$$\|\nabla u(x)\|_B \leq \frac{1}{\sigma} \sup_x \|\nabla g(x)\|_B. \tag{2.20}$$

We observe here that this estimate is still valid after the regularization procedure. When $B = \ell^2$, we recover the estimate (2.6).

As in [**36**], [**21**], we can also obtain, if g is C^∞, similar estimates for the higher derivatives and get for example the control of $\|\operatorname{Hess} u(x)\|_{\mathcal{L}(B)}$ using the control of $\|\operatorname{Hess} g(x)\|_{\mathcal{L}(B)}$, $\|\nabla g(x)\|_{B'}$ in a suitable space B' for which mp(σ) is also satisfied, and an additional assumption on the derivatives of order 3 on Φ. We refer for example to the estimate (4.3) in [**39**].

3. Around Brascamp-Lieb Inequalities

H. J.Brascamp and E. Lieb [**2**] have proved the following celebrated inequality which plays an important role in different contexts in the study of the Schrödinger equation.

THEOREM 3.1. *Let* $F(x) = \exp\big(-\Phi(x)\big)$, $x \in \mathbb{R}^m$, *with* Φ *in* C^2 *and strictly convex. We assume that* Φ *has a minimum and consequently* F *decays exponentially in all directions. Let* $h \in C^1(\mathbb{R}^m)$, *and let us assume that* $\operatorname{var} h < \infty$. *Then*

$$\operatorname{var} h \leq \langle \nabla h, (\Phi_{xx})^{-1} \nabla h \rangle \tag{3.1}$$

where ∇h *is the gradient of* h.

We want to give a new proof of this theorem in the spirit of Helffer-Sjöstrand [**21**]. For h in C^1, such that ∇h is bounded, we have seen in Section 2 that there exists a gradient vector field v s.t.:

$$h - \langle h \rangle = \nabla\Phi \cdot v - \operatorname{div} v, \tag{3.2}$$

where $\langle h \rangle = \int h \exp(-\Phi)\, dx / \int \exp(-\Phi)\, dx$ is the mean value of h. If we take the formula giving the variance, we get:

$$\begin{aligned} \operatorname{var}(h) &\stackrel{\text{def}}{=} \frac{\int (h - \langle h \rangle)^2 \exp -\Phi\, dx}{\int \exp -\Phi\, dx} \\ &= \frac{\int (h - \langle h \rangle)(v \cdot \nabla\Phi - \operatorname{div} v) \exp -\Phi\, dx}{\int \exp -\Phi\, dx}, \end{aligned} \tag{3.3}$$

and finally

$$\operatorname{var}(h) = \frac{\int v \cdot \nabla h \exp(-\Phi)\, dx}{\int \exp(-\Phi)\, dx}. \tag{3.4}$$

As in [**21**], we get by differentiation of (3.2):

$$\nabla h = (\nabla\Phi \cdot \nabla - \Delta) v + \operatorname{Hess} \Phi v. \tag{3.5}$$

Then we get:

$$\nabla h \cdot v = \big((\nabla\Phi \cdot \nabla - \Delta) v \cdot v\big) + (\operatorname{Hess} \Phi\, v) \cdot v. \tag{3.6}$$

Let us now integrate over $\mathbb{R}^m$ the formula (3.6) and we get, using the positivity of $\operatorname{Hess} \Phi$ and after an integration by parts:

$$\begin{aligned} \int \nabla h \cdot v \exp -\Phi\, dx &= \int \big((\nabla\Phi \cdot \nabla - \Delta) v \cdot v\big) \exp(-\Phi)\, dx \\ &\qquad + \int (\operatorname{Hess} \Phi\, v) \cdot v \exp(-\Phi)\, dx \\ &= \int |\nabla v|^2 \exp -\Phi\, dx + \int |(\operatorname{Hess} \Phi)^{1/2} v|^2 \exp -\Phi\, dx. \end{aligned}$$

So we have finally:

$$\begin{aligned} \int \nabla h \cdot v \exp(-\Phi)\, dx &= \int |\nabla v|^2 \exp(-\Phi)\, dx \\ &\qquad + \int |(\operatorname{Hess} \Phi)^{1/2} v|^2 \exp(-\Phi)\, dx. \end{aligned} \tag{3.7}$$

We now use Cauchy-Schwarz and obtain the following inequality:

$$\int |(\mathrm{Hess}\,\Phi)^{1/2} v|^2 \exp(-\Phi)\,dx \le \int |(\mathrm{Hess}\,\Phi)^{-1/2}\nabla h|^2 \exp(-\Phi)\,dx. \tag{3.8}$$

Using again Cauchy-Schwarz, we get

$$\left|\int \nabla h \cdot v \exp(-\Phi)\,dx\right| \le \int |(\mathrm{Hess}\,\Phi)^{-1/2}\nabla h|^2 \exp(-\Phi)\,dx$$

which is exactly the Brascamp-Lieb inequality.

Remark 3.2. One can not say that the proof of the Theorem 3.1 which is presented here is easier than the original one (this is probably the contrary) but we have replaced the induction over the dimension by a direct argument which is perhaps more enlightning.

4. Around FKG Inequalities

We have shown in [**21**], how one can recover a theorem due to Cartier (generalizing the celebrated FKG inequalities) using the maximum principle and we discussed possible applications of this technique for the Schrödinger operator. We learned more recently from I. Herbst that new generalizations of FKG inequalities containing partly the statements we got in [**21**] were already obtained by him in collaboration with L. Pitt [**22**]. We consider here a very specific case (actually not explicitly studied in [**22**]) and we shall discuss in this section the different approaches. Our idea is to analyze what has been done in the case of the Schrödinger operator and to extend the techniques and the results to the case of the Kac operator. We shall in particular present here an approach which is in the spirit of the proof by Brascamp-Lieb of the logconcavity of the first eigenfunction for the Schrödinger operator. We shall never try to give the most general assumptions on the potential (regularity ...) but we are more interested in explaining the methods.

THEOREM 4.1. *Let V be a C^∞ potential such that:*

$$V \ge C|x|^2 - C \text{ with } C > 0, \tag{4.1}$$

$$|D_x^\beta V| \le C_\beta (1+|x|)^{(2-|\beta|)_+} \tag{4.2}$$

and s.t. the following condition is satisfied:

$$\partial^2_{ij} V(x) \le 0 \text{ for } i \ne j, \quad \forall x \in \mathbb{R}^m. \tag{4.3}$$

Then, if u_1 is the first normalized eigenfunction of the Schrödinger operator

$$S_V = -\Delta + V. \tag{4.4}$$

or of the Kac operator

$$K_V = \exp\Big(-\frac{V}{2}\Big)\cdot \exp\Delta \cdot \exp\Big(-\frac{V}{2}\Big), \tag{4.5}$$

the correlation attached to the measure $(u_1)^2 dx$ satisfies:

$$\mathrm{Cor}(i,j) = \langle (x_i - \langle x_i\rangle)(x_j - \langle x_j\rangle)\rangle \ge 0 \tag{4.6}$$

We recall that the result of Cartier was the following:

THEOREM 4.2. *Let Φ be a C^∞ potential such that* (4.1), (4.2) *and* (4.3) *are satisfied; then the correlation attached to the measure* $\exp\bigl(-\Phi(x)\bigr)\,dx$ *satisfies* (4.6). *More generally, the functions $x \to x_i$ and $x \to x_j$ can be replaced by any monotone increasing functions f and g (say with bounded gradient) with respect to each of the variables*:

$$\bigl\langle (f - \langle f\rangle)(g - \langle g\rangle)\bigr\rangle \geq 0 \tag{4.7}$$

We shall actually prove Theorem 4.1 as a consequence of the theorem of Cartier. We emphasize that the assumptions (4.1) and (4.2) are much stronger than needed and as in the proof of Cartier's Theorem and as in the proof of the Brascamp-Lieb inequalities one can surely establish the theorems with much weaker assumptions.

A PROOF OF CARTIER'S THEOREM. We sketch here another proof inspired by discussions with I. Herbst and which combines the proof of [**22**] (cf. more recently [**5**]) and some remarks given in [**21**]. In this paper ([**22**, Theorem 1.3]), the positivity of the correlation is obtained by using the fact that the operator $\exp(-tA)$, where A is the selfadjoint operator on $L^2(\mathbb{R}^m, \exp(-\Phi)\,dx)$ introduced in (2.7):

$$A = -\Delta + \nabla\Phi\nabla$$

satisfies, if (4.3) is satisfied, the following "positivity preserving" property

($\mathcal{P}^+$) If f satisfies the condition $\nabla f \succeq 0$ in the sense that all the component of ∇f are positive (we shall write $f \in \mathcal{C}^+$), then $\exp(-tA)f$ has the same property.

In our case let us observe (see [**21**]), that, by doing the conjugation: $u = \exp\Phi/2 \cdot \tilde{u}$, we get the operator:

$$\tilde{A} = -\Delta + \left|\frac{\nabla\Phi}{2}\right|^2 - \Delta\Phi \text{ on } L^2(\mathbb{R}^m; dx).$$

This operator is essentially selfadjoint, with compact resolvent and clearly $\tilde{A}$ is positive and has zero as an eigenvalue with multiplicity 1. Coming back to our initial operator it is clear that we can invert A on the orthogonal of the first eigenfunction which is the constant function 1. Then we denote by $(A^{\text{red}})^{-1}$ the inverse of A reduced to the orthogonal to the zero eigenfunction, and vanishing on the constants. We observe that the property of positivity preservation of $\exp(-tA)$ is transmitted to $(A^{\text{red}})^{-1}$ considered as the limit as ϵ tends to 0^+ of $(A^{\text{red}} + \epsilon)^{-1}$. (Note that the projector on the first eigenvector of A has also the property). Then once we know that $(A^{\text{red}})^{-1}$ has this property, we get easily the property for the correlation in the following way. Let us study $\text{Cor}(f, g)$ with f, $g \in \mathcal{C}^+$. Let us write:

$$g - \langle g\rangle = A\bigl((A^{\text{red}})^{-1}g\bigr)$$

which is true because $g - \langle g\rangle$ is orthogonal to the zero eigenvector. Then $u = (A^{\text{red}})^{-1}g \in \mathcal{C}^+$ and we have:

$$\text{Cor}(f, g) = \langle \nabla f \cdot \nabla u\rangle \geq 0.$$

In [**21**], we proved that $(A^{\text{red}})^{-1}$ has the property in the case when Φ was strictly pseudoconvex using the maximum principle. $\square$

CORRELATION INEQUALITIES FOR THE KAC OPERATOR. We start from the following expression for $u_1(x)^2$

$$u_1(x)^2 = \lim_{p\to\infty} \mu_1^{-p}\big(K_V^{(p)}\big)(x,x), \tag{4.8}$$

where $K_V^{(p)}(x,y)$ is the distributional kernel of K_V^p, and we would like to use a convergence result for the moments. Under our strong assumptions above, one can actually prove that the convergence above is in $\mathcal{S}(\mathbb{R}^m)$, by using the following approach. We first remark that:

$$\mu_1^{-p}K_V^p = \pi_1 + \sum_{j\geq 2}\left(\frac{\mu_j}{\mu_1}\right)^p \pi_j \tag{4.9}$$

where π_j is the projector on the j-th eigenvector, and the convergence is *a priori* in $\mathcal{L}(L^2)$. We recall that μ_j is the decreasing sequence of eigenvalues and that we have:

$$\mu_2 < \mu_1. \tag{4.10}$$

We now observe the two inequalities:

$$\|\mu_1^{-p}K_V^p - \pi_1\|_{\mathcal{L}(L^2)} \leq \left(\frac{\mu_2}{\mu_1}\right)^p \tag{4.11}$$

and for any $k\in\mathbb{N}$, $\ell\in\mathbb{N}$

$$\|\mu_1^{-1}K_V - \pi_1\|_{\mathcal{L}(B^{-\ell},B^k)} \leq C_{k,\ell}, \tag{4.12}$$

where

$$B^k(\mathbb{R}^m) = \{u\in L^2(\mathbb{R}^m), x^\beta D_x^\delta u \in L^2(\mathbb{R}^m), \text{ for } |\beta|+|\delta|\leq k\}.$$

We then write

$$\mu_1^{-p}K_V^p - \pi_1 = (\mu_1^{-1}K_V - \pi_1)\left(\sum_{j=0}^{p-1}(\mu_1^{-1}K_V)^j\pi_1^{(p-1-j)}\right)$$

and get (for $p\geq 1$):

$$\begin{aligned}&\|\mu_1^{-p}K_V^p - \pi_1\|_{\mathcal{L}(B^{-l},B^k)}\\ &\qquad\leq p\|(\mu_1^{-1}K_V-\pi_1)\|_{\mathcal{L}(L^2,B^k)}\max\big(\|\pi_1\|_{\mathcal{L}(B^{-k},L^2)}, \|\mu_1^{-1}K_V\|_{\mathcal{L}(B^{-l},L^2)}\big).\end{aligned}$$

We then obtain:

$$\|\mu_1^{-p}K_V^p - \pi_1\|_{\mathcal{L}(B^{-\ell},B^k)} \leq p\widetilde{C}_{k,\ell}, \tag{4.13}$$

and by interpolation we get:

$$\|\mu_1^{-p}K_V^p - \pi_1\|_{\mathcal{L}(L^2,B^k)} \leq C_k(p)^{1/2}\left(\frac{\mu_2}{\mu_1}\right)^{p/2}, \tag{4.14}$$

or more generally for any $0<\delta<1$:

$$\|\mu_1^{-p}K_V^p - \pi_1\|_{\mathcal{L}(L^2,B^k)} \leq C_{k,\delta}(p)^{\delta}\left(\frac{\mu_2}{\mu_1}\right)^{p(1-\delta)}. \tag{4.15}$$

and using the adjoint we get finally:

$$\|\mu_1^{-2p} K_V^{2p} - \pi_1\|_{\mathcal{L}(B^{-k}, B^k)} \leq \widetilde{C}_{k,\delta}(p)^\delta \left(\frac{\mu_2}{\mu_1}\right)^{2p(1-\delta)}, \tag{4.16}$$

for any $0 < \delta < 1$.

This relation permits to get, as it is classical, the exponential convergence of the kernel of $\mu_1^{-2p} K_V^{2p}$ to the kernel of π_1 in $\mathcal{S}(\mathbb{R}^m \times \mathbb{R}^m)$ and in particular (taking the restriction to the diagonal) we get the:

LEMMA 4.3. *Suppose V is a C^∞ potential satisfying* (4.1)–(4.2), *then the function obtained by restriction to the diagonal of the distribution kernel of $\mu_1^{-2p} K_V^{2p}$. $(\mu_1^{-2p} K_V^{2p})(x, x)$ converges exponentially rapidly as $p \to \infty$ to $u_1(x)^2$ in $\mathcal{S}(\mathbb{R}^m)$.*

In particular, we can apply this result for the correlations and we get (at least for even p):

$$\int (\mu_1^{-p} K_V^{(p)})(x, x)\, dx \to \int u_1(x)^2\, dx \text{ as } p \to \infty, \tag{4.17}$$

$$\int x_i (\mu_1^{-p} K_V^{(p)})(x, x)\, dx \to \int x_i u_1(x)^2\, dx \text{ as } p \to \infty, \tag{4.18}$$

and

$$\int x_i x_j (\mu_1^{-2p} K_V^{(2p)})(x, x)\, dx \to \int x_i x_j u_1(x)^2\, dx \text{ as } p \to \infty. \tag{4.19}$$

So the study of the correlation for the measure $u_1(x)^2\, dx$ is naturally related to the study of particular correlation functions attached to the measure on $\mathbb{R}^{mp}$ of the form $(\exp -\Phi)\, dy_1 \dots dy_p$ with:

$$\Phi(y_1, \dots, y_p) = \sum_{j=1}^{p} \left(V(y_j) + \frac{1}{4}(y_{j+1} - y_j)^2 \right). \tag{4.20}$$

In order to apply Cartier's theorem to this measure we then have only to analyze the condition (4.3) for Φ. So we have to look to the derivative $X_j^i X_\ell^k \Phi$ where X_j^i corresponds to the derivative with respect to the variable y_{ji}. If $i = k$, then we have $j \neq \ell$, and the assumption on V gives the result. If $i \neq k$, then the only non vanishing terms correspond to $j = \ell$, and this derivative is equal to $-1/2 < 0$. We then obtain the theorem. □

Actually, we can do a little better and we shall prove more directly that $-\ln u_1$ satisfies indeed condition (4.3) if V does. We emphasize, as was observed in [**22**] the similarity with the way followed by Brascamp-Lieb in order to prove, if V is convex, the logconcavity of the first eigenfunction of the Schrödinger operator and we will treat in the case of the Kac operator the two problems similarly.

We first observe that it is sufficient to analyze the same property for the function $K^{(p)}(x, x)$ according to the Lemma (we can then take the limit $p \to \infty$). So we have to study, the function:

$$-\ln K^{(p)}(x, x)$$

and study if it satisfies condition (4.3), or if it is convex. The following lemma is due to Brascamp-Lieb (cf. Corollary 3.5 in [**2**]):

LEMMA 4.4. *If $F(x, y)$ is strictly positive, logconcave (with respect to the variables (x, y)) (and satisfies suitable integrability assumptions (essentially F is of Gaussian type)), then $G(x) = \int F(x, y)\, dy$ is logconcave.*

Similarly we shall get:

LEMMA 4.5. *If $f(x, y) = -\ln F(x, y)$ satisfies condition (4.3) (with respect to the variables (x, y)) (and satisfies suitable integrability assumptions), then $g(x) = -\ln\big(G(x)\big)$, with $G(x) = \int F(x, y)\, dy$, satisfies the same condition with respect to x.*

PROOF. Let us differentiate two times g with respect to the variables x_i, x_j:

$$(\partial_{x_i} g)(x) = \frac{1}{G(x)} \int (\partial_{x_i} f)(x, y) F(x, y)\, dy$$

and

$$\begin{aligned}(\partial^2_{x_i x_j} g)(x) = & \frac{1}{G(x)} \int (\partial^2_{x_i x} f)(x, y) F(x, y)\, dy \\ & - \frac{1}{G(x)} \int (\partial_{x_i} f)(x, y)(\partial_{x_j} f)(x, y) F(x, y)\, dy \\ & + \frac{1}{G(x)^2} \Big(\int (\partial_{x_i} f)(x, y) F(x, y)\, dy \Big) \Big(\int (\partial_{x_i} f)(x, y) F(x, y)\, dy \Big).\end{aligned}$$

We rewrite this relation on the following more classical form:

$$\begin{aligned}(\partial^2_{x_i x_j} g)(x) = & \langle (\partial^2_{x_i x_j} f)(x, \cdot) \rangle_x \\ & - \big\langle \big((\partial_{x_i} f) - \langle (\partial_{x_i} f) \rangle_x \big) \big((\partial_{x_j} f) - \langle (\partial_{x_j} f) \rangle_x \big) \big\rangle_x \end{aligned} \tag{4.21}$$

where $\langle u \rangle_x$ is the expectation of the function u with respect to the measure $d\mu_x = F(x, y)\, dy / G(x)$. We now observe that the first term of the r.h.s of (4.21)) is negative because

$$(\partial^2_{x_i x_j} f) \leq 0 \text{ if } i \neq j.$$

Now in order to examine the second term

$$\big\langle \big((\partial_{x_i} f) - \langle (\partial_{x_i} f) \rangle_x \big) \big((\partial_{x_j} f) - \langle (\partial_{x_j} f) \rangle_x \big) \big\rangle_x$$

we observe that we can apply the Cartier's Theorem because the function $y \to (\partial_{x_i} f)(x, y)$ is monotone with respect of each variable y_k due to the inequality $(\partial^2_{y_k x_i} f)(x, y) \leq 0$ which is satisfied by assumption. This proves the lemma. □

Remark 4.6. Using the Brascamp-Lieb inequalities and formula (4.21), we can recover Theorem 4.2 of Brascamp-Lieb [**2**] which gives the following inequality between $m \times m$ symmetric matrices:

$$(\partial^2_{x_i x_j} g)(x) \geq \big\langle (\partial^2_{x_i x_j} f)(x, \cdot) - \partial^2_{x_i y_k} f \cdot [(f''_{yy})^{-1}]_{k\ell} \partial^2_{y_\ell x_j} f \big\rangle_x \tag{4.22}$$

FKG in the case of Schrödinger. Here we follow the ideas of Brascamp-Lieb and remain at a relatively formal level. The details can be found in the book by B. Simon [**32**]. The result is here a consequence of two limit procedures.

The first one consists in writing:

$$u_1(x)^2 = \lim_{p \to \infty} \exp p\lambda_1 \big(\exp(-pS_V) \big)(x, x) \tag{4.23}$$

and *the second one* consists in writing the Feynmann-Kac formula, or more simply the Trotter formula:

$$\exp(-tS_V) = \lim_{n\to\infty}\left(\exp\left(-\frac{tV}{n}\right)\cdot\exp\left(\frac{t\Delta}{n}\right)\right)^n. \tag{4.24}$$

The control of the convergence in (4.23) is the same as for Kac, once we have verified that $\exp(-S_V)$ has the same properties as K_V. Under the strong conditions we have given on V it is for example a consequence of a global pseudodifferential theory (see for example [**11**]). For (4.24), the basic result given by the Trotter formula is that (4.24) is valid in the sense of the strong convergence (see for example Reed-Simon [**29**]). But more seems needed for pointwise convergence which is used by Brascamp and Lieb in their proof, and this is a actually a consequence of the proof of the Feynman-Kac formula (see for example [**9**]). Anyway one can also use one of the two following procedures in order to avoid this problem. The first one (which was suggested to us by JP Solovej) is to use regularization by gaussians, $\Omega(\tau,x_0,y_0)(x,y) = \text{const}\cdot\tau^{2m}\exp(-\tau)(x-x_0)^2\exp(-\tau)(y-y_0)^2$ and we observe that there are logconcave and ferromagnetic. We then first prove that $\langle\Omega(\tau,x_0,y_0),\exp(-tP)\Omega(\tau,x_0,y_0)\rangle$ is logconcave as a function of (x_0,y_0) (using the Trotter formula) (the convolution of two logconcave functions is logconcave (see [**32**])) and then using the fact that the distributional kernel of $\exp(-tP)$ is a temperate distribution and C^∞ (and actually in $\mathcal{S}$ under the strong assumptions we have done on V), we get the convergence of $\langle\Omega(\tau,x_0,y_0),\exp(-tP)\Omega(\tau,x_0,y_0)\rangle$ to $\exp(-tP)(x_0,y_0)$ (locally uniformly with respect to (x_0,y_0)) as τ tends to ∞. Here we were sketching the proof of the logconcavity but let us observe that the proof works equally well in order to get the "FKG" property.

The other approach in order to get the information is to observe that we need only to observe that for one gaussian Ω_0:

$$\lim_{t\to\infty}(\exp t\lambda_1\exp\big(-t(-\Delta+V)\big)\Omega_0 = c\cdot u_1$$

so we need only to know about the logconcavity (or the ferromagnetic character) of $\exp\big(-t(-\Delta+V)\big)\Omega_0$ which is easier to obtain from the Trotter formula. We refer to the book [**32**, p. 136–139] for detailed proofs around logconcavity. We can then conclude that $-\ln u_1$ is convex if V is convex and that it satisfies the condition (4.3) if V does.

Another proof based on the maximum principle is given in [**21**] but this proof assumes that V is strictly uniformly convex. We shall give in Section 5 a more precise control of the logconcavity, modelled on the control of the logconcavity as obtained for the Schrödinger case by [**33**] and [**35**].

5. Logconcavity for the Kac Operator with Convex Potential

There are at least two ways to obtain a precized control of the logconcavity of the first eigenvector. One will be to follow the proof we give in Section 4 for the logconcavity and the other consists to treat a more implicit equation, in the spirit of the proofs of [**35**], in the case of the Schrödinger operator. Let us observe that in this proof that we present now we use already the fact (in order to simplify the argument) that we know that the first eigenfunction is logconcave. We start of the equation satisfied by u_1:

$$u_1(x) = \mu_1^{-1}(K_V u_1)(x) \tag{5.1}$$

that is:

$$u_1(x) = \mu_1^{-1} \int \exp\Big(-\frac{V(x)}{2}\Big) \cdot \exp\Big(-\frac{(x-y)^2}{4}\Big) \cdot \exp\Big(-\frac{V(y)}{2}\Big) \cdot u_1(y)\, dy \tag{5.2}$$

which we rewrite with $g(x) = -\ln u_1(x)$:

$$\exp(-g(x)) = \text{const} \int \exp(-f(x,y))\, dy \tag{5.3}$$

with

$$f(x,y) = \frac{V(x)+V(y)}{2} + \frac{(x-y)^2}{4} + g(y) \tag{5.4}$$

We shall now try to apply the result given in (4.22):

$$(\partial^2_{x_i x_j} g)(x) \geq \big\langle (\partial^2_{x_i x_j} f)(x,\cdot) - \partial^2_{x_i y_k} f \cdot [(f''_{yy})^{-1}]_{k\ell} \partial^2_{y_\ell x_j} f \big\rangle_x. \tag{5.5}$$

Let us compute the different terms:

$$\begin{aligned}
(\partial^2_{x_i x_j} f)(x,y) &= \frac{\delta_{ij}}{2} + \frac{1}{2}\big(\operatorname{Hess} V(x)\big)_{ij},\\
(\partial^2_{x_i y_k} f)(x,y) &= -\frac{\delta_{ik}}{2},\\
(\partial^2_{y_i y_j} f)(x,y) &= \frac{\delta_{ij}}{2} + \frac{1}{2}\big(\operatorname{Hess} V(y)\big)_{ij} + (\operatorname{Hess} g)_{ij}(y).
\end{aligned}$$

We introduce now:

$$\begin{aligned}
\sigma &= \inf_x \lambda_{\min}(\operatorname{Hess} V(x)),\\
\phi &= \inf_x \lambda_{\min}(\operatorname{Hess} g(x)),
\end{aligned}$$

and we get:

$$2\phi \geq (\sigma+1) - \frac{1}{(1+\sigma+2\phi)} \tag{5.6}$$

This gives a minoration of ϕ, because the right hand side is strictly positive if $\sigma > 0$. If $\sigma = 0$, we observe that this gives: $2\phi^2 \geq 2\phi$ and $\phi = 0$ is solution.

Remark 5.1. It is interesting to ask if this estimate is optimal. For this we consider the quadratic case, and look for the very simple one dimensional case: $V(x) = \sigma x^2/2$ where the first eigenvalue is (see for example[**39**])

$$\exp\Big(-\frac{1}{2}\cosh^{-1}(\sigma+1)\Big)$$

and one first eigenvector is given by:

$$\exp\Big(-\frac{1}{4}\sqrt{\sigma^2+2\sigma}\,x^2\Big).$$

The phase is consequently:

$$\phi(x) = \frac{x^2}{2}\phi$$

with $\phi^2 = \sigma^2/4 + \sigma/2$. We observe that this is the case where the equality occurs in (5.6). We can indeed compute

$$2\phi = (\sigma + 1) - \frac{1}{1 + \sigma + 2\phi},$$
$$2\phi(1 + \sigma + 2\phi) = (\sigma + 1)(1 + \sigma + 2\phi) - 1,$$

and finally this gives

$$4\phi^2 = (\sigma + 1)^2 - 1 = \sigma^2 + 2\sigma.$$

So finally we have proved the following:

PROPOSITION 5.2. *If V is strictly convex and satisfies:*

$$\sigma = \inf_x \lambda_{\min}\big(\operatorname{Hess} V(x)\big) > 0,$$

then the first eigenfunction of the Kac operator u_1 is logconcave and more precisely $\Phi(x) = -\ln u_1$ satisfies:

$$\phi \geq \left(\frac{\sigma^2}{4} + \frac{\sigma}{2}\right)^{1/2} \tag{5.7}$$

with $\phi = \inf_x \lambda_{\min}\big(\operatorname{Hess}\Phi(x)\big)$.

We observe that this is coherent with the Schrödinger case as σ is small. Let us briefly recall the formal proof for the control of the convexity in the case of Schrödinger (see [**35**, Section 2, formulas 2.3–2.4]).

If $\Phi(x) = -\ln u_1(x)$, where $u_1(x)$ is the first eigenfunction we have:

$$|\nabla\Phi|^2 = V + \Delta\Phi - \sigma_1.$$

Taking the Hessian, we get (as in [**33**])

$$2\nabla\Phi \cdot \nabla \operatorname{Hess}\Phi - \Delta \operatorname{Hess}\Phi + 2 \operatorname{Hess}\Phi \circ \operatorname{Hess}\Phi = \operatorname{Hess} V \tag{5.8}$$

and we get, using the maximum principle (and modulo some regularization argument):

$$2\phi^2 \geq \sigma. \tag{5.9}$$

6. Thermodynamic Limit for the Kac Operator

If the problem of the existence of the thermodynamic limit has been investigated in detail in the case of the Schrödinger equation and the case of Laplace integrals, we do not know about general results concerning this limit for the Kac operator outside very particular examples like the motivating example introduced in Section 1 where the proof consists in coming back to the initial problem in statistical mechanics. We shall consider here a more general class where the existence of the thermodynamic limit can be easily performed and present two completely different approachs. The first one is a direct proof modeled on the proof given in the case of the Schrödinger operator in [**19**]. The second one is reminiscent of the origin of the problem and consists in studying a Laplace integral with more variables. The proof is then related to what was done in [**39**] and [**17**].

6.1. The direct approach. Let us consider our Kac operator:

$$K_{V^{(m)}} = K^{(m)} = \exp(-V(x)/2) \cdot \exp(\Delta^{(m)}) \cdot \exp(-V(x)/2). \tag{6.1}$$

with:

$$V(x) = \frac{1}{2}\sum_{k=1}^{m} x_k^2 + \sum_{k=1}^{m} W(x_k, x_{k+1}) \tag{6.2}$$

(with the convention that $m+1=1$) which is a compact operator on $L^2(\mathbb{R}^m)$. Here: $\Delta^{(m)} = \sum_{k=1}^{m}(\partial_{x_k})^2$. We just write here the following conditions on W:

$$|\nabla W| \leq C. \tag{6.3}$$

We introduce the potential:

$$\widetilde{W}(s,t) = \frac{1}{4}(s^2+t^2) + W(s,t)$$

so we can write also:

$$V(x) = \sum_{k=1}^{m} \widetilde{W}(x_k, x_{k+1})$$

and $\widetilde{W}$ has (after possibly the addition of a constant) the following properties:

$$\widetilde{W}(t,s) \geq 0, \tag{6.4}$$

$$\widetilde{W}(t,s) \geq C_1(t^2+s^2) - D_1, \tag{6.5}$$

$$\widetilde{W}(t,s) \leq C_2(t^2+s^2) - D_2, \tag{6.6}$$

for some constants C_1, C_2, D_1, D_2 with $C_1 > 0$. Let us denote by $\mu_1^{(m)}$ or more simply by $\mu(m)$ (in this section) the first eigenvalue of $K^{(m)}$ and $u^{(m)} = u_1^{(m)}$ denotes the strictly positive normalized first eigenfunction. Then we want to prove the:

PROPOSITION 6.1. *Under the assumptions* (6.3)–(6.5) *the sequence* $-\ln\big(\mu(m)\big)/m$ *is convergent to a limit* E_∞ *when* m *tends to* ∞ *moreover there exists a constant* C *such that*:

$$\left| E_\infty - \left(-\frac{\ln\big(\mu(m)\big)}{m} \right) \right| \leq C/m.$$

Remark 6.2 (*Comparison with Schrödinger*). We recall that similar results or stronger results were obtained in the case of the Schrödinger operator in [**19**] and [**36**]. As will be clear in the proof, the correspondance (which can be justified in the semi-classical context) is

$$\lambda_1(m) \leftrightarrow -\ln\big(\mu(m)\big)$$

where $\lambda_1(m)$ is the first eigenvalue of the Schrödinger operator.

MAJORATION, MINORATION. We first remark that, due to the normalization condition (6.5):

$$\mu(m) \leq 1, \tag{6.7}$$

and (6.6) implies:

$$\mu(m) \geq \exp(-C_3 m), \tag{6.8}$$

with $C_3 > 0$. This is obtained by a simple comparison with the quadratic case where everything can be computed explicitely (see Appendix A). It is now convenient to introduce the notation:

$$\lambda(m) = -\ln \mu(m).$$

As in the case of the Schrödinger equation, the following *a priori* estimate is crucial and is the object of the:

LEMMA 6.3. *There exists a constant C such that for all $m \geq 1$, we have for any j:*

$$\|x_j u^{(m)}\|^2 \leq C. \tag{6.9}$$

PROOF. We recall first the classical proposition (cf. for example [**30**]):

PROPOSITION 6.4. *Let A, B bounded selfadjoint operators, with bounded inverse such that: $0 < A \leq B$, then we have: $B^{-1} \leq A^{-1}$ and $\ln A \leq \ln B$.*

We shall apply this theorem with $A = K^{(m)}$ and $B = \exp -V$. These two operators are bounded but with unbounded inverse and we have to use a regularization procedure in order to apply the proposition. We simply consider the family, depending on $\epsilon > 0$:

$$A = K + \epsilon \text{ and } B = \exp -V + \epsilon.$$

We then obtain:

$$\ln(K + \epsilon) \leq \ln(\exp -V + \epsilon),$$

or:

$$(-\ln(\exp -V + \epsilon)\phi \mid \phi) \leq (-\ln(K + \epsilon)\phi \mid \phi)$$

for any ϕ in L^2.

We now apply this last inequality to $u^{(m)}$:

$$(-\ln(\exp -V + \epsilon)u^{(m)} \mid u^{(m)}) \leq -\ln(\mu(m) + \epsilon).$$

As $\epsilon \to 0$, we then obtain by monotone convergence (V and $u^{(m)}$ are positive):

$$0 \leq (Vu^{(m)} \mid u^{(m)}) \leq -\ln \mu(m),$$

and using (6.5):

$$2\sum_j \|x_j u^{(m)}\|^2 \leq \lambda(m) + Cm,$$

for some constant C.

We now observe that the potential is invariant by circular permutation and as usual the first eigenfunction as the same property. We then get (6.9). □

COMPARISON BETWEEN $\lambda(m+p)$ AND $\lambda(m) + \lambda(p)$. As usual, the existence of the thermodynamic limit can be deduced from the property of approximate additivity, that is from the existence of a constant C such that:

$$-C + \lambda(m) + \lambda(p) \leq \lambda(m+p) \leq C + \lambda(m) + \lambda(p), \quad \forall m, p. \tag{6.10}$$

In order to prove this property, we start from:

$$K^{(m+p)} = \exp\Big(-\frac{\delta V_{(m,p)}}{2}\Big) \cdot K^m \cdot \widehat{K}_{(m,p)} \cdot \exp\Big(-\frac{\delta V_{(m,p)}}{2}\Big) \tag{6.11}$$

where

$$\begin{aligned}\delta V_{(m,p)} = \widetilde{W}(x_{m+p}, x_1) + \widetilde{W}(x_m, x_{m+1}) - \widetilde{W}(x_{m+p}, x_{m+1})\\ - \widetilde{W}(x_1, x_m),\end{aligned} \tag{6.12}$$

$$\widehat{K}_{(m,p)} = \exp\Big(-\frac{V_{(m,p)}}{2}\Big) \cdot \exp(-\hat{\Delta}_{(m+1,m+p)}) \cdot \exp\Big(-\frac{V_{(m,p)}}{2}\Big), \tag{6.13}$$

$$V_{(m,p)} = \sum_{k=m+1}^{m+p-1} \widetilde{W}(x_k, x_{k+1}) + \widetilde{W}(x_{m+p}, x_{m+1}) \tag{6.14}$$

and

$$\hat{\Delta}_{(m+1,m+p)} = \sum_{k=m+1}^{m+p} (\partial_{x_k})^2. \tag{6.15}$$

It is clear that the spectrum of $\widehat{K}_{(m+1,m+p)}$ and K_p are the same. On the other hand, the variables occuring in K_p and $\widehat{K}_{(m+1,m+p)}$ are separate. For all ϕ in C_0^∞, we have:

$$(K^{(m+p)}\phi \mid \phi) \leq \mu(p) \cdot \mu(m) \cdot \big(\exp(\widetilde{W}(x_1, x_m) + \widetilde{W}(x_{m+p}, x_{m+1}))\phi \mid \phi\big), \tag{6.16}$$

hence (using Proposition 6.4):

$$\ln \mu(m+p) \leq \ln \mu(p) + \ln \mu(m) + \widetilde{W}(x_1, x_m) + \widetilde{W}(x_{m+p}, x_{m+1}). \tag{6.17}$$

We then apply (6.17) with $\phi = u^{(m+p)}$ and get:

$$\begin{aligned}\ln \mu(m+p) \leq \ln \mu(p) + \ln \mu(m)\\ + \int [\widetilde{W}(x_1, x_m) + \widetilde{W}(x_{m+p}, x_{m+1})] u^{(m+p)}(x)^2 \, dx\end{aligned}$$

and therefore

$$\ln \mu(m+p) \leq \ln \mu(p) + \ln \mu(m) + C \tag{6.18}$$

for some constant C independent of m. We have indeed:

$$\begin{aligned}\int [\widetilde{W}(x_1, x_m) + \widetilde{W}(x_{m+p}, x_{m+1})] u^{(m+p)}(x)^2 \, dx \leq C(\|x_m u^{(m+p)}\|^2\\ + \|x_{m+1} u^{(m+p)}\|^2 + \|x_1 u^{(m+p)}\|^2 + \|x_{m+p} u^{(m+p)}\|^2) + C\end{aligned}$$

and we can apply Lemma 6.3. In order to majorize $\lambda(m+p)$, we proceed formally on the following way by writing:

$$K^{(m)} \cdot \widehat{K}_{(m+1,m+p)} = \exp(\delta V_{(m,p)}/2) \cdot K^{(m+p)} \cdot \exp(\delta V_{(m,p)}/2).$$

We have:

$$\begin{aligned}K^{(m)} \cdot \widehat{K}_{(m+1,m+p)} &\leq \mu(m+p). \exp(\delta V_{(m,p)})\\ &\leq \mu(m+p) \cdot \exp\big(W(x_{m+p}, x_1) + W(x_m, x_{m+1})\big).\end{aligned}$$

Hence:

$$\ln K^{(m)} + \ln \widehat{K}_{(m+1,m+p)} \leq \ln \mu(m+p) + \big(W(x_{m+p}, x_1) + W(x_m, x_{m+1})\big).$$

Using this time, the function:

$$\hat{u}_{(m,p)}(x) = u^{(m)}(x_1, \dots, x_m).u^{(p)}(x_{m+1}, \dots, x_{m+p}),$$

we obtain:

$$\begin{aligned}
&\int \big(W(x_{m+p}, x_1) + W(x_m, x_{m+1})\big) \cdot \hat{u}_{(m,p)}(x)^2 \\
&\quad \leq C(2\|x_m \hat{u}_{(m,p)}\|^2 + 2\|x_{m+1}\hat{u}_{(m,p)}\|^2 + \|x_1 \hat{u}_{(m,p)}\|^2 + \|x_{m+p}\hat{u}_{(m,p)}\|^2) + C \\
&\quad \leq \hat{C}
\end{aligned}$$

(modulo a regularization argument). □

6.2. The indirect approach. We first recall some properties obtained in Section 4. We have proved as a consequence of Lemma 4.3 that:

$$-\ln \mu(m) = \lim_{p\to\infty} \frac{-1}{p} \ln \int K_V^{(p)}(x,x)\, dx^{(m)}. \tag{6.19}$$

We recall also that $\int K_V^{(p)}(x,x)\, dx^{(m)}$ is a Laplace integral associated to the phase (cf. (4.20)) $\Phi^{(p)}$ defined on $\mathbb{R}^{mp} = (\mathbb{R}^m)^p$ by:

$$\int_{\mathbb{R}^{mp}} \exp -\Phi^{(p)}(y_1, \dots, y_p)\, dy_1^{(m)} \dots dy_p^{(m)} \tag{6.20}$$

with:

$$\Phi^{(p)}(y_1, \dots, y_p) = \sum_{j=1}^{p} \Big(V(y_j) + \frac{1}{4}(y_{j+1} - y_j)^2 \Big). \tag{6.21}$$

Here a point x in $\mathbb{R}^{mp}$ can be written on different ways:

$$x = (x_1, \dots, x_m) = (y_1, \dots, y_p) \text{ with } y_j \in \mathbb{R}^m, x_i \in \mathbb{R}^p,$$

where

$$x_{ij} = y_{ji}.$$

And when we use the notation with the x_i, we shall write:

$$\Phi^{(p)}(y_1, \dots, y_p) = \tilde{\Phi}^{(m)}(x_1, \dots, x_m). \tag{6.22}$$

As in the preceding subsection, it is sufficient to prove (cf. (6.10)) the existence of C such that:

$$-C + \lambda(k) + \lambda(\ell) \leq \lambda(k+\ell) \leq C + \lambda(k) + \lambda(\ell), \quad \forall k, \ell, \tag{6.23}$$

where $\lambda(s) = -\ln \mu(s)$. According to (6.19), it is sufficient to prove the existence of C such that:

$$-C + \lambda(k,p) + \lambda(\ell,p) \leq \lambda(k+\ell,p) \leq C + \lambda(k) + \lambda(\ell,p), \quad \forall k, \ell, p, \tag{6.24}$$

where

$$\lambda(m,p) = \frac{-1}{p} \ln \int_{\mathbb{R}^{mp}} \exp(-\Phi^{(p)}(y_1, \dots, y_p))\, dy_1^{(m)} \dots dy_p^{(m)}. \tag{6.25}$$

We prefer to consider this integral in the form:

$$\lambda(m,p) = \frac{-1}{p} \ln \int_{\mathbb{R}^{mp}} \exp\big(-\tilde{\Phi}^{(m)}(x_1,\dots,x_m)\big)\, dx_1^{(p)} \dots dx_m^{(p)}. \tag{6.26}$$

We have then transformed the problem in a problem concerning a Laplace integral where p plays in some sense the role of a parameter. The proof given in [**21**] (Proposition 1.1) is very explicit with respect to the constants and we just follow the proof of this proposition with a more precized control of the constants. There are however two differences.

- We need a vector-valued version of this proposition.
- We are not exactly in the case of a sublinear perturbation of $\big(\sum_{i,j} x_{ij}^2\big)/2$ but with a perturbation of a nondiagonal quadratic form.

We shall prove the following proposition:

PROPOSITION 6.5. *Let us consider the family of potentials introduced in* (6.21)–(6.22) *that we index by* $m \in \mathbb{N}$ $\tilde{\Phi}^{(m)}(x)$ *with* $V^{(m)}(y) = y^2/2 + \Psi^{(m)}$ *where* $y \to \Psi^{(m)}(y)$ *is a* C^∞*-function on* $\mathbb{R}^m$ *satisfying the following properties*:

$$\Psi^{(m)}(0) = 0, \tag{6.27}$$

$$|\nabla \Psi^{(m)}|_{\ell^\infty} \leq C_1, \tag{6.28}$$

$$\big|\nabla\big(\Psi^{(m+n)} - (\Psi^{(m)} \oplus \Psi^{(n)})\big)\big|_{\ell^1} \leq C_2, \tag{6.29}$$

$$\big|\Delta\big(\Psi^{(m+n)} - (\Psi^{(m)} \oplus \Psi^{(n)})\big)\big| \leq C_3. \tag{6.30}$$

Then

$$-D + \lambda(k,p) + \lambda(\ell,p) \leq \lambda(k+\ell,p) \leq D + \lambda(k,p) + \lambda(\ell,p), \quad \forall k,\ell, \tag{6.31}$$

where D *is independent of* p *and explicitely computable from* C_1, C_2, C_3.

PROOF OF THE PROPOSITION. We first observe that:

$$\tilde{\Phi}^{(m)}(x_1,\dots x_m) = \sum_{i,j} x_{ij}^2 - \frac{1}{2}\sum_{i,j} x_{ij}x_{i\,j+1} + \sum_j \Psi^{(m)}(x_{1j},\dots,x_{mj}) \tag{6.32}$$

or

$$\tilde{\Phi}^{(m)}(x_1,\dots,x_m) = \sum_i \theta(x_i) + \tilde{\Psi}^{(m)}(x) \tag{6.33}$$

with

$$\theta(z) = \sum_j z_j^2 - \frac{1}{2}\sum_j z_j z_{j+1} \tag{6.34}$$

and

$$\tilde{\Psi}^{(m)}(x) = \sum_j \Psi^{(m)}(x_{1j},\dots,x_{mj}). \tag{6.35}$$

We now introduce an interpolating family parametrized by $(m,n) \in \mathbb{N}^2$ and $s \in [0,1]$:

$$\tilde{\Phi}^{(m,n)}(x,s) = (1-s)\tilde{\Phi}^{(m+n)}(x) + s\big(\tilde{\Phi}^{(m)}(x') + \tilde{\Phi}^{(n)}(x'')\big) \tag{6.36}$$

with $x = (x', x'') \in \mathbb{R}^{n+m}$.

We then have to study the corresponding:

$$\lambda(m,n,p,s) = \frac{1}{p} \ln \int_{\mathbb{R}^{m+n}} \exp(-\tilde{\Phi}^{(m,n)}(x,s))\, dx$$

and we observe that:

$$\lambda(m,n,p,0) = \lambda(m+n,p); \quad \lambda(m,n,p,1) = \lambda(m,p) + \lambda(n,p).$$

In order to control the difference $\lambda(m+n,p) - \big(\lambda(m,p)+\lambda(n,p)\big)$, we have consequently to estimate uniformly

$$\partial_s \lambda(m,n,p,s) = \frac{-1}{p}\left(\frac{\int_{\mathbb{R}^{m+n}} \left(\partial_s\left(\tilde{\Phi}^{(m,n)}(x,s)\right) \exp\left(-\tilde{\Phi}^{(m,n)}(x,s)\right)\right) dx^{(m+n)}}{\int_{\mathbb{R}^{m+n}} \exp(-\tilde{\Phi}^{(m,n)}(x,s))\, dx^{(m+n)}}\right). \tag{6.37}$$

Using the invariance by permutation we get:

$$\begin{aligned} &- \partial_s \lambda(m,n,p,s) \\ &= \left(\frac{\int_{\mathbb{R}^{m+n}} \left(\partial_s\left(\Phi^{(m,n)}(x_{11},\dots,x_{m1},s)\right) \exp\left(-\tilde{\Phi}^{(m,n)}(x,s)\right)\right) dx^{(m+n)}}{\int_{\mathbb{R}^{m+n}} \exp\left(-\tilde{\Phi}^{(m,n)}(x,s)\right) dx^{(m+n)}}\right). \end{aligned} \tag{6.38}$$

with

$$\partial_s \Phi^{(m,n)}(y,s) = \Psi^{(m+n)} - \big(\Psi^{(m)} \oplus \Psi^{(n)}\big).$$

Using assumption (6.27), we can write

$$\begin{aligned} &\left(\Psi^{(m+n)} - \big(\Psi^{(m)} \oplus \Psi^{(n)}\big)\right)(y) \\ &\qquad = \sum_{\ell=1}^{m+n}\left[y_\ell \int_0^1 \left(\partial_{y_\ell}\left(\Psi^{(m+n)} - \big(\Psi^{(m)} \oplus \Psi^{(n)}\big)\right)(ty)\right) dt\right]. \end{aligned} \tag{6.39}$$

An integration by part gives:

$$-\int_{\mathbb{R}^{m+n}} \big(\partial_s \Phi^{(m,n)}(x_{11},\dots,x_{m1},s)\big)\cdot \exp(-\tilde{\Phi}^{(m,n)}(x,s))\, dx = A+B+C \tag{6.40}$$

with

$$\begin{aligned} A &= -\int_{[0,1]}\int_{\mathbb{R}^{m+n}} \Delta\left(\Psi^{(m+n)} - \big(\Psi^{(m)} \oplus \Psi^{(n)}\big)\right)(tx_{\cdot 1}) \\ &\qquad \times \exp(-\tilde{\Phi}^{(m,n)}(x,s))t\, dx\, dt \\ B &= \int_{[0,1]}\int_{\mathbb{R}^{m+n}} \left(\nabla\left(\Psi^{(m+n)} - \big(\Psi^{(m)} \oplus \tilde{\Psi}^{(n)}\big)\right)(tx_{\cdot 1})\right) \\ &\qquad \times \left((1-s)\nabla\tilde{\Psi}^{(m+n)} + s\nabla\big(\tilde{\Psi}^{(m)} \oplus \tilde{\Psi}^{(n)}\big)\right)(x) \\ &\qquad \times \exp(-\tilde{\Phi}^{(m,n)}(x,s))\, dx\, dt \\ C &= \sum_{\ell=1}^{m+n}\int_{[0,1]}\int_{\mathbb{R}^{m+n}} \frac{1}{2} x_{\ell 2}\left(\partial_{y_\ell}\left(\Psi^{(m+n)} - \big(\Psi^{(m)} \oplus \Psi^{(n)}\big)\right)(tx_{\cdot 1})\right) \\ &\qquad \times \exp(-\tilde{\Phi}^{(m,n)}(x,s))\, dt\, dx, \end{aligned} \tag{6.41}$$

where $x_{\cdot 1} = (x_{11},\dots,x_{m1})$.

A new integration by part for C gives for the last term:

$$C = D + \frac{1}{4}(A + B + C)$$

with

$$D = \int_{[0,1]} \int_{\mathbb{R}^{m+n}} \frac{1}{4}\Big(\nabla\Big(\Psi^{(m+n)} - \big(\Psi^{(m)} \oplus \tilde{\Psi}^{(n)}\big)\Big)(tx_{\cdot 1})\Big)$$
$$\times \Big((1-s)\nabla\tilde{\Psi}^{(m+n)} + s\nabla(\tilde{\Psi}^{(m)} \oplus \tilde{\Psi}^{(n)})(x_{\cdot 2})\Big) \exp\big(-\tilde{\Phi}^{(m,n)}(x,s)\big)\, dx\, dt.$$

We finally get that:

$$-\partial_s \lambda(m,n,p,s) = \frac{4}{3}(A + B + D) \tag{6.42}$$

We eventually verify that by assumption A, B and C are bounded and this proves the proposition. □

6.3. Estimates on the correlations. We now come back to the correlations and we want to estimate the decay as $\operatorname{dist}(i,j)$ in $\mathbb{Z}/m\mathbb{Z}$ is large. In order to simplify the explanations we assume that $\Phi^{(p)}$ is even so that the mean value of $\langle x_i \rangle$ is equal to 0. We want to prove that under the assumptions (2.14)–(2.17) the correlation function:

$$\operatorname{Cor}^{(m)}(i,j) = \int x_i \cdot x_j u_1^{(m)}(x)^2\, dx$$

can be estimate, for some $\delta > 0$ by:

$$|\operatorname{Cor}^{(m)}(i,j)| \leq C \exp -\delta \operatorname{dist}(i,j). \tag{6.43}$$

The strategy is the same as in the preceding subsection and we use this time the property (4.19). It is then sufficient to prove the equivalent of (6.43) for the correlations attached to the measure $\exp\big(-\Phi^{(p)}(x)\big)\, dx$ but uniformly with respect to p. More precisely we have to control:

$$\frac{\int_{\mathbb{R}^{(m+p)}} x_{i1} x_{j1} \exp\big(-\Phi^{(p)}(x)\big) dx}{\int_{\mathbb{R}^{(m+p)}} \exp\big(-\Phi^{(p)}(x)\big)\, dx}.$$

If we solve the basic equation:

$$x_{i1} - \langle x_{i1} \rangle = (-\Delta + \nabla\Phi \cdot \nabla) u_i,$$

an immediate integration by parts gives (this trick was introduced in [**21**] in the same context):

$$\operatorname{Cor}^{(m,p)}(i,j) = \langle \partial_{x_{j1}} u_i \rangle.$$

We are then reduced to a control of $\sup_{x \in \mathbb{R}^{(mp)}} |\partial_{x_{j1}} u_i(x)|$ which is a consequence of the results we give in example 2.4 and of the estimate (2.20).

As a particular case we can for example get the following theorem:

THEOREM 6.6. *Let W be a C^3 function on $\mathbb{R}^2$ such that $\nabla^p W$ is bounded (for $p = 1, 2, 3$). Then if*

$$\Psi^{(m)}(x, \beta) = \beta \sum_{i=1}^{m} W(x_i, x_{i+1})$$

and if β is small enough, then there exists $\kappa_0 > 0$ and C_0 s.t. the correlation

$$\mathrm{Cor}^{(m)}(i,j) = \langle x_i x_j \rangle_m - \langle x_i \rangle_m \langle x_j \rangle_m$$

satisfies

$$|\,\mathrm{Cor}^{(m)}(i,j)| \leq C_0 \exp -\kappa_0 \,\mathrm{dist}_{\mathbb{Z}/m\mathbb{Z}}(i,j) \tag{6.44}$$

for all m and all pairs (i,j).

7. Estimate for the Splitting for the Kac Operator

In some problems in statistical mechanics on a lattice $\mathbb{Z}^2$, the mechanism of reduction to a one dimensional lattice permits to reduce the general questions about correlations or thermodynamic limit to corresponding spectral properties for the Kac operator K_V. Let us first recall that it was proved in [**36**], that in the case of the Schrödinger operator, the assumption that V is strictly convex uniformly in $\mathbb{R}^m$, that is satisfying for some $\sigma > 0$

$$\inf_x \lambda_{\min}\big(\mathrm{Hess}\, V(x)\big) = \sigma > 0, \tag{7.1}$$

permits to get a minoration of the splitting between the second eigenvalue λ_2 and the first eigenvalue λ_1:

$$\lambda_2 - \lambda_1 \geq \sqrt{2\sigma}. \tag{7.2}$$

This condition appears to be optimal in the case of the harmonic oscillator in the sense that we get equality. So it is natural to ask for the same type of question in the case of the Kac operator. Under condition (7.1) (and some conditions on the derivatives), the operator K_V is compact and its spectrum is given by a sequence of eigenvalues tending to 0 μ_j that we order by the condition: $\mu_j \geq \mu_{j+1}$. Moreover μ_1 is simple and so it is again interesting to study the quotient: μ_2/μ_1 and to have a universal estimate of $1 - \mu_2/\mu_1$ in the case when V is strictly convex.

We solved this question (which was to our knowledge open, at least in this general framework) recently in [**17**] but there are of course results of this type in quantum field theory for particular cases (see the book of Glimm and Jaffe [**9**]), as was mentioned to us by T.Spencer in 1991. The following theorem is proved in [**17**]:

THEOREM 7.1. *Let V be a C^∞ potential satisfying* (7.1) *and the condition*

$$|(D_x^\alpha V)(x)| \leq C_\alpha \text{ for } |\alpha| \geq 2,$$

then the quotient (μ_2/μ_1) satisfies:

$$\frac{\mu_2}{\mu_1} \leq \exp\left(-\frac{\cosh^{-1}(1+\sigma)}{2}\right). \tag{7.3}$$

Remark 7.2. This result is not optimal and we think it is possible to improve it in

$$\frac{\mu_2}{\mu_1} \leq \exp\big(-\cosh^{-1}(1+\sigma)\big), \tag{7.4}$$

which is what we hope from the study of the "harmonic" Kac operator (see Appendix A).

Remark 7.3. In the case of the specific problem posed by M. Kac in [**24**], the potential was:

$$\frac{1}{4}\sum_{k=1}^{m}(x_k)^2-\sum_{k=1}^{m}\ln\cosh\left(\sqrt{\frac{\nu}{2}}(x_k+x_{k+1})\right), \tag{7.5}$$

and the assumptions of the theorem are satisfied if

$$\sigma=\frac{1-4\nu}{2}>0 \tag{7.6}$$

In particular, the majoration given by Theorem 7.1 is independent of the dimension m.

We recall that for this Kac model the splitting appears in the computation of the correlation between two lines which is given in this Kac model (see [**24**]) and the introduction). Finally, let us mention that different questions on the splitting are solved, in general in the semiclassical context, in the case of the Schrödinger equation ([**27, 18, 33, 28, 34, 35, 20**]) and in the case of the Kac operator ([**24, 3, 12**]).

Sketch of the proof. We start from the well known property

$$\int(\mu_1)^{-p}(K_V^{(p)})(x,x)dx\to\int u_1(x)^2dx=1 \tag{7.7}$$

as p tend to ∞, where $(K_V^{(p)})(x,y)$ denotes the distribution kernel of $(K_V)^p$.

But an easy computation based on the functional calculus gives

$$-\ln\mu_1+\left(\frac{1}{p}\ln\int(K_V^{(p)})(x,x)\,dx\right)=\left(\frac{k}{p}\right)\left(\frac{\mu_2}{\mu_1}\right)^p\left(1+\mathcal{O}\big(\exp(-\delta p)\big)\right) \tag{7.8}$$

where $\delta>0$ and k is the multiplicity of μ_2.

Consequently a control, **by other means**, of the l.h.s. in (7.8) of the following type:

$$\left|-\ln\mu_1+\left(\frac{1}{p}\ln\int(K_V^{(p)})(x,x)\,dx\right)\right|\le C\exp(-\kappa p)$$

for some C and some $\kappa>0$, will give, by taking $p\to+\infty$ the following majoration:

$$\frac{\mu_2}{\mu_1}\le\exp(-\kappa). \tag{7.9}$$

In the spirit of what has been done in [**36, 19, 21**], we shall obtain a universal estimate of some (possibly) optimal κ, and as a consequence we shall get a universal estimate for the splitting. Let us recall that, in the case of the Schrödinger operator, a universal estimate was obtained for the splitting between the second eigenvalue λ_2 and the first eigenvalue λ_1 in [**35**] (by the maximum principle and a direct computation) and later in [**19**] as a consequence of the Brascamp-Lieb inequalities [**2**] (see formula (7.2)). Here we have to think to the correspondence $\lambda_j=-\ln\mu_j$ (which is asymptotically correct in the semiclassical limit) between the eigenvalues μ_j of the Kac operator and the eigenvalues λ_j of the Schrödinger operator. In order to prove the analogue of (7.2) for the Kac operator, that is formula (7.3), we shall use again (see Section 6) that $\int K_V^{(p)}(x,x)\,dx$ can be written as:

$$\int(K_V^{(p)})(x,x)\,dx=\int\exp\big(-\Phi^{(p)}(y_1,\dots,y_p)\big)\,dy_1\dots dy_p \tag{7.10}$$

with:

$$\Phi^{(p)}(y_1,\ldots,y_p) = \sum_{j=1}^{p}\Big(V(y_j) + \frac{1}{4}(y_{j+1}-y_j)^2\Big), \tag{7.11}$$

where $y_j \in \mathbb{R}^m$ and where we take the convention that $y_{p+1} = y_1$.

The proof given in [**17**] follows the strategy initiated in [**36**] and developed in [**21**]. We shall apply here the theory developed more recently in [**39**] which was presented in Section 2. For each coordinate of $y_i \in \mathbb{R}^m$, we shall use the notation $y_i = (y_{ij}; j = 1,\ldots,m)$. Because we can consider a subsequence for a particular problem we shall assume that p is even which simplifies some notations.

We introduce a t-dependent family:

$$\Phi^{(p)}(y_1,\ldots,y_p;t) = \sum_{j=1}^{p}\Big(V(y_j) + \frac{t}{4}(y_{j+1}-y_j)^2\Big). \tag{7.12}$$

We are interested in the control of the convergence of $\ln\delta^{(p)}(t)/p$ with:

$$\delta^{(p)}(t) = \int \exp\Big(-\Phi^{(p)}(y_1,\ldots,y_p;t)\Big)\, dy^{(p)}. \tag{7.13}$$

We first observe that for $t = 0$,

$$\delta^{(p)}(0) = \Big(\int \exp(-V(y))\, dy\Big)^p.$$

The sequence $\ln\delta^{(p)}(0)/p$ is consequently stationary and we conclude that it is natural to look for the logarithmic derivative with respect to t and to study the convergence of:

$$\begin{aligned}\rho(p,t) &= \langle (y_1-y_2)^2\rangle_\Phi \\ &= \frac{\int (y_1-y_2)^2 \exp\left(-\Phi^{(p)}(y_1,\ldots,y_p;t)\right) dy^{(p)}}{\int \exp\left(-\Phi^{(p)}(y_1,\ldots,y_p;t)\right) dy^{(p)}}\end{aligned} \tag{7.14}$$

as p tends to ∞.

Here we observe that by the structure of the phase $\Phi^{(p)}$ we have:

$$\rho(p,t) = \frac{\int (y_i-y_{i+1})^2 \exp\left(-\Phi^{(p)}(y_1,\ldots,y_p;t)\right) dy^{(p)}}{\int \exp\left(-\Phi^{(p)}(y_1,\ldots,y_p;t)\right) dy^{(p)}}. \tag{7.15}$$

We shall use this property and choose $i = p/2$. In order to study the convergence with respect to p, we introduce as in [**36**] or later in [**21**] a new intyerpolating family depending on p:

$$\Phi^{(p,p)}(y',y'';t,s) = s\Phi^{(2p)}(y',y'';t) + (1-s)\big(\Phi^{(p)}(y';t) + \Phi^{(p)}(y'';t)\big). \tag{7.16}$$

We are interested in a control of the convergence of $\rho(2p,t) - \rho(p,t)$, with $\rho(p,t)$ defined in (7.15). Similarly we introduce now:

$$\rho(p,p,t,s) = \frac{\int (y_i-y_{i+1})^2 \exp(-\Phi^{(p,p)}(y',y'';t,s))\, dy^{(2p)}}{\int \exp\left(-\Phi^{(p,p)}(y',y'';t,s)\right) dy^{(2p)}}. \tag{7.17}$$

We observe that:

$$\rho(2p,t) = \rho(p,p,t,1)$$

and

$$\rho(p,t) = \rho(p,p,t,0).$$

We have consequently to analyze the derivative with respect to s, of the expression above. We observe indeed that if we get a uniform control with respect to (t,s) of $(\partial_s \rho)(p,p,t,s)$ of the type:

$$|(\partial_s \rho)(p,p,t,s)| \leq C \exp(-\kappa p), \tag{7.18}$$

we get also:

$$|\rho(\infty,t) - \rho(p,t)| \leq \widetilde{C} \exp(-\kappa p), \tag{7.19}$$

where $\rho(\infty,t) = \lim_{p\to\infty} \rho(p,t)$ and $\widetilde{C}$ is some constant.

So we have to analyze:

$$(\partial_s \rho)(t,s) = \operatorname{Cor}\big((y_i - y_{i+1})^2, \partial_s \Phi\big) \tag{7.20}$$

where:

$$\partial_s \Phi = 2y_p y_1 + 2y_{2p} y_{p+1} - 2y_p y_{p+1} - 2y_{2p} y_1.$$

We observe that the quantity we have to estimate is a finite (independent of p) sum of correlations of the type $\operatorname{Cor}(f,g)$ with:

$$f = y_{ij} \cdot y_{kj}; \quad g = y_{\ell v} \cdot y_{nv}$$

where $|i-k| \leq 1$, $i = p/2$ and ℓ, n are near 1 or p in $\mathbb{Z}/(2p\mathbb{Z}) = \mathbb{Z}^{(2p)}$. We now look for a vector field $v^{(i,j;p)}$ defined on $E^{2p} \times [0,1]^2$: $(y,t,s) \to v^{(i,j;p)}(y,t,s) \in E^{2p}$ (with $E = \mathbb{R}^m$) solution of:

$$y_{ij} = \langle y_{ij} \rangle_{(t,s)} + v \cdot \nabla_x \Phi - \operatorname{div}_x v. \tag{7.21}$$

Let us compute $\operatorname{Cor}(f^{ikj}, g^{\ell nr})$ with: $f^{ikj} = y_{ij} \cdot y_{kj}$; $g^{\ell nr} = y_{\ell r} \cdot y_{nr}$.

We then write:

$$\begin{aligned}\operatorname{Cor}(f^{ikj}, g^{\ell nr}) = {} & \langle v^{ij}_{kj}(x) \cdot y_{\ell r} \cdot y_{nr} \rangle - \langle v^{ij}_{kj}(x) \rangle \cdot \langle y_{\ell r} \cdot y_{nr} \rangle \\ & + \langle v^{ij}_{\ell r}(x) \cdot y_{kj} \cdot y_{nr} \rangle + \langle v^{ij}_{nr}(x) \cdot y_{kj} \cdot y_{\ell r} \rangle.\end{aligned} \tag{7.22}$$

We shall prove later that there exists κ and C s.t.

$$\sup_x |v^{ij}_{\ell r}(x)| \leq C \exp(-\kappa p) \tag{7.23}$$

$\forall i, j, \ell, r$ such that $i = p/2$, $|i-j| \leq 1$, $\inf(|\ell - p|, |\ell - 1|, |\ell - 2p|) \leq 1$.

We then conclude that $\langle v^{ij}_{\ell r}(x) \cdot y_{kj} \cdot y_{nr} \rangle$ and $\langle v^{ij}_{nr}(x) \cdot y_{kj} \cdot y_{\ell r} \rangle$ have the same majoration using the property that:

$$|\langle v^{ij}_{\ell r}(x) \cdot y_{kj} \cdot y_{nr} \rangle| \leq \sup_x |v^{ij}_{\ell r}(x)| (\langle (y_{kj})^2 \rangle + \langle (y_{nr})^2 \rangle)$$

and we have:

$$\sup_x |v^{ij}_{\ell r}(x)| (\langle (y_{kj})^2 \rangle + \langle (y_{nr})^2 \rangle) \leq \text{const}$$

using either the Brascamp-Lieb inequality [**2**] or again the trick that:

$$\langle (y_{kj})^2 \rangle = \langle v^{kj}_{kj} \rangle.$$

This permits to deduce the control of the two last terms from the control of $\sup_x |v^{kj}_{kj}(x)|$.

Because i and k are neighbors, we can not use (7.23) for the two first terms in the left hand side of (7.22):

$$\langle v^{ij}_{kj}(x) \cdot y_{\ell r} \cdot y_{nr} \rangle - \langle v^{ij}_{kj}(x) \rangle \cdot \langle y_{\ell r} \cdot y_{nr} \rangle \tag{7.24}$$

which is the correlation $\operatorname{Cor}(v^{ij}_{kj}, g^{\ell nr})$.

Let w^{ikj} the solution of

$$\nabla v^{ij}_{kj} = (-\Delta + \nabla\Phi \cdot \nabla) w + (\operatorname{Hess}\Phi) w.$$

We shall have to prove that w^{ikj} has essentially the same properties as ∇v^{ij}_{kj}. On the other hand we have:

$$\operatorname{Cor}(v^{ij}_{kj}, g^{\ell nr}) = \langle w^{ikj}_{\ell r} \cdot y_{nr} \rangle + \langle w^{ikj}_{nr} \cdot y_{\ell r} \rangle. \tag{7.25}$$

If we prove that:

$$\sup_x |w^{ikj}|_{\ell r} \leq C \exp(-\kappa p) \tag{7.26}$$

under the same conditions as in (7.23), we shall have the following control for $\operatorname{Cor}(v^{ij}_{kj}, g^{\ell nr})$:

$$|\operatorname{Cor}(v^{ij}_{kj}, g^{\ell nr})| \leq C \exp(-\kappa p). \tag{7.27}$$

Finally we have proved, that if (7.23) and (7.26) are satisfied, then (7.18) is satisfied and Theorem 7.1 is proved for some κ. It remains to prove (7.23) and (7.26) with the computation of a very explicit κ. This is immediately seen by using the properties obtained in Example 2.5. We observe indeed that v^{ij} is a solution of (7.21) and that ∇y_{ij} is in the weighted space introduced in the presentation of this example for $\rho(\ell) = \exp \kappa \operatorname{dist}(\ell, i)$ where we take the distance in $\mathbb{Z}/p\mathbb{Z}$. If κ satisfies the condition (2.18), we then obtain by (2.20) that v^{ij} is in the same weighted space and we get finally the estimate (7.23). We can proceed similarly for (7.26). Let us finally observe that in order to get (7.9) for some κ_0, it is sufficient to have (7.27) for any $\kappa < \kappa_0$. The precise estimate of the decay will give us:

$$\kappa_0 = \frac{\cosh^{-1}(1+\sigma)}{2}. \quad \square$$

Remark 7.4. Coming back to our motivating example, we observe that we have solved the case when $\nu < 1/4$. The study of the minoration of the splitting and of the majoration of the splitting for $\nu > 1/4$ is still open. Let us recall that the announcement of a majoration of the splitting as $m \to \infty$ by $\exp(-\epsilon m)$ for some $\epsilon > 0$ was given in [**26**] but immediatly followed by the publication of an erratum. This problem is also open in the case of the Schrödinger operator (see however a partial answer in the semiclassical context in [**20**]).

Remark 7.5. We have learned from J. Sjöstrand that T. Spencer has obtained estimates for the splitting for another family of examples.

Acknowledgements. This survey was partially prepared during two stays in the Mittag-Leffler Institute where I have found very good working conditions and where I have stimulating discussions with I. Herbst, A. Jensen and J. P. Solovej. I want also to thank J. Sjöstrand who collaborates with me on this subject and who is at the origin of most of the techniques presented here. Finally it is also a pleasure for me to thank the organizers of the Vancouver Summer School J. Feldman, R. Froese, and L. Rosen for this very interesting meeting.

References

1. J. Avron, I. Herbst, and B. Simon *Schrödinger operators with magnetic fields III. Atoms in homogeneous magnetic fields*, Comm. Math. Phys. **79** (1981), 529–572.
2. H. J. Brascamp and E.Lieb, *On extensions of the Brunn-Minkovski and Prékopa-Leindler Theorems, including inequalities for Log concave functions, and with an application to diffusion equation*, J. Funct. Anal. **22** (1976), 366–389.
3. M. Brunaud and B. Helffer, *Un problème de double puits provenant de la théorie statistico-mécanique des changements de phase (ou relecture d'un cours de M. Kac)*, preprint LMENS, 1991.
4. P. Cartier, *Inégalités de corrélation en mécanique statistique*, Séminaire Bourbaki 25e année, 1972–73, no. 431, Lecture Notes in Math., vol. 383.
5. M.-F. Chen and F. Y.Yang, *On order preservation and positive correlations for multidimensional diffusions processes*, Probab. Theory Related Fields **95** (1993), no. 3, 421–428.
6. R. S. Ellis, *Entropy, large deviations, and statistical mechanics*, Grundlehren Math. Wiss., vol. 271, Springer, New York, 1985.
7. C. Fortuin, P. Kasteleyn, and J. Ginibre, *Correlation inequalities on some partially ordered sets*, Comm. Math. Phys. **22** (1971), 89–103.
8. J. Ginibre, *General formulation of Griffiths inequality*, Comm. Math. Phys. **16**, (1970), 318–328.
9. J. Glimm and A. Jaffe, *Quantum physics (a functional integral point of view)*, Springer Verlag, Second edition, 1987.
10. J. Glimm, A. Jaffe, and T. Spencer, *Phase transitions for Φ_2^4 models*, Comm. Math. Phys. **45** (1975), 203–216.
11. B. Helffer, *Théorie spectrale pour des opérateurs globalement elliptiques, Astérisque* **112**, 1–197.
12. ——, *On Schrödinger equation in large dimension and connected problems in statistical mechanics*, Proc. Conference in Differential Equations with Applications to Mathematical Physics (Atlanta 1992), (W. F. Ames, E. M. Harrell II, and J. V. Herod, eds.), Math. Sci. Engrg., vol. 192, Academic Press, San Diego, p. 153–166
13. ——, *Around a stationary phase theorem in large dimension*, J. Funct. Anal., (to appear).
14. ——, *Problèmes de double puits provenant de la théorie statistico-mécanique des changements de phase* II. *Modèles de Kac avec champ magnétique, étude de modèles près de la température critique*, preprint LMENS, 1992.
15. ——, *Décroissance exponentielle des fonctions propres pour l'opérateur de Kac: le cas de la dimension* > 1, Operator Calculus and Spectral Theory (Symposium on operator calculus and spectral theory, Lambrecht (Germany), 1991) (M. Demuth, B. Gramsch, and B.-W. Schulze, eds.) Oper. Theory Adv. Appl., vol. 57, Birkhäuser, Basel, 1992, p. 99–116.
16. ——, *Estimations sur les fonctions de corrélation pour des modèles du type de Kac*, Exposé au séminaire EDP de l'Ecole Polytechnique, 1993.
17. ——, *Universal estimate of the gap for the Kac operator in the convex case*, Report of the Mittag-Leffler Institut, 1993.
18. B. Helffer and J. Sjöstrand, *Multiple wells in the semi-classical limit* I, Comm. in Partial Differential Equations **9** (1984), no. 4, 337–408; II, Ann. Inst. H. Poincaré **42** (1985), no. 2, 127–212.
19. ——, *Semiclassical expansions of the thermodynamic limit for a Schrödinger equation*, Astérisque **210** (1992), 135–181.
20. ——, *Semiclassical expansions of the thermodynamic limit for a Schrödinger equation* II, Helv. Phys. Acta **65** (1992), 748–765.
21. ——, *On the correlation for Kac like models in the convex case*, Report mo. 9, 1992–93, Institut Mittag-Leffler, 1993. (to appear in J. Statist. Phys).
22. I. Herbst and L. Pitt, *Diffusion techniques in stochastic monotonicity and positive correlations*, Probab. Theory Related fields **87** (1991), 275–312.
23. M. Kac, *Statistical mechanics of some one-dimensional systems*, Studies in Mathematical Analysis and Related topics: Essays in Honor of Georges Polya, Stanford University Press,

Stanford, 1962, p. 165–169.
24. ——, *Mathematical mechanisms of phase transitions*, Brandeis lectures, Gordon and Breach, 1966
25. M. Kac and E. Helfand, *Study of several lattice systems with long-range forces*, J. Math. Phys. **4** (1963) 1078–1088.
26. M. Kac and C. Thompson, *On the mathematical mechanism of phase transition*, Proc. Nat. Acad. Sci. U.S.A. **55** (1966), 676–683; *Erratum*, **56** (1966) 1625.
27. ——, *Phase transition and eigenvalue degeneracy of a one dimensional anharmonic oscillator*, Stud. Appl. Math. **48** (1969), 257–264.
28. W. Kirsch and B. Simon, *Comparison theorems for the gap of Schrödinger operators*, J. Funct. Anal. **75** (1987), 396–410.
29. M. Reed and B. Simon, *Methods of modern mathematical physics*I. *Functional analysis*, Academic Press, 1972.
30. D. Ruelle, *Statistical mechanics*, Math. Physics monograph series, W. A. Benjamin, Inc., 1969.
31. B. Simon, *The $P(\phi)_2$ Euclidean quantum field theory*, Princeton Ser. Phys., 1974.
32. B. Simon, *Functional integration and quantum physics*, Pure Appl. Math., vol. 86, Academic Press, New York, 1979.
33. I. M. Singer, B. Wong, S. T. Yau and S. S. T. Yau, *An estimate of the gap of the first two eigenvalues of the Schrödinger operator*, Ann. Scuola Norm. Sup. Pisa Cl. Sci. (4) **12** (1985), 319–333.
34. J. Sjöstrand, *Potential wells in high dimensions* I, Ann. Inst. H. Poincaré Phys. Théor. **58** (1993), no. 1, 1–41.
35. ——, *Potential wells in high dimensions* II. *More about the one well case*, Ann. Inst. H. Poincaré Phys. Théor. **58** (1993), no. 1, 42–53.
36. ——, *Exponential convergence of the first eigenvalue divided by the dimension, for certain sequences of Schrödinger operator*, Astérisque **210** (1992), 303–326.
37. ——, *Schrödinger in high dimensions, asymptotic constructions and estimates*, Proc. French-Japanese Symposium on Algebraic Analysis and Singular Perturbations (CIRM, Luminy 1991), (to appear).
38. ——, *Evolution equations in a large number of variables*, Math. Nachr. (to appear).
39. —— *On ferromagnetic Laplace integrals and maximum principles*, 1993 (preliminary version)
40. A. D. Sokal, *Mean field bounds and correlation inequalities*, J. Statist. Phys. **28** (1982), 431–439.
41. C. Thompson, *Mathematical statistical mechanics*, Macmillan, New York, 1972.

Appendix A. Explicit Computations for the Harmonic Kac Operator

We want to compare:

$$\exp\left(\frac{-x^2}{2}\right)\cdot\exp\left(t\left(\frac{d}{dx}\right)^2\right)\cdot\exp\left(\frac{-x^2}{2}\right), \tag{A.1}$$

with

$$\exp -\left(-t\left(\frac{d}{dx}\right)^2 + x^2\right) \text{ for } t > 0. \tag{A.2}$$

We have by explicit computations (see [**41**] or [**17**]):

$$\begin{aligned}\exp\left(\frac{-x^2}{2}\right)\cdot\exp\left(t\left(\frac{d}{dx}\right)^2\right)\cdot\exp\left(\frac{-x^2}{2}\right)\\ = \exp\left[-\frac{\ln(z+\sqrt{z^2-1})}{4K\sqrt{z^2-1}}\cdot\left[-\frac{d^2}{d^2}+4K^2(z^2-1)x^2\right]\right]\end{aligned} \tag{A.3}$$

with: $K = 1/4t$ and $z = (2t+1)$.

We then obtain the explicit computation of the eigenvalues of

$$\exp\left(\frac{-x^2}{2}\right)\cdot\exp\left(t\left(\frac{d}{dx}\right)^2\right)\cdot\exp\left(\frac{-x^2}{2}\right)$$

as: $\exp\big(-(n-(1/2))\ln(z+\sqrt{z^2-1})\big)$ $(n\geq 1,\, n\in\mathbb{N})$.

In order to come back to our notation above we compute the formula for the operator $\exp(-\sigma x^2/4)\cdot\exp\big((d/dx)^2\big)\cdot\exp(-\sigma x^2/4)$ and we get for the eigenvalues μ_n:

$$\mu_n = \exp\left(-\left(\frac{2n-1}{2}\right)\cosh^{-1}(\sigma+1)\right), \tag{A.4}$$

and the splitting is given by:

$$\frac{\mu_2}{\mu_1} = \exp\big(-\cosh^{-1}(1+\sigma)\big). \tag{A.5}$$

So we see that we have lost in the proof sketched above a factor $1/2$ inside the exponential.

DÉPARTEMENT DE MATHÉMATIQUES ET INFORMATIQUE, ÉCOLE NORMALE SUPÉRIEURE, 45 RUE D'ULM, 75230 PARIS CEDEX, FRANCE

E-mail address: helffer@dmi.ens.fr

Centre de Recherches Mathématiques
CRM Proceedings and Lecture Notes
Volume 8, 1995

Localization Properties of Continuous Disordered Systems in d-dimensions

J.-M. Combes and P. D. Hislop

1. Introduction: Models and Results

The purpose of this article is to outline several recent results on localization for continuous random Schrödinger operators and the wave equation in random media in $d \geq 1$ dimensions ([**3**], [**5**], [**2**], [**4**]). Whereas almost sure (a.s.) spectral properties of lattice systems have been extensively studied (see the books [**1**] and [**11**] and references therein), not much is known about continuous systems in greater than one dimension (see, however, [**14**], [**17**] and [**19**]). We will present theorems on the a.s. existence of an interval of pure point spectrum for certain families of random additive and multiplicative perturbations of the Laplacian on $L^2(\mathbb{R}^n)$. We also obtain estimates on the a.s. decay of the corresponding eigenfunctions and the Lipschitz continuity of the integrated density of states for some cases. Some of the stochastic processes which we treat do not have lattice analogues, and they include the model of [**14**] as a special case.

By additive perturbations of the Laplacian, we mean random Schrödinger operators of the form

$$H_\omega = -\Delta + V_\omega \text{ on } L^2(\mathbb{R}^d), \tag{1.1}$$

where $V_\omega(x)$ is a stochastic process indexed by $\mathbb{R}^d$. Multiplicative perturbations have the form

$$H_\omega = -C_\omega^{-1} \Delta C_\omega^{-1} \text{ on } L^2(\mathbb{R}^d), \tag{1.2}$$

where $C_\omega = 1 + v_\omega$, for some non-negative stochastic process v_ω on $\mathbb{R}^d$. Such operators (1.2) arise in the study of acoustic and electromagnetic waves in random media. They were studied by Faris [**9**] in the lattice case (see also [**10**]). Whereas we are a long way from treating general processes V_ω and v_ω, we consider some families of stationary and non-stationary processes constructed from two basic processes which we call Anderson-like and Poisson-like. From these, we study stationary and non-stationary models for V_ω whereas we will consider only stationary v_ω here. In

---

1991 *Mathematics Subject Classification.* Primary: 82B44; Secondary: 81Q10, 35J10.

Lecture delivered by the second author at the Mathematical Quantum Theory conference at the University of British Columbia, Vancouver, Canada, 4-14 August 1993.

The research of the first author was supported by supported in part by CNRS.

The research of the second author was supported in part by NSF grants DMS93-07438 and INT 90-15895.

This is the final form of the paper.

the continuum, the potentials have some correlation, but we can also impose some strong correlation conditions in the Anderson-type processes.

The processes which we treat are determined by families of random variables on $\mathbb{R}^d$ and single-site distribution functions $u_i \geq 0$, $i \in \mathbb{Z}^d$ (some of our results apply to the case $u_i \leq 0$ but we restrict this discussion to the positive case for convenience, see also Klopp [**16**]). We take $\{\lambda_i(\omega)\}$ to be a family, in general, of correlated random variables on $\mathbb{Z}^d$ with probability space $\{\Omega, \mathbb{P}\}$. When the λ_i are i.i.d. (independent, identically distributed), $\{\Omega, \mathbb{P}\}$ has the familiar product structure. The conditional distribution of λ_i is assumed to be absolutely continuous. The Radon-Nikodym derivative $g_i(\lambda)$ is a bounded function supported on the interval $[0, \kappa_i]$. We assume $\sup_i \kappa_i \equiv M < \infty$ (except in the Gaussian case, see below). Next, we denote by $\{\xi_i(\omega')\}$ a family of vector-valued random variables with probability space $\{\Omega', \mathbb{P}'\}$ such that $\xi_i \in \Lambda_{R_i}(i)$, where $\Lambda_R(x)$ is a cube of side R centered at x. We require $R_i < 1/2$. Finally, we take $\{X_i(\omega')\}$ to be the points in $\mathbb{R}^d$ distributed according to the Poisson probability law with density ρ. For any Borel subset $S \subset \mathbb{R}$, this law is

$$\mathbb{P}_P\{\exists k \text{ points in } S\} = \frac{\left(|S|\rho\right)^k}{k!} e^{-|S|\rho},$$

where $|S|$ is the Lebesgue measure of S. The probability space for this Poisson process $\{\Omega_P, \mathbb{P}_P\}$ is also denoted by $\{\Omega', \mathbb{P}'\}$ to simplify notation.

The single-site distribution functions u_i must satisfy contain growth conditions for the Anderson-type and Poisson-type processes. In the Anderson case, we have

(u_A) $u_i \geq 0$ and $\exists c_0$, $c_1 > 0$, uniform in i, s.t. for some $\beta > 9d$,

$$c_0 \chi_{\Lambda_{3/2}(i)}(x) \leq u_i(x) \leq c_1 \left(1 + \|x - i\|\right)^{-\beta},$$

where $\chi_{\Lambda_l(x)}$ is the characteristic function of the cube $\Lambda_l(x)$.

In the Poisson case, u_i cannot vanish on any open set and must satisfy

(u_P) $u_i > 0$ and $\exists c_0$, $c_1 > 0$, uniform in i, and α, β, $9d \leq \beta < \alpha < \infty$ s.t.

$$c_0 \left(1 + \|x - i\|\right)^{-\alpha} \leq u_i(x) \leq c_1 \left(1 + \|x - i\|\right)^{-\beta}.$$

We can now give the form of our random potentials V_ω and processes v_ω. The Anderson-type potential is defined by

$$V^A_{\omega,\omega'}(x) \equiv \sum_{i \in \mathbb{Z}^d} \lambda_i(\omega) u_i\left(x - \xi_i(\omega') - i\right), \tag{1.3}$$

on the product probability space $\{\Omega \times \Omega', \mathbb{P} \times \mathbb{P}'\}$. In the case that the λ_i are i.i.d., $\xi_i = 0$ and $u = u_i$, this is a continuous version of the Anderson model (we can replace 3/2 by 1 in (u_A) in this case). The Poisson-type process is defined for λ_i i.i.d. and $u = u_i$ by

$$V^P_{\omega,\omega'}(x) \equiv \sum_{i \in \mathbb{Z}^d} \lambda_i(\omega) u\left(x - X_i(\omega')\right), \tag{1.4}$$

on the product space as above. We can take $v^X_{\omega,\omega'}$ in the multiplicative perturbation models to have the form of the right side of (1.3) for $X = A$ or (1.4) for $X = P$. The process V^P is ergodic but V^A is ergodic only if $\{\lambda_i\}$ and $\{\xi_i\}$ are i.i.d. and $u = u_i$. We will write $H^X_{\omega,\omega'}$, $X = A$ or P, for additive perturbations with potentials V^X

and similarly, for multiplicative perturbations with factors v^X. These operators are self-adjoint on $H^2(\mathbb{R}^d)$ (see the papers for details).

We will now give our main results in several theorems. For the stationary case treated in [**3**], we have for additive perturbations

THEOREM 1. *Let the density $\rho > 0$ be given and assume (u_X) , $X = A$ or P. Consider $H^X_{\omega,\omega'}$ as in (1.1). Then for $\delta_0 \equiv \|g\|^{-1}_\infty$ large enough, $\exists E_X \equiv E_X(\delta_0, u, \rho) > 0$ s.t. $\sigma(H^X_{\omega,\omega'}) \cap [0, E_X]$ is pure point $\mathbb{P} \times \mathbb{P}'$ almost surely. Furthermore, for $X = A$ and $\operatorname{supp} u$ compact, the corresponding eigenfunctions decay exponentially and the integrated density of states is Lipschitz continuous. For $X = A$ with $\operatorname{supp} u$ non-compact or for $X = P$, the corresponding eigenfunctions are bounded above by $c\|x\|^{-(\beta-d)/2}$, $\beta > 9d$ as in (u_X).*

As for multiplicative perturbations (1.2) with $C^X_{\omega,\omega'} = 1 + v^X_{\omega,\omega'}$, we have localization at high frequency. In [**5**], we prove

THEOREM 2. *Consider (1.2). Let the density $\rho > 0$ be given and assume u satisfies (u_X), $X = A$ or P. For δ_0 large enough, $\exists E_X \equiv E_X(u, \delta_0, \rho, d) \gg 0$ such that for any $E_X \leq E_0 < E_1 < \infty$ we have $\sigma(H^X_{\omega,\omega'}) \cap [E_0, E_1]$ is pure point with probability one, $X = A$ or P, provided δ_0 is large enough. The corresponding eigenfunctions decay a.s. as in Theorem 1.*

We plan to continue our investigation of random media and study the integrated density of states for multiplicative perturbations. We also intend to study the propagation of electromagnetic waves in random media. In this light, Theorem 2 is a first step.

For the model process $V^A_{\omega,\omega'}$, we can consider correlations between the random variables λ_i. As in [**8**], we can impose conditions on the conditional probability distribution of the λ_i (called in [**8**] "Wegner potentials" and stated above) and conditions on the correlations which replace independence. This condition can be formulated as follows:

(C) Let E be an event depending only on the random variables in $\Lambda_l(0)$ for l sufficiently large. Then for some $k \in \{2, 3, 4, \dots\}$ and some $\alpha > 1$, $\exists\, c > 0$, $\rho > 1$ (depending on k) such that if E_i represents the translation of event E to site $i \in \mathbb{Z}^d$, then for any set $\{i_1, \dots, i_k\} \subset \mathbb{Z}^d$ such that $|i_j - i_k| > 1/2l^\alpha$,

$$\mathbb{P}\left(\bigcap_{j=1}^{k} E_{i_j}\right) \leq c\mathbb{P}(E)^\rho.$$

We will also take the single-site potentials u_i as in (u_A) to have compact support. As the process $\{\lambda_i\}$ may now be unbounded, one checks using a Borel-Cantelli argument that $H^A_{\omega,\omega'}$ is bounded below by $-c\|x\|^2$ a.s., so it is a self-adjoint family. Although we can treat some non-stationary processes, we take $u_i = u$ here for simplicity. We have [**2**]

THEOREM 3. *Let $H^A_{\omega,\omega'}(\lambda) \equiv -\Delta + \lambda V^A_{\omega,\omega'}$ be an ergodic family of Schrödinger operators with $V^A_{\omega,\omega'}$ as in (1.3) and satisfying condition (C) and the other conditions discussed above. Then $\exists \lambda_0 > 0$ such that for $\lambda > \lambda_0 \exists$ an increasing function $E(\lambda)$ of λ s.t.*

$$\sigma\big(H^A_{\omega,\omega'}(\lambda)\big) \cap \big[-E(\lambda), E(\lambda)\big]$$

is pure point with probability one. The corresponding eigenfunctions are a.s. bounded by $c\|x\|^{-\sigma}$, *for some* $\sigma > 0$.

We can take, as examples, the coupling constants $\{\lambda_i\}$ to be a Gaussian process as in [**8**]. We refer to [**2**] for the details and description of other correlated processes.

Finally, we discuss some models with growing or decaying randomness. These models were inspired by the random lattice Hamiltonians studied in [**6**] and [**13**]. These are non-stationary processes obtained from $V^A_{\omega,\omega'}$ as follows. For $\alpha \in \mathbb{R}$, define

$$V^{NS}_{\omega,\omega'}(x) \equiv \sum_{i\in\mathbb{Z}^d} \lambda_i(\omega)u_i\big(x - i - \xi_i(\omega')\big)\big(1 + \|i\|\big)^{\alpha}, \tag{1.5}$$

where $\{\lambda_i\}$ are independent and distributed as described above, $\{\xi_i\}$ are as above, and u_i satisfies (u_A). (The factor $\big(1 + \|i\|\big)^{\alpha}$ can be replaced by more general increasing or decreasing functions of x). As stated below, for $0 < \alpha < d$, these models have only pure point spectrum a.s. and $\inf \sigma_{\text{ess}}(H^{NS}_{\omega,\omega'})$ depends on α in a stepwise fashion at least when $u = u_i$ has compact support. For $\alpha < -1$, on the otherhand, the spectrum is purely absolutely continuous. The a.s. spectral type for $-1 < \alpha < 0$ is presently unknown for $d \geq 2$, see [**6**] and [**18**] for the interesting spectral behavior which occurs for these long-range cases in one-dimension. The main result of [**4**] for the simple case of $\xi_i = 0$, $u_i = u$, and $\{\lambda_i\}$ i.i.d. and uniformly distributed on $[0, 1]$, is

THEOREM 4. *Let* $H^{NS}_{\omega} \equiv -\Delta + V^{NS}_{\omega}$, *with* V^{NS}_{ω} *given in* (1.5) *with the conditions above. If* $\alpha > d$, $\sigma(H^{NS}_{\omega})$ *is discrete a.s. Let* k *be such that* $d/(k+1) < \alpha \leq d/k$. *Then* $\exists a_k \geq 0$ *s.t.* $\sigma_{\text{ess}}(H^{NS}_{\omega}) = [a_k, \infty) = \sigma_{pp}(H^{NS}_{\omega})$ *a.s., with exponentially decaying eigenfunctions. Furthermore, the number of eigenvalues of* H^{NS}_{ω} *in* $[0, a_k]$ *is finite a.s. If* $u_i = \chi_{\Lambda_1(0)}$ $\forall i \in \mathbb{Z}^d$, *then* $a_k = \inf_{\Lambda_k} \sigma(-\Delta^D_{\Lambda_k})$, *where* Λ_k *is a connected region consisting of* k *unit cubes and* $-\Delta^D_{\Lambda_k}$ *is the Dirichlet Laplacian on* Λ_k.

In the following sections we will develop the main tools used in the proofs of Theorems 1–4. However, we will concentrate on the models of Theorems 1 and 2 for simplicity. In Section 2 we present an extension of the arguments of Simon-Wolff [**20**] and of Howland [**15**] necessary for the continuous models. Section 3 is devoted to Kotani's trick and Wegner's lemma for these models. Finally in Section 4 we present a brief discussion of resolvent estimates and the multi-scale analysis used in the proof of localization.

2. Perturbation of Singular Spectrum

Simon and Wolff [**20**] realized that the Aronszajn-Donoghue theory of the instability of the singular spectrum under rank one perturbations could be used effectively to control the continuous spectrum of the d-dimensional lattice Anderson model. This leads to a proof of localization when combined with the estimates of [**12**]. These ideas were further explored by Howland [**15**] who identified a class of perturbations, called H-finite by him, which preserves the pure point spectrum of a self-adjoint operator H_0. We present an extension of these results to a general class of relatively compact, positive perturbations. This result, Theorem 5, along with the spectral averaging method of Kotani, Corollary 8 of the next section, will allow us to control the continuous spectrum of the models introduced in Section 1.

For the case of additive perturbations, we consider a self-adjoint operator H_0 on a separable Hilbert space $\mathcal{H}$ with domain $D(H_0)$. Let V be a bounded self-adjoint operator admitting a factorization $V = CD^*$. We assume for $\lambda \in \mathbb{R}$, the additive perturbation

$$H_\lambda = H_0 + \lambda V, \tag{2.1}$$

is self-adjoint on $D(H_0)$. Let $R_\lambda(z) \equiv (H_\lambda - z)^{-1}$, whenever it exists. Consider the following assumptions on H_0 relative to an interval $I \subset \mathbb{R}$:

(A1) $\Gamma(z) \equiv D^* R_0(z) C$ is compact $\forall z \in \mathbb{C}$, $\operatorname{Im} z \neq 0$;
(A2) $\exists I_0 \subset I$, $|I_0| = |I|$ such that $\forall E \in I_0$,

$$\sup_{\varepsilon \neq 0} \| R_0(E + i\varepsilon) X \| < \infty,$$

where $X = C$ and D.

Let $\widetilde{\mathcal{H}} \equiv \left[f(H_\lambda) C\phi, f \in L^\infty(\mathbb{R}), \phi \in \widetilde{\mathcal{H}} \right]^{\text{cl}}$, where *cl* denotes the closure, and let $\widetilde{H}_\lambda$ denote the restriction of H_λ to $\widetilde{\mathcal{H}}$.

THEOREM 5. *Assume* (A1)-(A2); *then*

(1) $\sigma_{\text{ac}}(\widetilde{H}_\lambda) \cap I = \phi$ *and* $\sigma(\widetilde{H}_\lambda) \cap I_0$ *is pp* $\forall \lambda \in \mathbb{R} \setminus \{0\}$;
(2) *If, in addition,* $C = \pm D$, *then* $\sigma_{\text{sc}}(H_\lambda) \cap I = \phi$ *a.e.* $\lambda \in \mathbb{R}$ *with respect to Lebesgue measure; for such* λ, $\widetilde{H}_\lambda$ *is pp in* I *with finitely degenerate eigenvalues.*

Before indicating some of the ideas of the proof, we give the corresponding result for multiplicatively perturbed operators. Let $C_\lambda \equiv 1 + \lambda\eta$ be a bounded, positive, invertible operator for $\lambda \in I_{M_0} \equiv [-M_0, \infty)$ and consider the self-adjoint family H_λ of multiplicatively perturbed operators defined by

$$H_\lambda = C_\lambda^{-1} H_0 C_\lambda^{-1}. \tag{2.2}$$

We assume H_λ is self-adjoint on $D(H_0)$. Let $B_\lambda^2 \equiv 1 - C_\lambda^{-2}$, so B_λ is the positive square root of this positive operator. Let $\mathbb{R}^+$ be an interval disjoint from zero. Analogous to (A1)-(A2), we have two assumptions on H_0 relative to I:

(M1) $R_0(z) B_\lambda$ is compact $\forall z \in \mathbb{C}$, $\operatorname{Im} z \neq 0$, $\lambda \neq 0$
(M2) $\exists I_0 \subset I$, $|I_0| = |I|$, such that $\forall E \in I_0$ and $\lambda \neq 0$,

$$\sup_{\varepsilon \neq 0} \| R_0(E + i\varepsilon) B_\lambda \| \leq \infty.$$

Let $\widetilde{\mathcal{H}}$ and $\widetilde{H}_\lambda$ be defined as above. Note that $C_\lambda \widetilde{\mathcal{H}} = \widetilde{\mathcal{H}}$ and that B_λ and C_λ commute.

THEOREM 6. *Assume* (M1)-(M2). *Then* $\sigma_{\text{ac}}(\widetilde{H}_\lambda) \cap I = \phi \forall \lambda \in I_{M_0} \setminus \{0\}$. *Moreover,* $\sigma(\widetilde{H}_\lambda) \cap I \subset \bar{I}_0$ *and is pure point for Lebesgue a.e.* $\lambda \in I_{M_0}$.

To simplify notation, let H_λ^X, $X = A$ or M, denote the additive perturbation (2.1) for $X = A$ or the multiplicative perturbation (2.2) for $X = M$.

SKETCH OF THE PROOF OF THEOREMS 5 AND 6. (1) We first derive the analog of the Aronszajn formula for H_λ^X, $X = A$ and M. This relates boundary information about $R_0(E + i\varepsilon)$ to $R_\lambda(E + i\varepsilon)$, $E \in I_0$. In the A-case, let

$F_\lambda^A(z) \equiv D^* R_\lambda(z) C$ and in the M-case, let $F_\lambda^M(z) \equiv C_\lambda R_0(z) C_\lambda$. Also, in the M-case, note that

$$R_\lambda(z) = C_\lambda (H_0 - C_\lambda^2 z)^{-1} C_\lambda. \tag{2.3}$$

In the A-case, a simple calculation based on the second resolvent formula for $R_0(z)$ and $R_\lambda(z)$ leads to

$$F_\lambda^A(z)\big(1 + \lambda F_0^A(z)\big) = F_0^A(z), \tag{2.4}$$

for $\operatorname{Im} z \neq 0$. For multiplicative perturbations, we begin with (2.3) and write

$$R_\lambda(z) = C_\lambda \{(H_0 - C_\lambda^2 z)^{-1} - (H_0 - z)^{-1}\} C_\lambda + C_\lambda R_0(z) C_\lambda,$$

which, after multiplication on the right by B and some re-arranging, leads to

$$F_0^M(z) B = R_\lambda(z) B \big(1 - z B F_0^M(z) B\big). \tag{2.5}$$

In both (2.4) and (2.5), we have a factor of the form $1 + K^X(z)$ where $K^A(z) \equiv F_0^A(z)$ and $K^M(z) = -zBF_0^M(z)B$. By assumptions (A1) and (M1), respectively, these are compact. Furthermore, for $z = E + i\varepsilon$, $E \in I_0$ and $\varepsilon > 0$, we find that (A2) and (M2) imply that

$$\operatorname*{n-lim}_{\varepsilon \to 0} K^X(E + i\varepsilon) \equiv K^X(E),$$

exists and is compact. Hence, we can apply the Fredholm theorem to $1 + K^X(E)$ and conclude from (2.4) and (2.5) that $\text{n-lim}_{\varepsilon \to 0} F_\lambda^A(E + i\varepsilon)$ and $\text{n-lim}_{\varepsilon \to 0} R_\lambda(E + i\varepsilon)B$ exist and are finite for $E \in I_0$ unless $K^X(E)$ has an eigenvalue -1 with eigenfunction ξ. This immediately implies $\sigma_{\text{ac}}(\widetilde{H}_\lambda^X) \cap I = \phi$ since $|I \setminus I_0| = 0$. As for the possibility that $-1 \in \sigma_d\big(K^X(E)\big)$ for some $E \in I_0$, a calculation similar to the usual one appearing in the Birman-Schwinger argument shows that this implies that E is an eigenvalue of $\widetilde{H}_\lambda^X$. The corresponding eigenfunction ψ_E^X has the form $\psi_E^A = \text{s-lim}_{\varepsilon \to 0} R_0(E + i\varepsilon) C\xi$ and $\psi_E^M = \text{s-lim}_{\varepsilon \to 0} R_0(E + i\varepsilon) B C_\lambda \xi$. The de la Vallée Poisson Theorem then implies that the support of the sc spectral measure of $\widetilde{H}_\lambda^X$ lies in $I \setminus I_0$. These results hold for all $\lambda \in \mathbb{R} \setminus \{0\}$ when $X = A$, and for all $\lambda \in I_{M_0} \setminus \{0\}$, when $X = M$.

(2) To show that, in fact, $\sigma_{\text{sc}} \cap I = \phi$, a.e. λ, we need a result discussed in Section 3 on spectral averaging. Let E_λ be the spectral family for H_λ^X. For $X = A$ and $C = \pm D$, we show in Corollary 8 that

$$\int_{\mathbb{R}} \frac{d\lambda}{1 + \lambda^2} C E_\lambda(I \setminus I_0) C = 0,$$

from which it follows that $\sigma(\widetilde{H}_\lambda) \cap I$ is *pp* a.e. λ. In the multiplicative case, we show that for any $0 < M_1 < \infty$,

$$\int_0^{M_1} (\lambda - M_1)^2 C_\lambda E_\lambda(I \setminus I_0) C_\lambda \, d\lambda = 0,$$

so again we conclude that $\sigma(\widetilde{H}_\lambda) \cap I$ is *pp*, a.e. λ. This completes the sketch of the proof of Theorems 5 and 6. $\square$

We now indicate how to apply Theorems 5 and 6 to our model Hamiltonians $H^X_{\omega,\omega'}$. We fix $\omega' \in \Omega'$ and consider variations of $\omega \to \widetilde{\omega}$ for which only λ_0 changes. In the A-case, we obtain

$$H^X_{\tilde{\omega}} = H^X_\omega + \lambda V^X$$

where $\lambda \equiv \lambda_0(\tilde{\omega}) - \lambda_0(\omega)$ and

$$V^X(x) = \begin{cases} u(x - \xi_0(\omega')), & X = A \\ u(x - X_0(\omega')), & X = P. \end{cases} \tag{2.6}$$

If we write $H_\lambda \equiv H^X_{\tilde{\omega}}$, we obtain (2.1) if we take $C = D = (V^X)^{1/2}$ as in (2.6). In the M-case, we write (dropping the X superscript),

$$H^X_{\tilde{\omega}} = (1 + v_\omega)(1 + v_{\tilde{\omega}})^{-1} H_\omega (1 + v_{\tilde{\omega}})^{-1}(1 + v_\omega).$$

Letting $C_\lambda \equiv (1 + v_{\tilde{\omega}})(1 + v_\omega)^{-1}$, a simple calculation shows that

$$C_\lambda \equiv 1 + \lambda u (1 + v_\omega)^{-1},$$

where u stands for the right side of (2.6) and $\lambda \in [-M_0, M_0]$. Then we obtain (2.2) with $\eta \equiv u(1 + v_\omega)^{-1}$. Assuming (A2) and (M2) (it is easily seen that (A1) and (M1) are verified), we obtain the result that for a.e. λ, $\sigma(H^X_\lambda) \cap I$ is pure point. A probabilistic argument, as given in [**3**], then allows us to conclude that for a.e. (ω, ω'), $\sigma(H^X_{\omega,\omega'}) \cap I$ is pure point. The intervals $I = [0, E_X]$ and $I = [E_0, E_1]$ are established by resolvent estimates as outlined in Section 4.

3. Spectral Averaging: Wegner's Lemma & Kotani's Trick

The important fact that averages of the spectral measures E_λ of self-adjoint operators H_λ (as considered in Section 2) have very regular properties seems to have been discovered by Kotani. We have used this fact in the proof of Theorems 5 and 6 to eliminate the singular continuous spectrum for a.e. λ. Equally important in the theory of random operators are probabilistic estimates on the spectrum of local Hamiltonians which are usually called Wegner-type lemmas. These estimates are used in the multi-scale analysis to control the occurence of singular blocks. Both of these results can be derived from certain a priori estimates on the resolvent which we shall now present.

For the additive case, we consider a family $H_\lambda = H_0 + \lambda V$ of self-adjoint operators on $D(H_0)$. The important assumption is that there exists a constant $c_0 > 0$ and a bounded, self-adjoint operator B s.t.

$$0 \le c_0 B^2 \le V. \tag{3.1}$$

The assumption in the multiplicative case is that $H_\lambda = C_\lambda^{-1} H_0 C_\lambda^{-1}$, $\lambda \in J \equiv [-M_0, M_1]$, M_0, $M_1 < \infty$, and $C_\lambda = 1 + \lambda\eta \ge 1$. We assume also, as in (3.1), that there exists a constant $c_0 > 0$ and a bounded, self-adjoint operator χ s.t.

$$0 \le c_0 \chi^2 \le \eta. \tag{3.2}$$

The key technical lemma is

LEMMA 7. (1) Additive case. *Assume* (3.1). *For* $\operatorname{Im} z < 0$, $\operatorname{Im}\lambda \geq 0$, *define*

$$K(\lambda, z) \equiv B(H_\lambda - z)^{-1}B,$$

and for $t > 0$, $\varepsilon \geq 0$,

$$F_t(\varepsilon, z) = \int_{\mathbb{R}} \frac{d\lambda}{1+\lambda^2 t} K(\lambda + i\varepsilon, z).$$

Then for c_0 *as in* (3.1),

$$\|F_t(\varepsilon, z)\| \leq \pi c_0^{-1}. \tag{3.3}$$

(2) Multiplicative case. *Assume* (3.2). *For* $\operatorname{Im} z > 0$, $\operatorname{Im}\lambda \geq 0$ *define*

$$K(\lambda, z) \equiv \chi(H_\lambda - z)^{-1}\chi,$$

and for $\varepsilon \geq 0$,

$$F(\varepsilon, z) \equiv \int_{-M_0}^{M_1} (\lambda - M_1)^2(\lambda + M_0)^2 K(\lambda + i\varepsilon, z)\, d\lambda.$$

Then, for any $0 < \alpha < \infty$, *for all* ε *sufficiently small,* $\delta \geq 0$, $E > 0$ *and* c_0 *as in* (3.2),

$$\|F(\varepsilon, E + i\delta)\| \leq \frac{c(M_0, \alpha)}{c_0 E}, \tag{3.4}$$

where $c(M_0, \alpha)$ *is a positive constant independent of* (E, δ, ε).

Before discussing the proof of this lemma, we present a corollary which is a version of Kotani's trick and which was used in the proof of Theorems 5 and 6. We assume that (3.1) and (3.2) hold.

COROLLARY 8. (1) Additive case. *For any Lebesgue measurable subset* $L \subset \mathbb{R}$ *and for any* $g \geq 0$, $g \in L^\infty(\mathbb{R})$ *with compact support,*

$$\left\| \int_{\mathbb{R}} g(\lambda) B E_\lambda(L) B \, d\lambda \right\| \leq c_0^{-1} \|g\|_\infty |L|,$$

where c_0 *is as in* (3.1).

(2) Multiplicative case. *For any Lebesgue measurable subset* $L \subset \mathbb{R}^+$ *with* $E_0 = \inf\{x \in L\} > 0$ *and for any* $g \geq 0$, $g \in L^\infty(\mathbb{R})$ *with* $\operatorname{supp} g \subset [0, M_1]$,

$$\left\| \int_0^{M_1} g(\lambda)\chi E_\lambda(L)\chi \, d\lambda \right\| \leq c_0^{-1} E_0^{-1} c(\alpha, M_1) \|g\|_\infty |L|,$$

where $c(\alpha, M_1)$ *is as in* (3.4).

The proof of this corollary follows from Lemma 7 and Stone's formula. Finally, we present a form of Wegner's lemma suitable for our models. For a finite cube $\Lambda \subset \mathbb{R}^d$, let H_Λ^X, $X = A$ or P, be the restriction of $H_{\omega,\omega'}^X$ to Λ with Dirichlet boundary conditions and the random variables in $\mathbb{R}^d \setminus \Lambda$ fixed. For the case of multiplicative perturbations (1.2), let $H_{\omega,\omega'}^{X,\Lambda}$ be defined similarly. We write H_Λ^X for short.

THEOREM 9. (1) Additive case. *For any interval* $[0,\kappa]$ $\exists$ *constant* $C_W < \infty$ *depending only on* κ *and linearly on* $\|g\|_\infty$ *such that if* $\eta < 1$ *and* $[E_0-\eta, E_0+\eta] \subset [0,\kappa]$,

$$\mathbb{P}\Big\{\operatorname{dist}\big(\sigma(H_\Lambda^A), E_0\big) > \eta\Big\} \leq C_W|\Lambda|\eta,$$

and for any $\tau > 1 \exists \sigma > 0$ *s.t.*

$$\mathbb{P}\Big\{\operatorname{dist}\big(\sigma(H_\Lambda^P), E_0\big) > \eta\Big\} \leq C_W\big(|\Lambda|^\tau\eta + e^{-|\Lambda|^\sigma}\big),$$

provided $|\Lambda|$ *is large enough.*

(2) Multiplicative case. *For any interval* $(0,\kappa]$, $0 < \kappa$, $\exists$ *constant* $C_W < \infty$ *depending on* κ *and linearly on* $\|g\|_\infty$ *s.t. if* $[E_0-\eta, E_0+\eta] \subset (0,\kappa]$,

$$\mathbb{P}\Big\{\operatorname{dist}\big(\sigma(H_\Lambda^A), E_0\big) < \eta\Big\} \leq (E_0-\eta)^{-1}C_W|\Lambda|\eta,$$

and for any $\tau > 1$, $\exists \sigma > 0$ s.t.

$$\mathbb{P}\Big\{\operatorname{dist}\big(\sigma(H_\Lambda^P), E_1\big) < \eta\Big\} \leq (E_0-\eta)^{-1}C_W\big(|\Lambda|^\tau\eta + e^{-|\Lambda|^\sigma}\big)$$

provided $|\Lambda|$ *is large enough.*

SKETCH OF THE PROOF. Let $I_\eta \equiv [E_0-\eta, E_0+\eta]$ and let E_λ be the spectral family for H_λ. We fix $\omega' \in \Omega'$ and let $\mathbb{E}_1$ and $\mathbb{P}_1$ denote the expectation and probability with respect to those random variables on Ω in Λ. We consider the random variables associated with $\mathbb{Z}^d \cap (\mathbb{R}^d \setminus \Lambda)$ as frozen. We obtain estimates uniform in these variables. We have by Chebyshev's inequality

$$\begin{aligned}\mathbb{P}_1\Big\{\operatorname{dist}\big(\sigma(H_\lambda), E_0\big) < \eta\Big\} &\leq \mathbb{E}_1\Big(\operatorname{Tr}\big(E_\lambda(I_\eta)\big)\Big)\\ &\leq e^{E_0+\eta}\mathbb{E}_1\Big(\operatorname{Tr}\big(E_\lambda(I_\eta)e^{-H_\lambda}\big)\Big).\end{aligned} \tag{3.5}$$

For each $j \in \mathbb{Z}^d \cap \Lambda$, let H_{Λ_j} be the Neuman Laplacian on the unit cube Λ_j centered at j. We can then replace H_λ in (3.5) by $\bigoplus_j H_{\Lambda_j}$ (see [**3**]). In the additive case, the positivity of V allows us to replace H_{Λ_j} by $-\Delta_{N,j}$, the Neumann Laplacian on Λ_j. The multiplicative case is slightly more complicated. Note that $H_{\Lambda_j} = -C_\omega^{-1}\Delta_{N,j}C_\omega^{-1}$. Since $\eta \in C^2(\mathbb{R}^d)$, we have

$$H_{\Lambda_j} = -\nabla C_\omega^{-2}\nabla + F,$$

for a bounded function F depending on $\partial_j\eta$ and $\partial_i\partial_j\eta$. Due to the fact that $C_\omega^{-1} > 1$ a.s., we obtain

$$H_{\Lambda_j} \geq -\Delta_{N,j} + f_0,$$

for some finite constant f_0. These bounds allow us to write

$$\operatorname{Tr}\big(E_\lambda(I_\eta)e^{-H_\lambda}\big) \leq \sum_{j\in\mathbb{Z}^d\cap\Lambda} c_0\operatorname{Tr}\big(\chi_j E_\lambda(I_\eta)e^{-\Delta_{N,j}}\big), \tag{3.6}$$

where χ_j is the characteristic function on Λ_j. Let $\{\phi_{n,j}\}$ be a complete set of eigenfunctions of $-\Delta_{N,j}$ with eigenvalues $E_{n,j}$. We can expand the trace in (3.6) to obtain

$$\mathbb{E}_1\Big(\operatorname{Tr}\big(E_\lambda(I_\eta)e^{-H_\lambda}\big)\Big) \leq c_1\sup_n\Big\langle\phi_{n,j}, \mathbb{E}_1\big(\chi_j E_\Lambda(I_\eta)\chi_j\big)\phi_{n,j}\Big\rangle. \tag{3.7}$$

Finally, the expectation $\mathbb{E}_1(\chi_j E_\lambda(I_\eta)\chi_j)$ is uniformly bounded above as in Corollary 8 due to the fact that assumptions (3.1) and (3.2) hold with B and χ taken to be χ_j; see [**3**] for the details. □

Finally, we turn to the technical Lemma 7. The idea of the proof in the additive case is due to [**17**].

SKETCH OF THE PROOF OF LEMMA 7. One begins the proof for the A-case with inequality

$$-\operatorname{Im} K(\lambda, z) \geq c_0(\operatorname{Im}\lambda) K(\lambda, z)^* K(\lambda, z), \tag{3.8}$$

from which it follows that

$$\|K(\lambda, z)\| \leq \min\left((c_0 \operatorname{Im}\lambda)^{-1}, |\operatorname{Im} z|^{-1}\|B\|^2\right). \tag{3.9}$$

In the M-case, we first write

$$K(\lambda, z) = \chi C_\lambda (H_0 - C_\lambda^2 z)^{-1} C_\lambda \chi$$

and consider $\widetilde{R}_\lambda(z) \equiv (H_0 - C_\lambda^2 z)^{-1}$ and $\widetilde{K}(\lambda, z) \equiv \chi \widetilde{R}_\lambda(z)\chi$. A slightly long calculation leads to the analog of (3.8),

$$-\operatorname{Im} \widetilde{K}(\lambda + i\varepsilon, E + i\delta) \geq 2c_0 \varepsilon E \widetilde{K}^* \widetilde{K}$$

from which it follows that

$$\|\widetilde{K}(\lambda + i\varepsilon, E + i\delta)\| \leq \min\left(c_1\delta^{-1}, (2c_0 E\varepsilon)^{-1}\right).$$

From the definition of $\widetilde{K}$ and the bound $|C_\lambda| \leq 1 + M_1$, we obtain

$$\|K(\lambda + i\varepsilon, E + i\delta)\| \leq \min(1 + M_1)^2\left((2\varepsilon c_0 E)^{-1}, c_1\delta^{-1}\right). \tag{3.10}$$

Next, we use the analyticity of $K(\lambda, z)$ in the half-plane $\operatorname{Im}\lambda > 0$ in both cases. In the A-case, we evaluate the integral defining $F_t(\varepsilon, z)$ by a contour integral in $\operatorname{Im}\lambda > 0$ taken along a semi-circle of radius R. The limit $R \to \infty$ is controlled using bounds (3.9). In the M-case, we take a rectangular contour in $\operatorname{Im}\lambda > 0$ with sides $\{\lambda \mid -M_0 \leq \lambda \leq M_1\}$, $\Gamma_\pm \equiv \{M_\pm + is \mid 0 \leq s \leq \alpha\}$ for some $\alpha > 0$ and $M_+ \equiv M_1$, $M_- \equiv M_0$, and $\sum_\alpha \equiv \{s + i\alpha \mid -M_0 \leq \alpha \leq M_1\}$. The integral is evaluated using bounds (3.10). □

4. Resolvent Estimates and Multi-scale Analysis

As we have seen in Section 2, we need to verify hypotheses (A2) and (M2) on the resolvent of our random Hamiltonians for almost every realization of the potentials $V^X_{\omega,\omega'}$ and the coefficients $C^X_{\omega,\omega'}$, respectively. We do this using a multi-scale analysis based on the ideas of [**12**], [**7**] and [**21**]. The idea is to begin at a length scale l_0 and bound the resolvent of the finite volume Hamiltonian with a "good" probability. This requires an initial, a priori estimate which we obtain using quantum tunnelling estimates in the additive case and a Combes-Thomas type argument in the multiplicative case. We then iterate these estimates through a rapidly increasing sequence of length scales l_k controlling possible singularities (small divisors) using the Wegner-type lemma as given in Theorem 9. The details of this method are given in [**3**] and [**5**]. Here, we simply point out some features of the proof in the continuous case.

We need some notation. Let $|x| \equiv \max\{|x_i| \mid i = 1, \ldots, d\}$ and let $\|x\|$ be the Euclidean norm of x. For $l > 0$, denote by Λ_l the cube of side l centered at the origin,

$$\Lambda_l \equiv \{x \mid |x_i| < l/2, i = 1, \ldots, d\}.$$

Let $\tilde{\Lambda}_l$ denote the sub cube $\{x \subset \Lambda_l \mid d(x, \partial\Lambda_l) > \delta\}$, for a fixed $\delta > 0$. Let $\chi_l \in C^2(\Lambda_l)$ be a function such that $\chi_l \mid \tilde{\Lambda}_l = 1$, $\chi_l \geq 0$, and $\operatorname{supp} \nabla\chi_l \subset \Lambda_l \setminus \tilde{\Lambda}_l$. Let H be a random family of self-adjoint operators and let H_Λ denote the restriction of H to Λ with Dirichlet boundary conditions. We fix $E > 0$. We write $R_\Lambda(z) \equiv (H_\Lambda - z)^{-1}$ and, for $z = E + i\varepsilon$, $R_\Lambda(\varepsilon) \equiv (H_\Lambda - E - i\varepsilon)^{-1}$. Finally, for a sequence l_k, we will write $R_k(z)$, etc., for $R_{\Lambda_{l_k}}(z)$, etc. We first present a general theorem which shows how to pass from finite volume estimates to infinite volume estimates. Let l_k be an increasing length scale and let $\varepsilon_k > 0$ be a decreasing sequence $\varepsilon_k \to 0$. Let $f > 0$ be a non-decreasing function on $\mathbb{R}$ such that $f^{-1} \equiv 1/f \in L^2_{\text{loc}}(\mathbb{R})$. We consider two hypotheses:

[H1] $\exists c > 0$ and, for each k, a probability $\tilde{p}_k$ s.t. $(1 - \tilde{p}_k) \in l^1$, s.t. with probability $\tilde{p}_k$,

$$\sup_{\varepsilon > 0} \|W(\chi_k) R_k(\varepsilon) \chi_{k-1}\| \leq \varepsilon^2_{k+1},$$

where $W(\chi_k) \equiv [-\Delta, \chi_{\Lambda_{l_k}}]$.

[H2] For sequences (l_k, ε_k) as in **[H1]**, we have $\forall k \in \mathbb{N}$,

$$p_k \equiv \mathbb{P}\Big\{\operatorname{dist}\big(\sigma(H_k), E\big) \leq f(\varepsilon_k)\Big\} \in l^1.$$

THEOREM 10. *Let l_k, ε_k and f be as above.*

(1) Short-range case. *Let c be a function of compact support. If* **[H1]**-**[H2]** *hold, then for a.e.* ω $\exists d_\omega > 0$ *finite s.t.*

$$\sup_{\varepsilon > 0} \|(H - E - i\varepsilon)^{-1} c\| < d_\omega \|c\|_\infty.$$

(2) Long-range case. *Let c be a function satisfying* $|c(x)| < \delta(c)\big(1 + \|x\|\big)^{-\beta}$ *for some* $\beta > 0$ *s.t.* $\big(\varepsilon_{k+1}\big)\big(l_{k-1}\big)^{-\beta} \in l^1$. *If* **[H1]**-**[H2]** *hold, then for a.e.* ω $\exists d_\omega > 0$ *finite s.t.*

$$\sup_{\varepsilon > 0} \|(H - E - i\varepsilon)^{-1} c\| < d_\omega \delta(c).$$

The proof of this theorem relies on iteration of the geometric resolvent equation through scales l_k. The geometric resolvent equation is the following. Let $\Lambda \subset \Lambda' \subset \mathbb{R}^d$ and let $\chi_\Lambda \in C^2(\Lambda)$. As in **[H1]**, let $W(\chi_\Lambda) \equiv [-\Delta, \chi_\Lambda]$. An application of the second resolvent formula to R_Λ and $R_{\Lambda'}$ leads to

$$\chi_\Lambda R_{\Lambda'}(z) = R_\Lambda(z)\chi_\Lambda + R_\Lambda(z) W(\chi_\Lambda) R_{\Lambda'}(z).$$

In the iteration, localized resolvents as in **[H1]** are bounded by ε^2_{k+1} with probability $\tilde{p}_k$ whereas other resolvents $R_k(\varepsilon_k)$ are bounded $f(\varepsilon_k)^{-1}$ with probability $\geq 1 - p_k$ by **[H2]**. The result then follows by the first Borel Cantelli lemma.

Consequently, to verify (A2) and (M2), we must verify **[H1]**-**[H2]**. However **[H1]** still requires estimating the resolvents $R_k(z)$ for $k \in \mathbb{N}$. We replace **[H1]** by a hypothesis **[H1]** $(\boldsymbol{\gamma_0, l_0})$ involving $H_{\Lambda_{l_0}}$ for an initial length scale l_0 and a form of **[H2]**, called **[W]**, which again controls the small divisors. The proof that {**[H1]**

(γ_0, l_0), [W]} $\implies$ {[H1], [H2]} involves the multi-scale analysis. We state these two new hypotheses and the main lemma here:

[H1] (γ_0, l_0) For some $\gamma_0 > 0$, $l_0 > 1$, $\exists \xi > 2d$ s.t.

$$\mathbb{P}\left\{\sup_{\varepsilon>0}\|W(\chi_{l_0})R_{\Lambda_{l_0}}(E+i\varepsilon)\chi_{l_0/3}\| \leq e^{-\gamma_0 l_0}\right\} \geq 1 - l_0^{-\xi}.$$

[W] There exists $C_W < \infty$, $\tau \geq 1$, $\sigma > 0$ s.t. for all sufficiently large cubes Λ and all $\eta > 0$,

$$\mathbb{P}\Big\{\operatorname{dist}\big(\sigma(H_\Lambda), E\big) < \eta\Big\} \leq C_W\big(|\Lambda|^\tau \eta + e^{-|\Lambda|^\sigma}\big). \tag{4.1}$$

The particular form of the probability estimate on the right side of (4.1) is necessary to accommodate the Poisson potential, see [**3**] for the details.

LEMMA 11. *Let $\gamma_0 > 0$ and assume* [**W**]. *There exists a minimum length scale $l^* \equiv l^*(\gamma_0, d, C_W, \tau)$ such that if* [**H1**] (γ_0, l_0) *holds for $l_0 > l^*$, then for $l_k \equiv l_0^{(3/2)^k}$, $\xi > 2d$, $\exists 0 < \kappa < \infty$ such that for each $k \in \mathbb{N}$*

$$\mathbb{P}\Big\{\|W_{l_k}R_{l_k}(E+i\varepsilon)\chi_{l_{k/3}}\| \leq e^{-1/2\kappa\gamma_0 l_k}\Big\} \geq 1 - l_k^{-\xi}.$$

Given Lemma 11, we select l_k as there, set $\varepsilon_{k+1} \equiv (l_k)^{-(3+\delta_1)\tau d}$ for some $0 < \delta_1 \ll 1$, and take $f(\varepsilon) = \varepsilon^{1/2-\delta}$ for some $0 < \delta \ll \delta_1$. We then verify the hypotheses of Theorem 10. This leads to the following.

COROLLARY 12. *Let $\gamma_0 > 0$ and assume* [**W**]. *There exists a minimum length scale $l^* = l^*(\gamma_0, C_W, \tau, d)$ such that if* [**H1**] (γ_0, l_0) *holds for $l_0 > l^*$ and if the function c satisfies $|c(x)| \leq \delta(c)\big(1+\|x\|\big)^{-\beta}$ for $\beta > 9d\tau/2$, then for a.e. $\omega \exists d_\omega > 0$ finite s.t.*

$$\sup_{\varepsilon>0}\|(H-E-i\varepsilon)^{-1}c\| \leq d_\omega \delta(c). \tag{4.2}$$

Consequently, since [**W**] is proved in Section 3 for our models, once we prove [**H1**] (γ_0, l_0), this corollary implies localization by Theorem 5 and the discussion following it.

Although bounds of the form (4.2) are sufficient to prove localization, they do not give decay estimates on the eigenfunctions. We remark that, given the exponential bound in [**H1**] (γ_0, l_0), we can improve Theorem 10 and Corollary 12 yielding estimates strong enough to give eigenfunction decay. The upper bounds depend upon whether $\operatorname{supp} c$ is compact or not. For the models, this implies that the eigenfunctions of $H^A_{\omega,\omega'}$ with a single-site potential u of compact support are a.s. bounded above by an exponential. For the Poisson case and the Anderson case with $\operatorname{supp} u$ non-compact, we have only an upper bound of the form $\|x\|^{-\beta}$, for $\beta > 9d$. It is not known if either of these bounds are optimal. We refer to [**3**] and [**5**] for the details.

NOTE ADDED IN PROOF. A more general form of the spectral averaging theorem is given in Combes, Hislop, Mourre, "Spectral Averaging, Perturbation of Singular Spectrum, and Localization", submitted to Trans. AMS. □

References

1. R. Carmona and J. Lacroix, *Spectral theory of random Schrödinger operators*, Birkhäuser, Boston, 1990.
2. J.-M. Combes, S. DeBièvre, and P. D. Hislop, *Localization for continuous Schrödinger operators with correlated potentials*, preprint (1993).
3. J.-M. Combes and P. D. Hislop, *Localization for continuous random Hamiltonians in d-dimensions*, J. Funct. Anal. **124** (1994), 149–180.
4. J.-M. Combes, P. D. Hislop, and V. Jaksic, *Spectral properties of random Schrödinger operators with unbounded potentials: continuous models*, preprint (1993).
5. J.-M. Combes, C. Shubin Christ, P. D. Hislop, and A. Tip, *Localization for the wave equation in random media*, SIAM J. Math. Anal., (submitted).
6. F. Delyon, B. Simon, and B. Souillard, *From power pure point spectrum to continuous spectrum in disordered systems*, Ann. Inst. H. Poincaré **42** (1985), 283.
7. H. von Dreifus and A. Klein, *A new proof of localization in the Anderson tight binding model*, Commun. Math. Phys. **124** (1989), 285–299.
8. ——, *Localization for random Schrödinger operators with correlated potentials*, Commun. Math. Phys. **140** (1991), 133–147.
9. W. Faris, *Localization for a random discrete wave equation in random media*, IMA vol. 7, (G. Papanicolaou, ed.), Springer-Verlag, New York, 1987.
10. —— *A localization principle for multiplicative perturbations*, J. Funct. Anal. **67** (1986), 105–114.
11. A. Figotin and L. Pastur, *Spectra of random and almost-periodic operators*, Berlin, Springer-Verlag 1992.
12. J. Fröhlich and T. Spencer, *Absence of diffusion in the Anderson tight binding model for large disorder or low energy*. Commun. Math. Phys. **88** (1983) 151–184.
13. Y. A. Gordon, V. Jaksic, S. Molcanov, and B. Simon, *Spectral properties of random Schrödinger operators with unbounded potentials*, Commun. Math. Phys. **157** (1993), 23–50.
14. H. Holden and F. Martinelli, *On absence of diffusion near the bottom of the spectrum for a random Schrödinger operator on* $L^2(\mathbb{R}^v)^+$, Commun. Math. Phys. **93** (1984), 197–217.
15. J. Howland, *Perturbation theory of dense point spectra*, J. Funct. Anal **74** (1987), 52–80.
16. F. Klopp, *Localization for some continuous random Schrödinger operators*, Commun. Math. Phys., (to appear).
17. S. Kotani and B. Simon, *Localization in general one dimensional systems*, II, Commun. Math. Phys. **112** (1987), 103–120.
18. S. Kotani and N. Ushiroya, *One-dimensional Schrödinger operators with random decaying potentials*, Commun. Math. Phys. **115** (1989), 247–266.
19. F. Martinelli and E. Scoppola. *Introduction to the mathematical theory of localization*, La Rivista del Nuovo Cimento **10 (3)** (1987).
20. B. Simon and T. Wolff, *Singular continuous spectrum under rank one perturbation and localization for random Hamiltonians*. Comm. Pure Appl. Math. **39** (1986), 75–90.
21. T. Spencer, *Localization for random and quasi-periodic potentials*, J. Stat. Phys. **51** (1988), 1009–1019.

DÉPARTMENT DE MATHÉMATIQUES, UNIVERSITÉ DE TOULON ET DU VAR, 83130 LA GARDE, FRANCE

E-mail address: combes@cptsg2.univ-mrs.fr

MATHEMATICS DEPARTMENT, UNIVERSITY OF KENTUCKY, LEXINGTON, KY 40506-0027, USA
E-mail address: hislop@ms.uky.edu

Centre de Recherches Mathématiques
CRM Proceedings and Lecture Notes
Volume 8, 1995

Cluster WKB

T. A. Osborn and F. H. Molzahn

ABSTRACT. A connected graph description of the Schrödinger propagator's WKB expansion is constructed for quantum systems with arbitrary smooth Hamiltonians. Use of variables describing fluctuations about a classical trajectory with fixed endpoints yields a family of perturbed $(\hat{q}, \hat{p})$-quadratic Hamiltonians. The Dyson series for this problem is summed into exponential form by exposing the graph structure present in each perturbative order. The coefficients in this cluster representation are classical transport integrals of derivatives of the perturbation's Weyl symbol, coupled by a Green function of the phase space Jacobi operator.

1. Introduction and Summary

A natural cluster structure of perturbation theory is exposed by employing classical dynamics. In this fashion a constructive approach to the WKB approximation is developed. This contribution provides an overview of the results obtained in Ref. [**22**].

The time evolution of a quantum system with Hamiltonian $\widehat{H}(t)$ (for each $t \in \mathbb{R}$, a self-adjoint operator acting in Hilbert space $\mathcal{H}$) is described by the time evolution operator $U(t,s)$. It is a unitary operator valued solution of the Schrödinger evolution equation

$$i\hbar \frac{\partial}{\partial t} U(t,s) = \widehat{H}(t)\, U(t,s) \tag{1.1}$$

with initial value $U(s,s) = I$ (the identity on $\mathcal{H}$). Thus $\psi_t = U(t,s)\psi_s$ is the state vector evolved from initial data $\psi_s \in \mathcal{H}$ at time s. In (1.1) $2\pi\hbar$ is Planck's constant.

In this work we consider the case $\mathcal{H} = L^2(\mathbb{R}^d, \mathbb{C})$, corresponding to a scalar wavefunction with configuration space $\mathbb{R}^d$. Then $U(t,s)$ is represented by its integral kernel $K(t,s,x,y)$,

$$U(t,s)\psi_s(x) = \int_{\mathbb{R}^d} dy\, K(t,s,x,y)\psi_s(y), \quad x \in \mathbb{R}^d. \tag{1.2}$$

We now summarize the results of a derivation of an $\hbar \downarrow 0$ formal asymptotic expansion for the 'propagator' K. Let $Q \equiv (t,s,x,y)$ denote the standard pairs of space and time arguments. Then it is found (under certain conditions to be elaborated

1991 *Mathematics Subject Classification.* Primary: 81Q20; Secondary: 81S30.
This is the final form of the paper.

upon in Section 5) that K has the cluster expansion

$$K(Q) \overset{\hbar\downarrow 0}{\sim} (2\pi\hbar)^{-d/2} e^{i\Phi(Q)} \big|D_c(Q)\big|^{1/2} e^{(i/\hbar)S_c(Q)} F(Q,\hbar), \tag{1.3a}$$

$$F(Q,\hbar) = \exp\bigg(\sum_{j=1}^{\infty} \frac{1}{j!}\mathcal{L}_j(Q,\hbar)\bigg). \tag{1.3b}$$

Equation (1.3a) contains an essential singularity at $\hbar = 0$. The phase Φ is $\hbar$-independent and determined by the requirement that $K(Q) \to \delta(x-y)$ as $t \to s$. In (1.3b) the cluster integrals $\mathcal{L}_j$ vanish at $\hbar = 0$. The leading coefficients S_c and D_c are, respectively, the classical action and Van Vleck determinant, for a classical mechanical system associated with $\widehat{H}(t)$ as follows.

Let $H_{\mathrm{w}} \leftrightarrow \widehat{H}$ denote the standard [**2, 11, 23, 26, 27**] Wigner-Weyl symbol $\leftrightarrow$ operator correspondence. The Weyl symbol $H(t)_{\mathrm{w}}$ is a function on the classical phase space $\mathbb{R}^{2d} \simeq \mathbb{R}^d_q \times (\mathbb{R}^d)^*_p$, defined in terms of the kernel $\langle\cdot|\, \widehat{H}(t)\, |\cdot\rangle : \mathbb{R}^d \times \mathbb{R}^d \to \mathbb{C}$ by the Wigner transform

$$H(t)_{\mathrm{w}}(q,p) = \int_{\mathbb{R}^d} d^d v\, e^{-ip\cdot v/\hbar} \left\langle q + \frac{v}{2}\right| \widehat{H}(t) \left| q - \frac{v}{2}\right\rangle. \tag{1.4}$$

The symbol $H(t)_{\mathrm{w}}$ is real valued because $\widehat{H}(t)$ is self-adjoint. In typical applications it is the case that $H(t)_{\mathrm{w}}(z) = H_c(z,t) + \mathcal{O}(\hbar)$. Here H_c is the $\hbar$-independent 'classical Hamiltonian' associated with $\widehat{H}$. The classical Hamiltonian induces a classical trajectory $z_c = (q_c\, p_c) \colon \mathbb{R} \to \mathbb{R}^{2d}$ solving Hamilton's canonical equations

$$\frac{d}{d\tau} z_c(\tau) = J\nabla H_c(z_c(\tau),\tau). \tag{1.5}$$

Here ∇H_c is the $2d$-dimensional phase space gradient, and J is the matrix $\left[\begin{smallmatrix} 0 & \delta \\ -\delta & 0 \end{smallmatrix}\right]$ associated with the classical Poisson bracket $\{z^i, z^j\} = J^{ij}$. Assume that in some region of the parameters Q there exists a unique trajectory $z_c(\tau;Q)$ which is a solution of (1.5) with the boundary conditions

$$q_c(s) = y, \quad q_c(t) = x \quad (s < t). \tag{1.6}$$

Hamilton's principal function is the action of this trajectory,

$$S_c(Q) \equiv \int_s^t d\tau\, \big[p_c(\tau;Q)\cdot \dot{q}_c(\tau;Q) - H_c(z_c(\tau;Q),\tau)\big], \tag{1.7}$$

while

$$D_c(Q) \equiv \det\bigg(-\frac{\partial^2 S_c(Q)}{\partial x\, \partial y}\bigg). \tag{1.8}$$

Formula (1.3a), with the attendant definitions (1.7)–(1.8), displays the WKB asymptotic form of the propagator, familiar from the semiclassical literature [**3, 4, 7, 9, 16, 17, 20, 24**]. Note that this form has usually been taken as an ansatz for the small $\hbar$ structure of K. Upon substitution of the ansatz into the Schrödinger equation, recursive techniques may be employed to compute the $\hbar$ expansion coefficients of $F(Q,\hbar)$. We want to emphasize that in the result being presented here, (1.3) does *not* have the status of an ansatz. Rather it is the outcome of a formal derivation which additionally provides explicit nonrecursive formulae for the $\hbar$ expansion of $F(Q,\hbar)$. The only other literature, of which we are aware, which attempts to directly construct the WKB approximation of the propagator in terms of trajectories [**7, 12, 14, 18**] is based on the phase space path integral representation of K. These works did not develop the analysis of the coefficient structure sufficiently far so as to uncover the cluster structure found in (1.9).

The explicit structure of F will now be described. In order to simplify the presentation, we shall restrict consideration in the remainder of this paper to the case of a Weyl-quantized Hamiltonian. Effectively, this means that $\widehat{H}(t)$ has an $\hbar$-free symbol: $H_{\mathrm{w}} = H_c$. This is the case in the commonly occurring situation where H_c is given and $\widehat{H}(t)$ is obtained from it by Weyl's quantization prescription. (The case in which $H(t)_{\mathrm{w}}$ has $\hbar$ dependence of $\mathcal{O}(\hbar)$ poses no great difficulty. See Ref. [**22**].)

Under these circumstances the function F is found to admit the exponential cluster representation (1.3b) with the jth cluster integral $\mathcal{L}_j$ defined by

$$\text{(1.9)}\quad \mathcal{L}_j(Q,\hbar) = (i\hbar)^{-j} \sum_{C\in\mathcal{C}_j} \int_{[s,t]^j} d^j\tau \Big(\prod_{\beta\in EC} [e^{i\hbar b_\beta} - 1] \Big) \\ \times \prod_{l=1}^{j} e^{(i\hbar/2) b_{\{l,l\}}} H_c^{(3)}(0, z_c(\tau_l;Q), \tau_l).$$

Some of the notation in (1.9) requires explanation.

The set $\mathcal{C}_j$ consists of all simple connected labeled graphs, [**28**] or clusters, on j vertices. A cluster in $\mathcal{C}_j$ is thus a pair $C = (\bar{\jmath}, EC)$ composed of vertex set $\bar{\jmath} \equiv \{1,\ldots,j\}$, and edge set EC containing the unordered pairs $\beta = \{k,l\}$ of linked vertices $k,\ l \in \bar{\jmath}$. In (1.9) the vertex l is associated with the symbol $H_c^{(3)}(\cdot,\tau_l)$, which is the remainder in the quadratic order Taylor expansion of the classical Hamiltonian about the point $z_c(\tau;Q)$, viz.

$$\text{(1.10)}\quad H_c^{(3)}(\zeta, z_c, \tau) = H_c(\zeta + z_c, \tau) - H_c(z_c,\tau) - \zeta\cdot\nabla H_c(z_c,\tau) \\ - \frac{1}{2}\zeta\cdot\nabla\nabla H_c(z_c,\tau)\zeta.$$

Each link β is associated with a differential operator

$$\text{(1.11)}\qquad b_\beta(\tau) = \mathbb{D}_k \cdot G_c(\tau_k,\tau_l;Q)\mathbb{D}_l,$$

where $\mathbb{D}_k$ is the phase space gradient acting upon the kth symbol $H_c^{(3)}(\cdot, z_c, \tau_k)$ in (1.9). Finally, G_c is the Green function of the phase space Jacobi operator

$$\text{(1.12)}\qquad \mathcal{J}_c(\tau) \equiv \frac{d}{d\tau} - J\nabla\nabla H_c(z_c(\tau;Q),\tau),$$

where (as in (1.10)) $\nabla\nabla H_c$ is the Hessian matrix. Thus G_c is a $2d\times 2d$ matrix valued function obeying $\mathcal{J}_c(\tau)G_c(\tau,\sigma;Q) = J\delta(\tau-\sigma)$ and certain homogeneous boundary conditions related to (1.6); for details see Section 3.

A critical property of the cluster integrals is that

$$\text{(1.13)}\qquad \mathcal{L}_j(Q,\hbar) = \mathcal{O}\big(\hbar^{\lceil j/2\rceil}\big), \quad j \geq 1,$$

where $\lceil a \rceil$ denotes the largest integer $\geq a$. This feature has two important implications.

1. Combining (1.13) with (1.3b) verifies that the action phase $e^{iS_c/\hbar}$ carries all the essential singularity of K as $\hbar \downarrow 0$, in accordance with the standard WKB ansatz.
2. The higher $\hbar$-expansion coefficients of $\log K$ can be graphically determined as finite sums of H_c's derivatives integrated along the full dynamical trajectory $z_c(\cdot; Q)$. This follows after expanding $e^{i\hbar b_\beta}$ and $e^{(i\hbar/2)b_{\{l,l\}}}$ in (1.9) as series in $\hbar$ and examining the terms that contribute to a given $\hbar$ order.

To develop point 2 more fully, note that (1.13) leads to a re-organization of $\log F$ in powers of $\hbar$ of the form

$$\sum_{j=1}^{\infty} \frac{1}{j!} \mathcal{L}_j(Q, \hbar) = \sum_{N=1}^{\infty} (i\hbar)^N \left(\sum_{j=1}^{2N} W_j^N(Q) \right), \tag{1.14}$$

where W_j^N is the portion of $\mathcal{L}_j/j!$ proportional to $(i\hbar)^N$. The W_j^N coefficients are easily extracted from (1.9), and may be represented via multi-graphs with loops. An edge β in the underlying cluster C replicates into a multiplet of edges in the multi-graph according to the exponent of b_β, and similarly powers of $b_{\{l,l\}}$ lead to loops attached to vertex l. Because $H_c^{(3)} = \mathcal{O}(\zeta^3)$, it must be struck by at least three $\mathbb{D}$'s before setting $\zeta = 0$. This leads to a selection of multi-graphs in which each vertex has incidence three or greater. For the $(i\hbar)^1$ corrections, the relevant graphs are shown in the figure.

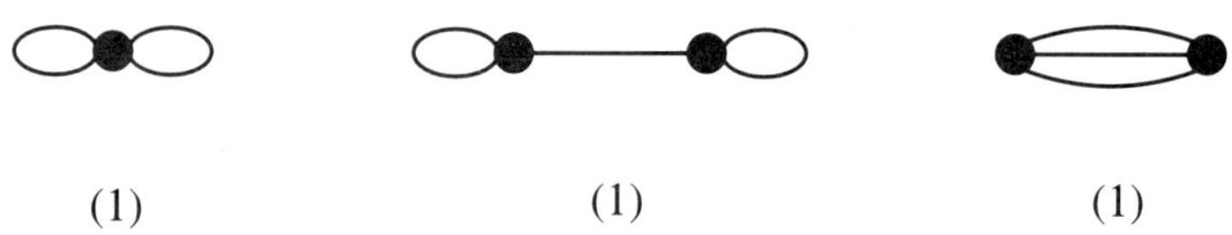

FIGURE 1. Graphs for $i\hbar W_j^1$.

The related coefficient formulae are

$$W_1^1(Q) = \frac{1}{2!\, 2^2} \int_s^t d\tau\, (\mathbb{D} \cdot G_c(\tau, \tau)\mathbb{D})^2\, H_c[\tau], \tag{1.15a}$$

$$\begin{aligned} W_2^1(Q) = \int_s^t d\tau_1 \int_s^t d\tau_2\, \mathbb{D}_1 \cdot G_c(\tau_1, \tau_2)\mathbb{D}_2 \Big\{ &\frac{1}{2!\,3!} (\mathbb{D}_1 \cdot G_c(\tau_1, \tau_2)\mathbb{D}_2)^2 \\ &+ \frac{2}{(2!\,2)^2} (\mathbb{D}_1 \cdot G_c(\tau_1, \tau_1)\mathbb{D}_1)(\mathbb{D}_2 \cdot G_c(\tau_2, \tau_2)\mathbb{D}_2) \Big\} H_c[\tau_1] H_c[\tau_2]. \end{aligned} \tag{1.15b}$$

Note that here $H_c^{(3)}$ has been conveniently replaced by $H_c[\tau] \equiv H_c(z_c(\tau; Q), \tau)$, with $\mathbb{D}_j$ acting on the z_c-argument of $H_c[\tau_j]$. This is possible because the third derivatives of the two symbols are identical.

The topics of the succeeding four sections are respectively: 2) the definition of the fluctuation problem about $z_c(\tau; Q)$ via a unitary transformation, 3) Jacobi Green functions, 4) quantum perturbation theory in terms of classical trajectories, and 5) the cluster re-summation method.

2. Classically Guided Propagators

Classical motion can be systematically introduced into the representation of the exact propagator by using the Heisenberg phase space translation operators [**13, 15, 25, 26**]. The result is a unitarily equivalent statement of the original quantum problem, which is characterized in terms of successively higher order fluctuations about the classical motion.

The Heisenberg phase space translations are the family of unitary operators on $\mathcal{H}$, parameterized by $z_0 = (q_0, p_0) \in \mathbb{R}^{2d}$,

$$T(z_0) \equiv \exp\left(\frac{i}{\hbar} z_0 \cdot J^{-1} \hat{z}\right). \tag{2.1}$$

Here $\hat{z} = (\hat{q}, \hat{p})$ denotes the canonical position and momentum operators. The basic properties of $T(z_0)$ required in our derivation are unitarity, z-translation, differentiability and their Dirac kernel representation. In order, these are given by

$$T(z_0)^{-1} = T(-z_0) = T(z_0)^{\dagger}, \tag{2.2a}$$

$$T(z_0)^{\dagger} \hat{z} T(z_0) = \hat{z} + z_0 I, \tag{2.2b}$$

$$i\hbar \frac{d}{dt} T(z(t)) = -T(z(t)) \dot{z}(t) \cdot J^{-1}\left(\hat{z} + \frac{1}{2} z(t)\right), \tag{2.2c}$$

$$\langle x | T(z_0) | y \rangle = \exp\left(\frac{i}{2\hbar} p_0 \cdot (x+y)\right) \delta(x - y - q_0). \tag{2.2d}$$

In the expressions above † is the adjoint and $z(\cdot)$ is any C^1 curve in phase space.

Now incorporate the classical trajectory $z_c(\tau) \equiv z_c(\tau; Q)$ into the quantum evolution. Take ψ_τ to be the time dependent state whose evolution is governed by $\widehat{H}(\tau)$. The transformed state $\phi_\tau = T(z_c(\tau))^{\dagger} \psi_\tau$ has the companion evolution operator $U_{z_c}(\tau, s)$ defined by

$$U(\tau, s) = T(z_c(\tau))\, U_{z_c}(\tau, s)\, T(z_c(s))^{\dagger}. \tag{2.3}$$

The unitary modification in (2.3) of the original evolution operator $U(\tau, s)$ has a dual motivation. First, the Heisenberg operators account for rigid-shape displacement of wave packets along z_c. Second, U_{z_c} will be represented by a perturbation expansion in fluctuations about z_c. The zeroth order representation of U_{z_c} will account for all quadratic fluctuations and perturbative terms will describe third and higher order effects. It is these higher order fluctuations that are eventually described by the cluster expansion (1.3b).

Further insight into the unitary transformation (2.3) follows from its kernel equivalent. Employing (2.2d) one finds

$$K(Q) = \exp\left\{\frac{i}{2\hbar}\left[p_c(t;Q) \cdot x - p_c(s;Q) \cdot y\right]\right\} K_{z_c(\cdot;Q)}(t, s, 0, 0). \tag{2.4}$$

The two coordinate arguments of K_{z_c} are originally $x - q_c(t)$ and $y - q_c(s)$. Both of these vanish as a consequence of the boundary condition (1.6). The momenta $p_c(t;Q)$ and $p_c(s;Q)$ are the endpoint values for the trajectory $z_c(\tau)$. Clearly, determining the diagonal kernel $K_{z_c(\cdot;Q)}(t, s, 0, 0)$ will give $K(Q)$.

Next consider the Hamiltonian operator generating evolution $U_{z_c}(\tau, s)$. From (2.2c) and (2.3) it directly follows that

$$\widehat{H}_{z_c}(\tau) = \dot{z}_c(\tau) \cdot J^{-1}\Big(\hat{z} + \frac{1}{2}z_c(\tau)\Big) + T\big(z_c(\tau)\big)^{\dagger}\, \widehat{H}(\tau)\, T\big(z_c(\tau)\big). \tag{2.5}$$

Given $\widehat{H}_{z_c}(\tau)$, the task at hand is to obtain a perturbative expansion about $z_c(\tau)$. This is achieved by computing the Weyl symbol of $\widehat{H}_{z_c}(\tau)$, Taylor expanding it about z_c and finally mapping the result back into operator form. Specifically, the Wigner transform of (2.5) reads

$$\widehat{H}_{z_c}(\tau)_{\mathrm{w}}(\zeta) = \dot{z}_c(\tau) \cdot J^{-1}\Big(\zeta + \frac{1}{2}z_c(\tau)\Big) + H_c(\zeta + z_c(\tau), \tau). \tag{2.6a}$$

Its power series expansion about the value 0 of fluctuation variable ζ has simple low order terms. The ζ-constant term is a scalar function of time; the ζ-linear term vanishes as a consequence of Hamilton's equation (1.5); and the ζ-quadratic term is the Hessian form $\zeta/2 \cdot \nabla\nabla H_c(z_c(\tau), \tau)\zeta$. The corresponding $\hat{z}$-expanded representation of $\widehat{H}_{z_c}$ is thus the sum of three distinct operators:

$$\widehat{H}_{z_c}(\tau) = P(\tau)I + \widehat{H}_0(\tau) + V(\tau) \tag{2.6b}$$

where

$$P(\tau) = H_c(z_c(\tau), \tau) + \frac{1}{2}\dot{z}_c(\tau) \cdot J^{-1}z_c(\tau) \tag{2.6c}$$

$$\widehat{H}_0(\tau) = \frac{1}{2}\hat{z} \cdot \nabla\nabla H_c(z_c(\tau), \tau)\hat{z} \tag{2.6d}$$

$$V(\tau) = H_c^{(3)}\Big(\frac{1}{i}\nabla_u, z_c(\tau), \tau\Big)e^{iu\cdot\hat{z}}\Big|_{u=0}. \tag{2.6e}$$

The operator $\widehat{H}_0(\tau)$ is all of the $\hat{z}$-quadratic terms in $\widehat{H}_{z_c}$. The operator $V(\tau)$ consists of terms of order $\hat{z}^3$ and higher. The right side of (2.6e) corresponds to Weyl quantization of $H_c^{(3)}$, cf. (1.10).

The operator $\widehat{H}_0 + V$ is the dynamically non-trivial part of $\widehat{H}_{z_c}$ and it defines its own propagator, $\widetilde{K}_{z_c}$, obeying

$$i\hbar\frac{\partial}{\partial\tau}\widetilde{K}_{z_c}(\tau, \sigma, x', y') = \big[\widehat{H}_0(\tau) + V(\tau)\big]\widetilde{K}_{z_c}, \tag{2.7a}$$

$$\lim_{\tau\to\sigma}\widetilde{K}_{z_c}(\tau, \sigma, x', y') = \delta(x' - y'). \tag{2.7b}$$

Here care has been taken to make the argument list of $\widetilde{K}_{z_c}$ distinct from the original space-time pair $Q = (t, s, x, y)$ which acts as a parameter determining $z_c(\tau; Q)$.

A straightforward calculation gives the complete bridge between the kernels $K(Q)$ and $\widetilde{K}_{z_c}$. The trivial term $P(\tau)$ contributes a phase to K which, when combined with the phase in (2.4), leads to

$$K(Q) = e^{iS_c(Q)/\hbar}\,\widetilde{K}_{z_c(\cdot;Q)}(t, s, 0, 0), \tag{2.8}$$

where S_c is the action (1.7).

In equation (2.7a) the Hamiltonian is partitioned into a leading quadratic term $\widehat{H}_0(\tau)$ plus a perturbative correction $V(\tau)$. As is well known [**8**] quadratic Hamiltonian problems are exactly solvable. Thus perturbation problem (2.7) has the attractive feature that the reference problem's propagator has a known analytical form. The precise sense in which the terms in the potential $V(\tau)$ are small relative to $\widehat{H}_0(\tau)$ is clarified by the linked graph propagator expansions found in Section 1. There it was seen that the physical scale parameter—or effective coupling constant—making the effects of $V(\tau)$ small is $\sqrt{\hbar}$.

3. Jacobi Fields and Phase Space Green Functions

This section provides a brief overview of the solutions of the Jacobi equation [**6, 19**] in a phase space setting. Most of the literature [**5**] concerned with this topic is based on the second-order d-dimensional Jacobi operator associated with the Euler-Lagrange equation in configuration space. Here the focus is on the Jacobi fields associated with Hamilton's equation of motion.

Jacobi fields describe the stability of classical motion in the neighbourhood of a given reference trajectory. Let ϵ be a parameter, not occurring in the Hamiltonian H_c, which indexes a family of trajectories, say $z_c(\tau;\epsilon)$. Differentiating (1.5) with respect to ϵ gives

$$\mathcal{J}(\tau)\frac{\partial z_c}{\partial \epsilon}\bigg|_{\epsilon=0} = 0, \tag{3.1a}$$

$$\mathcal{J}(\tau) \equiv \frac{d}{d\tau} - JH''(\tau), \tag{3.1b}$$

where the Hessian $\nabla\nabla H_c\big(z_c(\tau;0),\tau\big)$ defines the $2d\times 2d$ symmetric matrix function $H''(\tau)$. The function $\partial z_c/\partial\epsilon$ immediately yields the nearby orbit expansion

$$z_c(\tau;\epsilon) = z_c(\tau;0) + \epsilon\frac{\partial z_c}{\partial \epsilon}\bigg|_{\epsilon=0} + \mathcal{O}(\epsilon^2).$$

Any solution of (3.1a) is called a Jacobi field, and $\mathcal{J}(\tau)$ is referred to as the Jacobi operator along $z_c(\tau;0)$.

The central role of the phase space Green function is to construct solutions of the inhomogeneous Jacobi equation

$$\mathcal{J}Z = f, \quad f \in C^0 \equiv C^0\big([s,t],\mathbb{R}^{2d}\big). \tag{3.2}$$

Before proceeding to the Green function a more detailed specification of $\mathcal{J}(\tau)$ is required. Consider a generic form of $\mathcal{J}(\tau)$ that results if $H''(\tau)$ is taken to be an arbitrary symmetric matrix valued C^1 function of $\tau\in[s,t]$—not necessarily the Hessian of H_c along z_c. Define the domain of $\mathcal{J}$ to be

$$\mathcal{D}\mathcal{J} \equiv \Big\{Z \in C^1\big([s,t],\mathbb{R}^{2d}\big) \mid \pi_q Z(s) = 0 = \pi_q Z(t)\Big\}$$

where π_q denotes the projection of a $2d\times m$ matrix onto its first d rows, e.g. $\pi_q z_c = q_c$. The generic Jacobi operator $\mathcal{J}$ is the linear map $\mathcal{J}:\mathcal{D}\mathcal{J}\to C^0$. Invertibility of this map is possible only if the homogeneous Jacobi equation has no non-trivial solution. The homogeneous solutions are described by the following terminology. The time t is said to be *conjugate* to s if there exists $Z\in\mathcal{D}\mathcal{J}$ with $Z\neq 0$ and $\mathcal{D}\mathcal{J}=0$.

On the function space C^0, there is a skew scalar product which we denote by Ω,

$$\Omega(Z \mid W) \equiv \int_s^t d\tau\, Z(\tau) \cdot J^{-1} W(\tau). \tag{3.3}$$

Two basic properties of $\mathcal{J}$ follow at once: (a) If t is not conjugate to s, then $\mathcal{J}: \mathcal{DJ} \to C^0$ is injective; (b) $\mathcal{J}$ is skew-adjoint with respect to Ω, i.e.

$$\Omega(Z \mid \mathcal{J}W) = -\Omega(\mathcal{J}Z \mid W) \quad \forall\, Z, W \in \mathcal{DJ}. \tag{3.4}$$

When t is not conjugate to s, the phase space Green function $G(\tau, \sigma)$ provides a 'kernel' for the map $\mathcal{J}^{-1}: C^0 \to \mathcal{DJ}$. Define $G: [s,t]^2 \to \mathbb{R}^{2d \times 2d}$ to be the solution of

$$\mathcal{J}(\tau) G(\tau, \sigma) = J\, \delta(\tau - \sigma) \tag{3.5a}$$

satisfying the boundary conditions

$$\pi_q G(s, \sigma) = 0 = \pi_q G(t, \sigma) \quad \forall \sigma \in (s, t). \tag{3.5b}$$

Then the inhomogeneous problem (3.2) has a solution $Z \in \mathcal{DJ}$ given by the formula

$$Z(\tau) = \int_s^t d\sigma\, G(\tau, \sigma) J^{-1} f(\sigma). \tag{3.6}$$

Since $\mathcal{J}$ is skew-adjoint it readily follows that the Green function has the symmetry property (where $^\top$ denotes matrix transpose)

$$G(\tau, \sigma) = G(\sigma, \tau)^\top. \tag{3.7}$$

Fortunately, the Green function may be represented via an explicit combination of solutions to Hamilton's equation (1.5), and it is in this form that it appears in the study of the perturbation series for the problem (2.7). The notation $z_c(\tau; Q)$ denoted a trajectory obeying the two-point boundary condition (1.6). Of equal importance is the class of trajectories with specified initial data. The classical phase flow $g_c(\tau): \mathbb{R}^{2d} \to \mathbb{R}^{2d}$ is the family of maps (indexed by $\tau \in \mathbb{R}$) which arises when an initial value problem is posed for (1.5). Specifically, if the initial condition $z_c(t) = \zeta$ is imposed at time t, then in terms of this solution z_c to (1.5) the phase flow is defined by

$$g_c(\tau)(\zeta) = g_c(\tau \mid \zeta) = z_c(\tau). \tag{3.8}$$

The notations $g_c(\tau)$ and $g_c(\tau \mid \cdot)$ will be used interchangeably depending upon convenience. The Green function is built from both $g_c(\tau \mid \zeta)$ and $z_c(\tau; Q)$. Upon setting $H''(\tau) = \nabla\nabla H_c(z_c(\tau; Q), \tau)$ and requiring that t is not conjugate to s along $z_c(\cdot; Q)$, it may be shown that

$$\begin{aligned} G_c(\tau, \sigma; Q) = {} & -\Theta(\tau - \sigma) \frac{\partial g_c(\tau \mid x, p_c(t; Q))}{\partial p} \frac{\partial z_c(\sigma; Q)}{\partial x}^\top \\ & - \Theta(\sigma - \tau) \frac{\partial z_c(\tau; Q)}{\partial x} \frac{\partial g_c(\sigma \mid x, p_c(t; Q))}{\partial p}^\top, \end{aligned} \tag{3.9}$$

where Θ is the Heaviside unit step function. It is this Green function that enters the WKB expansion coefficients, cf. (1.15).

4. Perturbation Theory with Phase Space Variables

Quantum systems defined by perturbed quadratic Hamiltonians have propagators which admit exponential cluster expansions. This section restates the kernel valued Dyson series in terms of the classical trajectories of the quadratic portion of the full Hamiltonian. This revised representation is the starting point of the linked-graph summation method. These topics were treated in Ref. [**1**] for the special case of $\hat{p}$-quadratic (rather than $\hat{z}$-quadratic) unperturbed Hamiltonians, which have geodesic trajectories.

We begin with a summary and comparison of classical and quantum dynamics for quadratic systems. As in Section 3 we define all quadratic systems of interest via the real symmetric coupling matrix $H''(\tau)$. The quadratic quantum Hamiltonian and its Weyl symbol classical counterpart are given, respectively, by

$$\widehat{H}_0(\tau) = \frac{1}{2}\hat{z}\cdot H''(\tau)\hat{z}, \tag{4.1a}$$

$$H_0(z,\tau) = \frac{1}{2}z\cdot H''(\tau)z. \tag{4.1b}$$

Subscript 0 will indicate quantities associated with quadratic systems. The trajectory of H_0 satisfying two-point boundary condition (1.6) is $z_0(\tau;Q)$, while $S_0(Q)$ is its action.

The propagator $K_0(Q) \equiv \langle x|\, U_0(t,s)\, |y\rangle$ generated by $\widehat{H}_0(\tau)$ has an explicit representation given in terms of classical quantities,

$$K_0(Q) = (2\pi\hbar)^{-d/2} e^{-i(\pi/4)\,\mathrm{sign}\, H_{22}(s) + i\pi\mu_0/2} \big|D_0(t,s)\big|^{1/2} e^{iS_0(Q)/\hbar}. \tag{4.2}$$

The right side of this formula is just the 0th order WKB approximation [**3, 16, 20**], which for quadratic Hamiltonians is free of any higher $\hbar$-correction terms. The action S_0 assumes the homogeneous quadratic form

$$S_0(Q) = x\cdot A(t,s)x + x\cdot B(t,s)y + y\cdot C(t,s)y, \tag{4.3}$$

where the $d\times d$ matrices A, B, C may be constructed from Jacobi fields. This form of the action implies that the Van Vleck determinant $D_0(Q) = \det(-\partial^2 S_0(Q)/\partial x\partial y)$ is independent of x and y. The phase factor in $K_0(Q)$, composed of the signature of $H_{22}(s)$ [$\equiv \partial^2 H_0(z,s)/\partial p^2$] and Maslov index μ_0 along $z_0(\tau;Q)$, ensures that the propagator becomes $\delta(x-y)$ as $t\to s$ and induces a continuous evolution operator. If t is not conjugate to s then A, B and C are non-singular and $D_0(Q)$ is both finite and non-zero.

The detailed linkage between classical and quantum motion will be examined in terms of the interaction picture for the position-momentum observable $\hat{z}$. The time dependent operator

$$\hat{z}(\tau,t) \equiv U_0(\tau,t)^\dagger \hat{z} U_0(\tau,t) \tag{4.4a}$$

is the solution of the Heisenberg equation of motion

$$i\hbar\frac{d}{d\tau}\hat{z}(\tau,t) = \left[\hat{z}(\tau,t), U_0(\tau,t)^\dagger \widehat{H}_0(\tau) U_0(\tau,t)\right] \tag{4.4b}$$

obeying the initial condition $\hat{z}(t,t) = \hat{z}$. Because $\widehat{H}_0(\tau)$ is $\hat{z}$-quadratic, (4.4b) reduces to a linear equation for $\hat{z}(\tau,t)$. The resultant solution reads

$$\hat{z}(\tau,t) = g_0(\tau)\hat{z}. \tag{4.5}$$

The quantity $g_0(\tau)$ is the classical flow for $H_0(z,\tau)$ which is 'pinned' at time t, $g_0(t) = \mathrm{Id}$. Because H_0 is quadratic, the flow is a linear map of $\mathbb{R}^{2d}$ represented by a symplectic matrix.

The identity (4.5), for quadratic Hamiltonian dynamics, relates the classical and quantum initial value problems. In the ensuing development it is the two-point boundary value problem with data Q that is required. There arises a new position-momentum operator defined via the similarity transformation

$$\widehat{Z}(\tau;Q) \equiv K_0(Q)^{-1}\hat{z}(\tau,t)K_0(Q). \tag{4.6}$$

The description of $\widehat{Z}(\tau;Q)$ in terms of classical motion now takes the form

$$\widehat{Z}(\tau;Q) = z_0(\tau;Q) + \frac{\partial g_0(\tau)}{\partial p}\hat{p}. \tag{4.7}$$

When acting on a wave function $\psi(x)$, the quantity $z_0(\tau;Q)$ is a multiplication operator with respect to the variable x. One important consequence of (4.7) is the non-equal time commutator

$$\frac{1}{i\hbar}\big[\widehat{Z}^j(\tau;Q), \widehat{Z}^k(\tau';Q)\big] = \big\{g^j(\tau), g^k(\tau')\big\}\big(x, \nabla S_0(Q)\big). \tag{4.8}$$

The $\tau = \tau'$ value of the Poisson bracket on the right of (4.8) is J^{jk}. For non-quadratic systems the right side of (4.8) is no longer exact, but it is the leading term of an asymptotic expansion in $\hbar$.

With this summary of quadratic dynamics in place, we turn to the task of characterizing perturbation theory in terms of phase space trajectories. The general evolution problem to be considered is

$$i\hbar\frac{\partial}{\partial t}\widetilde{U}(t,s) = \big(\widehat{H}_0(t) + V(t)\big)\widetilde{U}(t,s), \quad \widetilde{U}(s,s) = I, \tag{4.9}$$

where $V(\tau)$ is a sufficiently smooth perturbation of the quadratic operator $\widehat{H}_0(\tau)$. The principal application occurs when $H''(\tau)$ is the Hessian $\nabla\nabla H_c(z_c(\tau;Q),\tau)$ and $V(\tau)$ the higher order fluctuation potential (2.6e).

Iteration of the operator valued integral equation equivalent to (4.9) leads to the standard time ordered Dyson series

$$\widetilde{U}(t,s) = \sum_{N=0}^{\infty}\frac{1}{(i\hbar)^N}\int_{\Delta_N^>} d^N\tau\, V(\tau_1,t)\cdots V(\tau_N,t)U_0(t,s). \tag{4.10}$$

The doubly argumented $V(\tau,t)$ is the interaction picture operator

$$V(\tau,t) \equiv U_0(\tau,t)^\dagger V(\tau)U_0(\tau,t), \tag{4.11}$$

whereas the integration region is the open hypertriangle $\Delta_N^> \equiv \big\{\tau \in [s,t]^N \mid t > \tau_1 > \tau_2 > \cdots > \tau_N > s\big\}$. Observe that in (4.11) the operator $V(\tau,t)$ coincides with its Schrödinger value $V(t)$ at time $\tau = t$. Next, pass from (4.10) to its kernel analog by taking Dirac matrix elements with respect to $\langle x|\cdots|y\rangle$,

$$\widetilde{K}(Q) = \sum_{N=0}^{\infty}\frac{1}{(i\hbar)^N}\int_{\Delta_N^>} d^N\tau\, V(\tau_1,t)\cdots V(\tau_N,t)K_0(Q), \tag{4.12}$$

where the operators $V(\tau_l,t)$ are understood to be functions of $\hat{z} = (\hat{q},\hat{p})$ wherein $\hat{q}$ acts via multiplication by x and $\hat{p}$ is $(\hbar/i)\nabla_x$.

As a final preparatory step, the unperturbed propagator $K_0(Q)$ will be pushed all the way to the left in (4.12). This is accomplished by conjugating each perturbation operator $V(\tau_l, t)$ in (4.12) with $K_0(Q)$. To this end define

$$V(\tau; Q) \equiv K_0(Q)^{-1} V(\tau, t) K_0(Q). \tag{4.13}$$

Note that this expression is non-singular for all t not conjugate to s. Thus the factored form of $\widetilde{K}$ is

$$\widetilde{K}(Q) = K_0(Q) F(Q), \tag{4.14a}$$

$$F(Q) = \sum_{N=0}^{\infty} \frac{1}{(i\hbar)^N} \int_{\Delta_N^{>}} d^N\tau \, V(\tau_1; Q) \cdots V(\tau_N; Q) 1. \tag{4.14b}$$

The scalar valued function $F(Q)$, referred to as the configuration function, is the multiplier required to convert the quadratic kernel K_0 into the full propagator $\widetilde{K}$.

To proceed further a detailed understanding of the integrand in (4.14b) is required. Begin with $V(\tau, t)$. If the perturbation $V(\tau)$ is represented via its Weyl symbol $V(\tau)_{\mathrm{w}}$, then

$$V(\tau, t) = V(\tau)_{\mathrm{w}} \left(\frac{1}{i} \nabla_u \right) \exp\bigl[iu \cdot \hat{z}(\tau, t) \bigr] \Big|_{u=0}. \tag{4.15}$$

Going one step further and conjugating (4.15) with K_0 leads to

$$V(\tau; Q) = V(\tau)_{\mathrm{w}} \left(\frac{1}{i} \nabla_u \right) \exp\bigl[iu \cdot \widehat{Z}(\tau; Q) \bigr] \Big|_{u=0}. \tag{4.16}$$

In this way one has an explicit expression of $V(\tau; Q)$ in terms of the 'phase space coordinate operators' $\widehat{Z}(\tau; Q)$.

Utilizing these ingredients, the integrand of the Nth term in (4.14b) reads

$$\mathcal{I}_N = \left(\prod_{l=1}^{N} V(\tau_l)_{\mathrm{w}} \left(\frac{1}{i} \nabla_{u^l} \right) \right) e^{iu^1 \cdot \widehat{Z}(\tau_1; Q)} \cdots e^{iu^N \cdot \widehat{Z}(\tau_N; Q)} 1 \Big|_{u=0}. \tag{4.17a}$$

The expression above simplifies by $\hat{q}$–$\hat{p}$ normal ordering the exponential product of operators. A complete normal ordering in closed form is possible because the non-equal time commutator (4.8) is a known c-number function, and because in (4.7) z_0 is (x, y)-linear and the $\hat{p}$ dependence is also linear. In these circumstances repeated application of the Baker-Campbell-Hausdorff formula leads, after some manipulation, to

$$\mathcal{I}_N = \exp\left\{ \frac{i\hbar}{2} \sum_{k,l=1}^{N} \nabla_{u^k} \cdot G_0(\tau_k, \tau_l) \nabla_{u^l} \right\} \prod_{l=1}^{N} V(\tau_l)_{\mathrm{w}} \left(z(\tau_l; Q) + u^l \right) \Big|_{u=0}. \tag{4.17b}$$

In this expression G_0 is the phase space Green function for the Jacobi operator of Hamiltonian H_0 along the trajectory $z_0(\cdot; Q)$. One clarification of (4.17b) is essential. Observe that one encounters the Green function $G_0(\tau_l, \tau_l)$ with coincident time arguments. In fact, this diagonal value of G_0 is not uniquely determined by the classical Jacobi Green function problem (3.5). Recall that $G_0(\tau, \sigma)$ constructs inhomogeneous solutions to the H_0-Jacobi equation by an integration over σ (see (3.6)), so altering the value of G_0 on a set of measure zero cannot change the inhomogeneous solution. The value of $G_0(\tau, \tau)$ is fixed by defining $\Theta(0) = 1/2$ in

(3.9). With this convention, the two expressions for $\mathcal{I}_N$ in (4.17a) and (4.17b) are in full agreement.

The last stage of simplification involves removing the time ordering restriction $\Delta_N^>$. The integrand $\mathcal{I}_N$ given by (4.17b) is a symmetric function of $(\tau_1, \dots, \tau_N)$ in the hypercube $[s,t]^N$. This symmetry follows from that of the Green function (3.7) and the commutativity of the scalar valued symbols $V(\tau_l)_{\mathrm{w}}$ (see Ref. [**21**]). Thus the ordered integration region $\Delta_N^>$ may be replaced by an integral over $[s,t]^N$ and a compensating factor of $1/N!$. We also make a notational modification in order to remove the somewhat elaborate u-derivative notation found in (4.17b). Instead, let us employ the phase space gradient $\mathbb{D}_k$ which is factor-specific in that it acts only on the perturbation symbol $V(\tau_k)_{\mathrm{w}}$, i.e.

$$\mathbb{D}_k \prod_{l=1}^{N} V(\tau_l)_{\mathrm{w}} = [\nabla V(\tau_k)_{\mathrm{w}}] \prod_{l \neq k} V(\tau_l)_{\mathrm{w}}.$$

Putting together all these changes, the desired form of the configuration function series reads

$$\begin{aligned}(4.18) \quad F(Q) = \sum_{N=0}^{\infty} \frac{(i\hbar)^{-N}}{N!} \int_{[s,t]^N} d^N\tau \prod_{1 \le k < l \le N} &\exp\{i\hbar \mathbb{D}_k \cdot G_0(\tau_k, \tau_l)\mathbb{D}_l\} \\ &\times \prod_{l=1}^{N} e^{(i\hbar/2)\mathbb{D}_l \cdot G_0(\tau_l,\tau_l)\mathbb{D}_l} V(\tau_l)_{\mathrm{w}}\big(z(\tau_l;Q)\big)\end{aligned}$$

The preceding analysis has rendered the perturbative series for $F(Q)$ in terms of two elementary classical quantities: the $H_0(z,\tau)$ phase space trajectory $z_0(\tau;Q)$, and the Green function $G_0(\tau,\tau')$ describing linearized deviation about $z_0(\tau;Q)$. Result (4.18) is useful in its own right for efficiently computing the leading terms in perturbation theory.

5. Cluster Exponentiation

The final aspect of our phase space analysis rests on identifying the graph combinatorial structure present at each order of perturbation, and then utilizing this structure to exactly sum the Dyson series for the configuration function.

From the outset the basic goal has been to obtain a formally exact solution of the fluctuation problem (2.6)–(2.7). So, in order to apply the results of Section 4 we must set $H''(\tau) = \nabla\nabla H_c(z_c(\tau;Q),\tau)$ and define the perturbation $V(\tau)$ by (2.6e). It is seen from the kernel relationship (2.8) linking $K(Q)$ to $K_{z_c(\cdot;Q)}(t,s,0,0)$ that a coordinate-restricted form of the series (4.18) is required. The variables (t,s,x,y) of Section 4 need to be relabeled (t,s,x',y') and then the endpoint conditions $x' = y' = 0$ are imposed. This greatly simplifies the dynamical functions associated with quadratic Hamiltonians. Specifically one has

$$(5.1a) \qquad z_0(\tau;t,s,0,0) = 0,$$

$$(5.1b) \qquad S_0(t,s,0,0) = 0,$$

$$(5.1c) \qquad D_0(t,s;Q) = D_c(Q),$$

$$(5.1d) \qquad G_0(\tau,\sigma;Q) = G_c(\tau,\sigma;Q).$$

Statement (5.1a) says that the quadratic trajectory has collapsed to a point. This fact forces the action S_0 to vanish. The equivalence of the quadratic Van Vleck determinant and Green function with those of the full H_c-classical dynamics holds because both satisfy the same equations of motion and the same boundary conditions.

Formulae (5.1) imply a corresponding simplification of the diagonal quadratic propagator from (4.14a), namely

$$K_0(t,s,0,0;Q) = (2\pi\hbar)^{-d/2}\left|D_c(Q)\right|^{1/2} e^{i\Phi(Q)}, \tag{5.2a}$$

$$\Phi(Q) \equiv \frac{\pi}{2}\mu(Q) - \frac{\pi}{4}\operatorname{sign}\frac{\partial^2 H_c}{\partial p^2}(z_c(s;Q),s). \tag{5.2b}$$

The Q dependence of K_0 arises from the fact that the Hessian defining $\widehat{H}_0(\tau)$ is Q dependent. Similar remarks apply to $D_0(t,s;Q)$ and $G_0(\tau,\sigma;Q)$. Formula (4.18) involves the gradients of $V(\tau)_{\mathrm{w}}(\zeta) = H_c^{(3)}(\zeta, z_c(\tau;Q),\tau)$ (in view of (2.6e)). They are

$$\mathbb{D}^\alpha V(\tau)_{\mathrm{w}}(0) = \nabla_\zeta^\alpha H_c^{(3)}(\zeta, z_c(\tau;Q),\tau)\big|_{\zeta=0}. \tag{5.3}$$

Next, consider the combinatorial nature of the Nth order perturbation expression for $F(t,s,0,0,\hbar)$. In order to clarify the structure present it is helpful to define

$$a_{kl} \equiv \left[e^{i\hbar\mathbb{D}_k\cdot G_c(\tau_k,\tau_l;Q)\mathbb{D}_l} - 1\right], \quad f_l \equiv e^{(i\hbar/2)\mathbb{D}_l\cdot G_c(\tau_l,\tau_l;Q)\mathbb{D}_l} H_c^{(3)}(0, z_c(\tau_l;Q),\tau_l).$$

With this notation the Nth contribution to (4.18) reads

$$\frac{1}{N!}\frac{1}{(i\hbar)^N}\int_{[s,t]^N} d^N\tau \prod_{1\le k<l\le N}(1+a_{kl})\, f_1\cdots f_N.$$

The multiple integral here is the sum of $2^{N(N-1)/2}$ integrals arising from independent choices of 1 or a_{kl} in each factor of $\prod_{k<l}$. Each of these integrals may be associated with an N-vertex graph: the function f_l is represented by the vertex labeled l; the product of operators a_{kl} defines an edge set with at most one link between each pair of vertices. If such a graph is disconnected it factors into a product of lower dimensional integrals which are connected. Examining the first few terms of the series (4.18) reveals that the low order simple connected graphs are

$$\mathcal{L}_1(Q,\hbar) = \frac{1}{i\hbar}\int_s^t d\tau\, f_1, \quad \mathcal{L}_2(Q,\hbar) = \frac{1}{(i\hbar)^2}\int_{[s,t]^2} d^2\tau\, a_{12} f_1 f_2,$$

$$\mathcal{L}_3(Q,\hbar) = \frac{1}{(i\hbar)^3}\int_{[s,t]^3} d^3\tau\,(a_{12}a_{13} + a_{12}a_{23} + a_{13}a_{23} + a_{12}a_{13}a_{23}) f_1 f_2 f_3.$$

The general j^{th} order cluster integral is given by (1.9).

In terms of this graph characterization the configuration function series (4.18) reads

$$F(Q,\hbar) = \sum_{N=0}^{\infty}\sum_m^{(N)} \frac{1}{m!}\prod_{j=1}^{N}\left(\frac{\mathcal{L}_j(Q,\hbar)}{j!}\right)^{m_j}. \tag{5.4}$$

The summation $\sum^{(N)}$ is over all N-multi-indices $m = (m_1, \dots, m_N)$ which partition N, i.e. $\sum_{j=1}^{N} jm_j = N$. For such a partition, m_j describes the number of j-vertex clusters in an Nth order graph. The summation of (5.4) into exponential form results simply by applying the Faa di Bruno formula [**10**] to the composite function $\exp \circ (\sum_{j\geq 1} \mathcal{L}_j / j!)$. The result is (1.3b).

Having completed the formal construction of the cluster expansion it is appropriate to summarize the various restrictive assumptions made in the course of the derivation. In each application of the WKB cluster method these assumptions need to be verified.

1. The existence of a single trajectory $z_c(\cdot; Q)$ satisfying the classical fixed endpoint boundary value problem is required. Of course, the existence of at least one trajectory satisfying (1.5)–(1.6) is a generic ingredient in any WKB description of $K(Q)$.
2. It is assumed that the Weyl symbol $\widehat{H}(\tau)_{\mathrm{w}}$ of the quantum Hamiltonian is sufficiently smooth in its variables z, τ, $\hbar$. With respect to z our formal calculations assume analyticity.
3. It is required that the fluctuation evolution problem, (2.7) and (4.9), admit a perturbative expansion based on the leading quadratic terms of the Hamiltonian $\widehat{H}_{z_c(\cdot;Q)}$. This means that the associated Dyson series must be either a convergent or an asymptotic series in the parameter $\hbar$.
4. A crucial assumption in our propagator study is that the final time t not be conjugate to s relative to $\mathcal{J}(\tau; Q)$.

References

1. A. O. Barvinsky and T. A. Osborn, *A phase-space technique for the perturbation expansion of Schrödinger propagators*, Univ. of Manitoba preprint MANIT-92-01, 1992.
2. M. V. Berry, *Semi-classical mechanics in phase space: A study of Wigner's function*, Philos. Trans. Roy. Soc. London Ser. A **287** (1977), 237–271.
3. M. V. Berry and K. E. Mount, *Semiclassical approximations in wave mechanics*, Rep. Prog. Phys. **35** (1972), 315–397.
4. G. D. Birkhoff, *Some remarks concerning Schrödinger's wave equation*, Proc. Nat. Acad. Sci. U.S.A. **19** (1933), 339–344.
5. C. DeWitt-Morette, *The semiclassical expansion*, Ann. Physics **97** (1976), 367–399.
6. C. DeWitt-Morette, A. Maheshwari, and B. Nelson, *Path integration in phase space*, Gen. Relativity Gravitation **8** (1977), 581–593.
7. ——, *Path integration in non-relativistic quantum mechanics*, Phys. Rep. **50** (1979), 255–372.
8. V. V. Dodonov and V. I. Man'ko, *Phase space eigenfunctions of multidimensional quadratic Hamiltonians*, Phys. A **137** (1986), 306–316.
9. D. Elworthy, A. Truman, and K. Watling, *The semi-classical expansion for a charged particle on a curved space background*, J. Math. Phys. **26** (1985), 984–990.
10. I. S. Gradshteyn and I. M. Ryzhik, formula 0.430.2, *Table of Integrals, Series, and Products*, Academic, New York, 1980, p. 19.
11. A. Grossmann, G. Loupias, and E. M. Stein, *An algebra of pseudodifferential operators and quantum mechanics in phase space*, Ann. Inst. Fourier (Grenoble) **18** (1968), 343–368.
12. J. R. Klauder and I. Daubeschies, *Quantum-mechanical path integrals with Wiener measure for all polynomial Hamiltonians. II*, J. Math. Phys. **26** (1985), 2239–2256.
13. J. R. Klauder and E. C. G. Sudarshan, *Fundamentals of quantum optics*, Benjamin, New York, 1968.
14. F. Langouche, D. Roekaerts, and E. Tirapegui, *WKB expansion for arbitrary Hamiltonians*, Nuovo Cimento A **64** (1981), 357–376.
15. R. G. Littlejohn, *The semiclassical evolution of wave packets*, Phys. Rep. **138** (1986), 193–291.
16. V. P. Maslov, *The quasi-classical asymptotic solutions of some problems in mathematical physics*, U.S.S.R. Comput. Math. and Math. Phys. **1** (1962), 123–141.

17. V. P. Maslov and M. V. Fedoriuk, *Semiclassical approximation in quantum mechanics*, Reidel, Dordrecht, 1981.
18. M. M. Mizrahi, *On the semiclassical expansion in quantum mechanics for arbitrary Hamiltonians*, J. Math. Phys. **18** (1977), 786–790.
19. ——, *Phase space path integrals, without limiting procedure*, J. Math. Phys. **19** (1978), 298–307.
20. ——, *On the WKB approximation to the propagator for arbitrary Hamiltonians*, J. Math. Phys. **22** (1981), 102–107.
21. F. H. Molzahn, *Exponential cluster solutions to quantum transport equations*, J. Phys. A **25** (1992), 4913–4940; J. Phys. A **26** (1993), 2275–2276 (Corrigendum).
22. F. H. Molzahn and T. A. Osborn, *A phase space fluctuation method for quantum dynamics*, Ann. Physics **230** (1994), 343–394.
23. J. E. Moyal, *Quantum mechanics as a statistical theory*, Math. Proc. Cambridge Philos. Soc. **45** (1949), 99–124.
24. T. A. Osborn and F. H. Molzahn, *Structural connections between two semiclassical approximations: The WKB and Wigner–Kirkwood approximations*, Phys. Rev. A **34** (1986), 1696–1707.
25. M. E. Taylor, *Noncommutative harmonic analysis*, Math. Surveys Monographs, vol. 22, American Mathematical Society, Providence, RI, 1986.
26. H. Weyl, *Quantenmechanik und gruppentheorie*, Z. Phys. **46** (1927), 1–46.
27. E. P. Wigner, *On the quantum correction for thermodynamic equilibrium*, Phys. Rev. **40** (1932), 749–759.
28. R. J. Wilson, *Introduction to graph theory*, Chap. 2, Academic, New York, 1975.

Department of Physics, University of Manitoba, Winnipeg, MB, Canada, R3T 2N2

E-mail address: tosborn@ccm.umanitoba.ca

Department of Physics, University of Manitoba, Winnipeg, MB, Canada, R3T 2N2

E-mail address: molzahn@cc.umanitoba.ca

Centre de Recherches Mathématiques
CRM Proceedings and Lecture Notes
Volume 8, 1995

The Selberg Zeta Function and Scattering Poles for Kleinian Groups

Peter A. Perry

ABSTRACT. This paper discusses joint work with S. J. Patterson.

Hyperbolic $n+1$-dimensional space H^{n+1} is a simply connected space of constant curvature that may be modelled by the unit ball in R^{n+1} with metric $(1-|z|^2)^{-2}|dz|^2$ or by the upper half-plane in R^{n+1} with metric $x_{n+1}^{-2}|dx|^2$. The isometries of H^{n+1} are Möbius transformations.

We consider nonsimply connected manifolds M of constant negative curvature and infinite volume obtained as quotients of H^{n+1} by certain discrete groups of isometries called Kleinian groups. The *Selberg zeta function* describes the length spectrum of closed geodesics on M, while the *scattering operator* describes the continuous spectrum of the Laplace-Beltrami operator on M. We show that the zeros of the Selberg zeta function are determined by the topology M, the eigenvalues of the Laplacian on M, and the poles of the scattering operator. We use this information to derive a Selberg trace formula for Kleinian groups.

1. Introduction

The relationship between the poles of the scattering operator and the geometry of the underlying scattering system has been the subject of intensive investigation in both acoustical and Schrödinger scattering. In recent years, attention has also focussed on 'geometric scattering' where the underlying dynamics is given by the Schrödinger or wave equation on a Riemannian manifold: the goal is to understand how geometric information is reflected in the scattering operator associated to a non-compact Riemannian manifold.

Just as acoustical and Schrödinger scattering theories require strong control on the spatial asymptotic behavior of the potential or obstacle, so geometric scattering theory requires strong control on the asymptotic geometry of the manifold. Although it is often not possible to construct a comparison scattering theory in the usual sense, it is desirable that the asymptotic geometry be described, 'locally' at least, by a known space such as a simply connected symmetric space. Although there are challenging problems connected with scattering on simply connected manifolds, we will consider non-simply connected ones with nontrivial closed geodesics. We will study a class of Riemannian manifolds, constructed as quotients of a simply connected space by discrete groups of isometries, for which this connection can be specified explicitly through the Selberg trace formula.

1991 *Mathematics Subject Classification.* Primary: 11F72, 58G25; Secondary: 35P25.
Supported in part by NSF grant DMS-9343893.
This is an announcement-details will be published elsewhere

Hyperbolic $n+1$-dimensional space $\mathbb{H}^{n+1}$ is a simply connected space of constant negative curvature which can be modelled as the open unit ball in $\mathbb{R}^{n+1}$ with the Poincaré metric

$$ds^2 = \frac{4\,|dz|^2}{(1-|z|^2)^2}.$$

The isometries of $\mathbb{H}^{n+1}$ are Möbius transformations. Manifolds of constant negative curvature with nontrivial topology and geometry can be constructed as quotients of $\mathbb{H}^{n+1}$ by discrete groups Γ of isometries. These quotient manifolds M have a natural Laplace operator Δ whose spectrum carries geometric information about M. In this lecture, we consider the relationship between the scattering operator $S(s)$ for Δ and the Selberg zeta function $Z(s)$ for a class of quotient manifolds M called convex co-compact hyperbolic manifolds. These manifolds are the union of a compact manifold-with-boundary and finitely many ends diffeomorphic to cylinders $\mathbb{R} \times N$ where N is a compact manifold without boundary of dimension n. The Selberg zeta function is the dynamical zeta function associated to geodesic flow on M. Meromorphy of the scattering operator follows from [**2, 9, 19, 24, 29**]; see also the lectures of Agmon in this volume for related results for manifolds of finite hyperbolic volume, and see [**3, 4, 7, 10, 11, 12, 14, 15, 16, 17, 18, 19, 25**] for other related work on the spectral theory of the Laplacian on M. Here we use the meromorphy of the scattering operator to prove a trace formula which enables us to characterize the zeros of the zeta function in terms of scattering poles and topological data of M. Using results of dynamics and function theory, we will derive an analogue of the Selberg trace formula for convex co-compact hyperbolic manifolds. This formula gives a precise relationship between scattering poles and lengths of closed geodesics.

2. Hyperbolic Space, Kleinian Groups, and Hyperbolic Manifolds

Real hyperbolic $n+1$-dimensional space $\mathbb{H}^{n+1}$ is such a globally symmetric space of constant sectional curvature -1. It may be modelled either as the unit ball $\mathbb{B}^{n+1}$ in $\mathbb{R}^{n+1}$ with metric

$$ds^2 = \frac{4\,|dx|^2}{(1-|x|)^2},$$

where $|dx|^2$ denotes the usual Euclidean metric, as the upper half-space $\mathbb{R}^{n+1}$ with metric

$$ds^2 = x_{n+1}^{-2}\,|dx|^2,$$

or as the hyperboloid

$$x_1^2 + \cdots + x_n^2 - x_{n+1}^2 = 1 \tag{1}$$

in $\mathbb{R}^{n+1}$ with metric induced from (1).

A geometric notion which is important for scattering theory is the notion of 'boundary at infinity' or 'geometric boundary.' Let M be a complete Riemannian manifold, and consider the unit speed geodesics $\gamma(t)$ of M which escape to infinity as $t \to \infty$. Call two such geodesics γ_1 and γ_2 *equivalent* if $\operatorname{dist}(\gamma_1(t), \gamma_2(t))$ is bounded as $t \to \infty$. The geometric boundary of M, denoted $\partial_\infty M$, is the set of equivalence classes of all geodesics that escape to infinity under this equivalence relation; this set corresponds to the set of scattering directions. In the case of

Euclidean $\mathbb{R}^{n+1}$, the geometric boundary is S^n; it is easiest to see in the ball model that the geometric boundary of $\mathbb{H}^{n+1}$ is also S^n.

The isometries of $\mathbb{H}^{n+1}$ are Möbius transformations that preserve $\mathbb{B}^{n+1}$ (ball model), Möbius transformations that preserve $\mathbb{R}^{n+1}$ (upper half-space model), or transformations of $SO(n,1)$, the linear transformations of determinant one preserving the quadratic form respectively (hyperboloid model).

There are three kinds of isometries of $\mathbb{H}^{n+1}$. *Hyperbolic transformations* have two fixed points at infinity and are conjugate in the upper half-space model to the composition of a dilation $x \mapsto e^{\ell} x$ and an $SO(n)$ rotation of the variables $x_1, \ldots,$ x_n. The quantity ℓ is called the *length* of the transformation. The length and the eigenvalues of the $SO(n)$ twisting matrix are conjugacy invariants. *Parabolic transformations* have a single fixed point at infinity and are conjugate in the upper half-space model to Euclidean motions of $\mathbb{R}^n$ that leave x_{n+1} unchanged. *Elliptic motions* are non-Euclidean rotations with a single fixed point either in $\mathbb{H}^{n+1}$ or on its boundary at infinity.

If γ is any hyperbolic isometry and $x \in \mathbb{H}^{n+1}$, then its differential $\gamma'(x)$ is a positive scalar multiple of an orthogonal transformation; we denote this scalar multiple by $|\gamma'(x)|$ and call it the *conformal dilation of* γ at x. Isometries of $\mathbb{H}^{n+1}$ extend to conformal (angle-preserving) maps on $\partial_\infty \mathbb{H}^{n+1}$. For boundary points x', we also denote by $|\gamma'(x')|$ the conformal dilation of the induced transformation on $\partial_\infty \mathbb{H}^{n+1}$.

A group Γ of isometries of $\mathbb{H}^{n+1}$ is *discrete* if it is discrete as a topological subgroup of the isometries of $\mathbb{H}^{n+1}$: in the hyperboloid model, this simply means that Γ is topologically discrete as a subgroup of $SO(n,1)$. Such a discrete group Γ acts *properly discontinuously* on $\mathbb{H}^{n+1}$, i.e., for any element $\gamma \in \Gamma$ and any bounded open subset $U \subset \mathbb{H}^{n+1}$, at most finitely many Γ-images of U intersect U. The elements of Γ also act on $\partial_\infty \mathbb{H}^{n+1}$ (as is most easily seen in the ball model); the *domain of discontinuity* $\Omega(\Gamma)$ is the subset of $\partial_\infty \mathbb{H}^{n+1}$ on which Γ acts properly discontinuously, and its complement in $\partial_\infty \mathbb{H}^{n+1}$ is called the *limit set* $\Lambda(\Gamma)$. A *fundamental domain* $\mathcal{F}$ for Γ in $\mathbb{H}^{n+1}$ is an open subset of $\mathbb{H}^{n+1}$ so chosen that for any $z \in \mathbb{H}^{n+1}$, $z = \gamma(z')$ for some $z' \in \mathcal{F}$, and z' is unique unless $z' \in \partial\mathcal{F}$.

The group Γ is called *geometrically finite* if there is a finite-sided fundamental domain $\mathcal{F}$, and the 'sides' are geodesic hyperplanes. In the upper half-space model, geodesic hyperplanes are either Euclidean hyperplanes parallel to the x_{n+1}-axis or hemispheres that intersect the Euclidean hyperplane $x_{n+1} = 0$ normally. Since we are interested in scattering theory, we consider geometrically finite groups Γ whose fundamental domains have infinite hyperbolic volume: we call such groups *co-infinite volume* discrete groups. We also impose two further conditions. Let $C(\Gamma)$ denote the hyperbolic convex hull of $\Lambda(\Gamma)$ in $\mathbb{H}^{n+1}$. We require that Γ be *convex co-compact*, i.e., for any fundamental domain $\mathcal{F}$, $\mathcal{F} \cap C(\Gamma)$ is a compact subset of $\mathbb{H}^{n+1}$. Finally, we require that Γ be *torsion-free*, i.e., that Γ contain no elements of finite order.

If Γ is geometrically finite, co-infinite volume, convex co-compact, and torsion free, then the quotient $\mathbb{H}^{n+1}/\Gamma$ is a complete Riemannian manifold M of infinite volume, and may be written $K \cup E$ where K is a compact convex manifold with smooth boundary and E is diffeomorphic to $(0,1) \times N$ where N is a compact Riemannian manifold without boundary, and may be multiply connected. The manifold K is the 'core' of M and E contains the end or ends of M. Heuristically, a scattering process consists of a wave propagating in from an end of M, scattering in the compact core K, and moving out again into the ends of M .

The geometric boundary of M is diffeomorphic to the smooth boundary of K. In fact, K can be chosen so that the diffeomorphism from E to $(0,1) \times N$ and the pullback of the hyperbolic metric to $(0,1) \times N$ calculated explicitly. The geometric boundary $B = \partial_\infty M$ is also diffeomorphic to $\Omega(\Gamma)/\Gamma$. Since the transformations of Γ act conformally on $\Omega(\Gamma)$, the geometric boundary inherits a natural conformal structure: that is, B admits an atlas of charts whose transition maps are conformal maps of $\mathbb{R}^n$. These maps are none other than elements of Γ acting on $\Omega(\Gamma)$.

Finally, it will be useful to have another description of M. There is a finite cover of M by charts of two types: (i) interior charts isometric to geodesic balls in hyperbolic space $\mathbb{H}^{n+1}$, and (ii) boundary charts isometric to the hemisphere $\{x \in \mathbb{R}_+^{n+1} : |x| < 1\}$ (upper half-space model).

3. The Laplace Operator, Spectral Theory, and the Scattering Operator

Let M be the quotient of $\mathbb{H}^{n+1}$ by a geometrically finite, convex co-compact, torsion-free discrete group of infinite co-volume. The Laplace-Beltrami operator on M is known to have at most finitely many eigenvalues in the interval $[0, n^2/4)$ and purely absolutely continuous spectrum of infinte multiplicity on the interval $[n^2/4, \infty)$ (see Lax and Phillips [**16**]). Thus the resolvent operator $R(s) = (-\Delta - s(n-s))^{-1}$ is a bounded-operator valued function in the half-plane $\mathrm{Re}(s) > n/2$ with at most finitely many poles corresponding to eigenvalues of Δ, and has boundary values between suitable weighted spaces on the line $\mathrm{Re}(s) = n/2$ which corresponds to the continuous spectrum. The resolvent also has a meromorphic continuation into the half-plane $\mathrm{Re}(s) < n/2$ whose poles correspond to the poles of the scattering operator (see [**2, 9, 17, 18, 19, 20, 24**]). The change of variables $s = n/2 - ik$ gives a parameterization of the resolvent's analyticity properties more familiar to workers in Schrödinger operators, where the resolvent is a bounded operator for $\mathrm{Im}(k) > 0$, has boundary values on the real k-axis, and has a meromorphic continuation to the half-plane $\mathrm{Im}(k) < 0$. On the other hand the parameterization in terms of s connects more closely with the usual definition of the Selberg zeta function.

The eigenfunctions that span the continuous spectral subspace of the Laplace operator are denoted $e(x, b; s)$ where $x \in M$, $b \in B$ and $s \in \mathbb{C}$. They represent waves originating at a point b at spatial infinity and scattering in the manifold. Because of the special geometric situation, these eigenfunctions are not ordinary smooth functions on the boundary but rather obey special transformation laws which we now describe. An *automorphic form of complex weight* s is a smooth function f on $\Omega(\Gamma)$ such that for every $\gamma \in \Gamma$, the identity

$$f(\gamma(x)) \, |\gamma'(x)|^s = f(x)$$

holds. Such forms can also be identified as smooth sections of a complex line bundle over B obtained as the quotient of $\Omega(\Gamma) \times \mathbb{C}$ with respect to the equivalence relation $(x, z) \sim (x', z')$ if $x' = \gamma(x)$ and $|\gamma'(x)|^s z' = z$ for some $\gamma \in \Gamma$. We denote the space of all such smooth functions by $\Gamma(s)$. Forms of complex weight n, interpreted as sections of a complex line bundle over B, are actually densities on B and can be invariantly integrated on B. Thus the spaces $\Gamma(s)$ and $\Gamma(n-s)$ are dual under the dual pairing $\langle f, g\rangle = \int_B f(x) g(x) \, dx$. For $\mathrm{Re}(s) = n/2$, complex conjugation maps

$\Gamma(s)$ to $\Gamma(n-s)$ and we may define an inner product on $\Gamma(s)$ by the formula

$$(f,g) = \int_B f(x)\overline{g(x)}\, dx.$$

Completing $\Gamma(s)$ in this inner product gives a Hilbert space $\mathcal{H}(s)$.

The functions $e(x,b;s)$ are, for fixed s and x, forms of complex weight s. As functions of s they are initially defined in the half-plane $\mathrm{Re}(s) > n/2$ and admit a meromorphic continuation to the half-plane $\mathrm{Re}(s) < n/2$. The *scattering operator* $S(s)$ is a unitary operator from $\mathcal{H}(n-s)$ to $\mathcal{H}(s)$ with the property that

$$e(x,b;s) = S(s)\big(e(x,\cdot,n-s)\big)(b)$$

where the notation means that the scattering operator acts on the boundary variable of $e(x,b;n-s)$.

The scattering operator, initially defined as a unitary map from $\mathcal{H}(n-s)$ to $\mathcal{H}(s)$ for $\mathrm{Re}(s) = n/2$, admits a meromorphic extension to a map from $\Gamma(n-s)$ to $\Gamma(s)$ for all but a discrete set of s. Unlike its counterpart in Schrödinger theory, it is not even a bounded operator for $\mathrm{Re}(s) > n/2$! Instead it is a pseudodifferential operator whose Schwarz kernel has leading singularity

$$k(x,y;s) \sim |x-y|^{-2s}$$

modulo smooth remainders.

Intuition from other scattering problems suggests that the poles of the meromorphically continued scattering operator should determined by closed geodesics on M. In order to show that this is the case we need to define the Selberg zeta function.

4. Closed Geodesics and the Selberg Zeta Function

Associated to each closed geodesic γ on M is an element of the fundamental group of M. These elements can be viewed as elements of the discrete group Γ. Denoting the corresponding group element by γ also, we let $\ell(\gamma)$ denote the length of γ and we let $\alpha_1(\gamma), \dots, \alpha_n(\gamma)$ denote the eigenvalues of the $SO(n)$ twisting matrix associated to γ. A closed geodesic is called *primitive* if it is not the composition (in the homotopy sense) of other closed geodesics.

The Selberg zeta function is defined as follows. Let $\{\gamma\}$ be a listing of the primitive closed geodesics of M. It is not difficult to see that the lengths obey the crude estimate

$$\#\{\gamma : \ell(\gamma) < x\} \le C\exp(nx)$$

so that the infinite product

$$Z(s) = \prod_{\{\gamma\}} \prod_{k_1,\dots,k_n \ge 0} \Big(1 - \alpha_1(\gamma)^{k_1} \dots \alpha_n(\gamma)^{k_n} \exp\big(-(s+k_1+\dots+k_n)\ell(\gamma)\big)\Big)$$

converges for $\mathrm{Re}(s) > n$. Note that when $n=1$ and M is a hyperbolic surface, this formula reduces to the simpler form

$$Z(s) = \prod_{\{\gamma\}} \prod_{k=0}^{\infty} \Big(1 - \exp\big(-(s+k)\ell(\gamma)\big)\Big).$$

Selberg introduced the zeta function in the context of hyperbolic surfaces of finite volume. Just as the Riemann zeta function has analytic properties reflecting the distribution of primes, so the Selberg zeta function has analytic properties reflecting the distribution of lengths of primitive geodesics. In the compact case, the zeta function also obeys an analogue of the Riemann hypothesis: its zeros are explicit zeros at the integers $s = 1, 0, -1, \ldots$ together with zeros on the line $\mathrm{Re}(s) = 1/2$ determined by eigenvalues of the Laplace operator on the surface. In the finite-volume, non-compact case, its zeros are either of the above type or are determined by poles of the scattering operator, a finite matrix-valued function of s.

We will show that the Selberg zeta function for our manifolds of infinite volume also has an analytic continuation to $\mathbb{C}$ and that its zeros are determined by the scattering poles of $S(s)$ and the underlying topology of M. From this will follow a Hadamard product representation of the zeta function and a Selberg-type trace formula that relates the length spectrum of closed geodesics to the eigenvalues of the Laplacian and scattering poles.

5. The Main Theorem

We can now state the main result.

THEOREM 5.1. *Let Γ be a geometrically finite, convex co-compact, torsion-free discrete group with infinite co-volume, and let $M = H^{n+1}/\Gamma$. The Selberg zeta function $Z(s)$ extends to an entire function of s with the following zeros:*

(a) Spectral zeros σ *with* $\mathrm{Re}(\sigma) > n/2$ *and $s(n-s)$ an eigenvalue of the Laplacian on M.*

(b) Scattering zeros ρ *with* $\mathrm{Re}(\rho) < n/2$ *and ρ a pole of the scattering operator $S(s)$.*

(c) Topological zeros *at $s = -k, 0, -1, -2, \ldots$ with multiplicities*

$$(2m + 2k - 1)\frac{(2m + k - 2)!}{(2m-1)!k!}\frac{(k + 2m - 2)!}{k!}\chi(M)$$

where $\chi(M)$ is the Euler characteristic of M.

Using methods of dynamics, Patterson and Perry have shown that the zeta function is an entire function of order at most $n + 1$. This result, together with the theorem above, imply:

THEOREM 5.2. *Under the hypotheses of Theorem 5.1,*

$$Z(s) = \exp\bigl(P(s)\bigr)G(s)^{\chi(M)}\prod_{\sigma}\Bigl(1 - \frac{s}{\sigma}\Bigr)\prod_{\rho}E\Bigl(\frac{s}{\rho}, n + 1\Bigr)$$

where $E(z, k) = (1 - z)\exp(z + z^2/2 + \cdots + z^k/k)$, $P(s)$ is a polynomial of degree at most $n + 1$ and $G(s)$ is a known entire function.

The coefficients of the polynomial $P(s)$ are not known. Nonetheless, we can use the Hadamard product representation above and the Euler product that defines the zeta function for $\mathrm{Re}(s) > n$ to obtain the following trace formula that provides a quantitative link between closed geodesics and scattering poles.

THEOREM 5.3 (SELBERG TRACE FORMULA). *Let $\hat{h} \in C_0^{\infty}(0, \infty)$. The for-*

mula

$$(2)\quad \sum_{\sigma} h(r_\sigma) + \sum_{\{\gamma\}} \sum_{m=1}^{\infty} \frac{e^{-m\ell(\gamma)(n/2)}}{\det(I - K(\gamma)^m e^{-m\ell(\gamma)})} \hat{h}(\ell(\gamma)) = \chi(M) \int H(t) h(t)\, dt + \sum_{\rho} h(r_\rho)$$

holds.

Here $K(\gamma)$ is the $SO(n)$ matrix associated to γ and $H(t) = G'(n/2-it)/G(n/2-it)$. Convergence of the sum over poles follows from the fact that, as zeros of an entire function of bounded order, the distribution function $N(x) = \#\{r_\rho : |r_\rho| \leq x\}$ grows at most polynomially in x.

6. Relative Trace Formulas and a 'Trace-Class Rigidity Theorem'

In two dimensions (where $n = 1$) it is possible to write down a more refined version of the trace formula due to Patterson [**25**]. We begin with the observation that, for the abelian group of dilations Γ of the form

$$\{(x_1, x_2) \mapsto \mu^n (x_1, x_2) : n \in \mathbb{Z}\}$$

where $\mu > 1$, a fundamental domain for Γ is the region

$$\{(x_1, x_2) : 1 \leq \sqrt{x_1^2 + x_2^2} \leq \mu\}$$

and the quotient $\mathbb{H}^2/\Gamma$ is a hyperbolic cylinder C_μ whose boundary at infinity is two circles and which has a single closed geodesic of length $\ell = \log \mu$. Given a hyperbolic surface M of infinite area of the type considered above, M is the union of a compact convex surface K bounded by N closed geodesics of length ℓ_i and N half-cylinders. The boundary at infinity of M is a finite union N circles. Let $S(s)$ be the scattering operator on M, let M_2 be the union of two copies of M and let

$$S_2(s) = \begin{pmatrix} S(s) & 0 \\ 0 & S(s) \end{pmatrix}$$

be the scattering operator on M_2. Let M_1 be the union of N hyperbolic cylinders C_{μ_i} with $\mu_i = e^{\ell_i}$. Then $S_2(s)$ and $S_1(s)$ can be viewed as acting on the same spaces (automorphic forms on the union of N circles) and it can be shown that $S_2(s) S_1(s)^{-1} = I + K(s)$ where $K(s)$ is a trace-class operator-valued function of s. As such the determinant $\Phi(s) = \det(I + K(s))$ is well-defined. Letting $Z_2(s) = Z_M(s)^2$ and $Z_1(s) = \prod_{i=1}^{N} Z_{C_i}(s)$ we then can show that

$$\frac{d}{ds} \log\left(\frac{Z_2(s)}{Z_1(s)}\right) = 4\pi \frac{\Gamma(s)\Gamma(1-s)}{\Gamma(s-1/2)\Gamma(1/2-s)} \chi(M) + \frac{d}{ds} \log \det(I + K(s))$$

where $\chi(M)$ is the Euler characteristic of M. From this and the order estimates on the zeta functions follows the trace formula

$$\sum_{\sigma} h(r_\sigma) + \sum_{\{\gamma\}} w_\gamma \sum_{m=1}^{\infty} \frac{e^{-m\ell(\gamma)(n/2)}}{\det(I - K(\gamma)^m e^{-m\ell(\gamma)})} \hat{h}(\ell(\gamma)) = 4\pi\chi(M) \int \frac{\Gamma(1/2-it)\Gamma(1/2+it)}{\Gamma(-it)\Gamma(it)} h(t)\, dt + \sum_{\rho} h(r_\rho)$$

where w_γ is a weight of either 1 or 2 depending on whether γ is a closed geodesic contained in the interior of K or is one of the geodesics that bound K, and r_ρ are zeros of the determinant.

This formula suggests the interesting possibility of 'renormalizing' the scattering operators of higher-dimensional hyperbolic manifolds by comparing with the scattering operators of hyperbolic cylinders. The following theorem shows that this is difficult if not impossible to do in higher dimensions.

THEOREM 6.1 ("TRACE-CLASS RIGIDITY"). *Let Γ_1 and Γ_2 be discrete groups acting on hyperbolic three-dimensional space $\mathbb{H}^3$ with domains of discontinuity $\Omega(\Gamma_1)$ and $\Omega(\Gamma_2)$. Let $M_1 = \mathbb{H}^3/\Gamma_1$ and $M_2 = \mathbb{H}^3/\Gamma_2$. Suppose that there is a diffeomorphism $\psi\colon \Omega(\Gamma_1) \to \Omega(\Gamma_2)$ which induces an isomorphism of Γ_1 and Γ_2. Let $S_1(s)$ and $S_2(s)$ be the scattering operators for Γ_1 and Γ_2, and let $S_{\mathrm{rel}}(s) = S_1(s) - \psi^* S_2(s)$. Suppose that $S_{\mathrm{rel}}(s)$ is a trace-class operator for* $\mathrm{Re}(s) = 1$. *Then Γ_1 and Γ_2 are conjugate in the group of isometries of $\mathbb{H}^3$, and M_1 and M_2 are isometric.*

The proof uses the explicit leading singularity of the scattering operator and an isomorphism theorem of Marden [**21**].

The trace formulae presented here provide a starting point for more detailed investigations into the zeta function and the distribution of scattering poles. Of particular interest are: (1) sharp upper and lower bounds on the asymptotic distribution of scattering poles, (2) study of determinants of Laplacians and their behavior under deformations, and (3) 'isospectral' problems involving metric perturbations of hyperbolic manifolds. A first upper bound on the distribution of scattering poles was proved by Perry [**31**] using the observation that the set of scattering poles is contained in the set of zeros of the zeta function and the zeta function is an entire function of bounded order. More recently, Sjöstrand and Zworski [**33**] obtained lower bounds on the number of scattering poles and Guillopé and Zworski obtained optimal upper bounds in two dimensions together with some results in higher dimensions [**13**].

References

1. S. Agmon, *Spectral theory of Schrödinger operators on Euclidean and non-Euclidean spaces*, Comm. Pure Appl. Math. **39** (1986), Number S, Supplement.
2. ——, *On the spectral theory of the Laplacian on non-compact hyperbolic manifolds*, Journées 'Équations aux derivées partielles', Exp. **XVII**, École Polytechnique, Palaiseau, 1987.
3. Y. Colin de Verdière, *Théorie spectral des surfaces de Riemann d'aire infinie*, Astérisque **132** (1985), 259–275.
4. J. Elstrodt, *Die Resolvante zum Eigenwertproblem der automorphen Formen in der hyperbolischen Ebene* I, Math. Ann. **203** (1973), 295–330; II, Math. Z. **132** (1973), 99–134; III, Math. Ann. **208** (1974), 99–132.
5. ——, *Die Selbergsche Spurformel für kompakte Riemannsche Flächen*, Jber. d. Dt. Math.-Verein **83** (1981), 45–77.
6. C. Epstein, (private communication).
7. J. D. Fay, *Fourier coefficients of the resolvent for a Fuchsian group*, J. Reine Angew. Math. **293/294** (1977), 143–203.
8. J. Fischer, *An approach to the Selberg trace formula via the Selberg zeta function*, Lecture Notes in Math. **1253**, Springer, New York, 1987.
9. R. Froese, P. Hislop, and P. Perry, *The Laplace operator on hyperbolic three-manifolds with cusps of non-maximal rank*, Invent. Math. **106** (1991), 295–333.
10. L. Guillopé, *Sur la distribution des longeurs des géodesiques fermées d'une surface compacte á bord totalement géodesique*, Duke Math. J. **53** (1986), 827–848.

11. ——, *Théorie spectrale de quelques variétés à bouts*, Ann. Sci. École Norm. Sup. **22** (1989), 137–160.
12. ——, *Fonctions zêta de Selberg et surfaces de géometrie finie*, Adv. Stud. Pure Math. **21** (1992), 33–70.
13. L. Guillopé and M. Zworski, *Upper bounds on the number of resonances for non-compact surfaces*, 1993 (preprint).
14. P. Lax and R. S. Phillips, *Translation representations for solutions of the non-Euclidean wave equation* I, Comm. Pure Appl. Math. **32** (1979), 617–667; II, **34** (1981), 347–358.
15. ——, *The asymptotic distribution of lattice points in Euclidean and non-Euclidean spaces*, J. Funct. Anal. **46** (1982), 280–350.
16. ——, *Translation representation for automorphic solutions of the wave equation in non-Euclidean spaces* I, Comm. Pure Appl. Math. **37** (1984), 303–324; II, 779–813; III, **38** (1985), 197–208.
17. N. Mandouvalos, *The theory of Eisenstein series and spectral theory for Kleinian groups*, Thesis, University of Cambridge, 1983.
18. ——, *The theory of Eisenstein series for Kleinian groups*, The Selberg Trace Formula and Related Topics, Contemp. Math., vol. 53, Amer. Math. Soc., Providence, RI, 1986, 357–370.
19. ——, *Spectral theory and Eisenstein series for Kleinian groups*, Proc. London. Math. Soc. **57** (1988), 209–238.
20. ——, *Scattering operator, Eisenstein series, inner product formula, and "Maass-Selberg relations' for Kleinian groups*, Mem. Amer. Math. Soc., vol. 78, Amer. Math. Soc., Providence, RI, 1989, no. 400.
21. A. Marden, *The geometry of finitely-generated Kleinian groups*, Ann. Math. **99** (1974), 383–462.
22. G. D. Mostow, *Quasiconformal mappings in n-space and the rigidity of hyperbolic space forms*, Inst. Hautes Études Sci. Publ. Math. **34** (1968), 53–104.
23. ——, *The rigidity of locally symmetric spaces*, Actes, Congrès intern. Math. **2** (1970), 187–197.
24. R. Mazzeo and R. Melrose, *Meromorphic extension of the resolvent on complete spaces with asymptotically constant negative curvature*, J. Funct. Anal. **75** (1987) 260–310.
25. S. J. Patterson, *The Laplace operator on a Riemann surface* I, Compositio Math. **31** (1975), 83–107; II, **32** (1976), 71–112; III, **33** (1976), 227–259.
26. ——, *The Selberg zeta function of a Kleinian group*, Number Theory, Trace Formulas, and Discrete Groups: Symposium in Honor of Atle Selberg, (Oslo, Norway, July 14–21, 1987) Academic Press, New York, 1989, pp. 409–442.
27. ——, *On Ruelle's zeta function*, Math. Gött., Schriftenreihe des Sonderforschungsbereichs Geometrie und Analysis **36** (1989) (preprint).
28. S. J. Patterson and P. A. Perry, *Divisor of the Selberg zeta function for Kleinian groups*, (in preparation).
29. P. A. Perry, *The Laplace operator on a hyperbolic manifold* II. *Eisenstein series and the scattering matrix*, J. Reine Angew. Math. **398** (1989), 67–91.
30. ——, *The Selberg zeta function and a local trace formula for Kleinian groups*, J. Reine Angew. Math. **410** (1990), 116–152.
31. ——, *The Selberg zeta function and scattering poles for Kleinian groups*, Bull. Amer. Math. Soc. **24** (1991), 327–333.
32. A. Selberg, *Harmonic analysis and discontinuous groups in weakly symmetric Riemannian spaces with applications to Dirichlet series*, J. Indian Math. Soc. **20** (1956), 47–87.
33. J. Sjöstrand and M. Zworski, *Lower bounds on the number of scattering poles*, 1993 (preprint).

Department of Mathematics, University of Kentucky, Lexington, Kentucky 40506-0027, U. S. A.

E-mail address: perry@ms.uky.edu

Centre de Recherches Mathématiques
CRM Proceedings and Lecture Notes
Volume 8, 1995

Charge Transport in Quantum Hall Systems and the Index of Projectors

R. Seiler

1. Introduction

Charge transport in non-dissipative systems has often a natural geometric interpretation. This became clear in recent years mainly through the analysis of quantum Hall systems in the framework of quantum mechanics. More recently the same phenomenon emerged in quantum field theory as well.

The quantum field theory point of view has been discussed by J. Fröhlich in the first week of this summer school, [**12**]. In this lecture I shall report on some recent work by J. Avron, B. Simon and myself [**3, 4**] and talk about certain aspects of the integer quantum Hall effect in the framework of quantum mechanics.

Let me remind you of the experimental setup and result. In a two dimensional electron layer the relation between Hall current and voltage is observed (Figures 1, 2) in the presence of a very large magnetic field in the transversal direction to the plane and at very small temperature.

For small current the functional dependance is linear (Ohm's law). Conductivity is an integer in units of e^2/h with very high precision. For more details I refer you to one of the several books for instance the one by Girvin and Prange [**18**].

Hall conductivity has been related to integers in two different manners. J. Bellissard — using technics of A. Connes — identified Hall conductivity as a Fredholm index [**6**]. Thouless, Kohomoto, Nightingale and de Nijs realized that the Hall conductivity is a topological integer [**19**] which turned out to be a Chern number. In both cases the authors considered a one particle model with random respectively periodic potential.

The Chern number approach has been extended to the many body case including interactions [**5, 15, 16**]. For the Fredholm index approach this is not yet the case.

In all approaches to the quantum Hall effect a key role is played by the gap condition and by gauge invariants. The gap condition infers that the groundstate is separated from the rest of the spectrum by a quantity which is large compared with the termal energy kT.

The objective of this lecture is to present the Fredholm index approach from a different point of view and to show how this approach is related to the Chern number approach. No technical details will be given.

1991 *Mathematics Subject Classification.* Primary: 81-06; Secondary 81Q99.
Summer School in Mathematical Quantum Theory, Vancouver, August 1993
This is the final form of the paper.

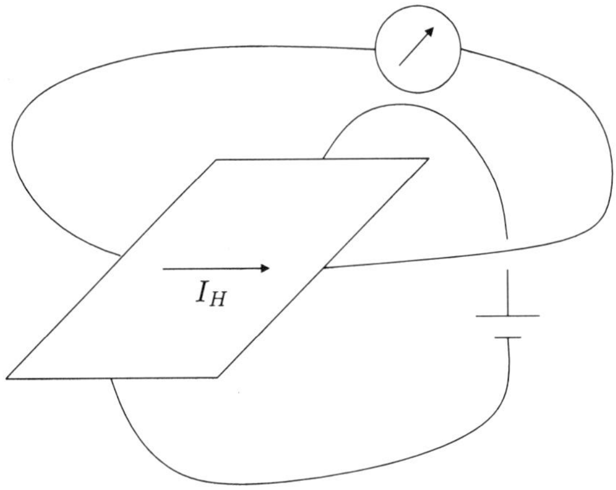

FIGURE 1. Experimental setup.

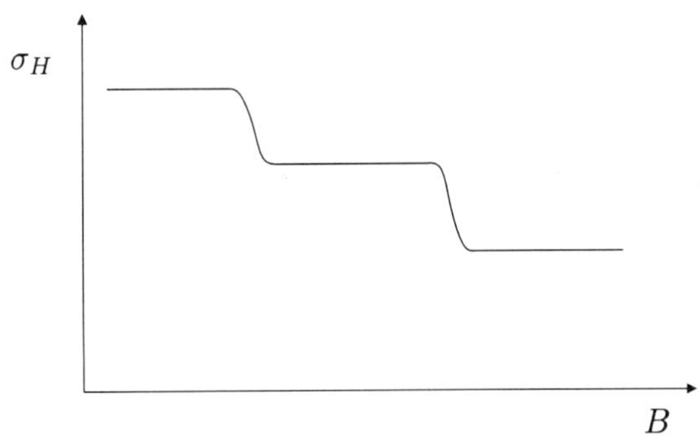

FIGURE 2. Hall conductivity as a function of B.

2. Charge Deficiency

The framework is that of non interacting nonrelativistic charged fermions moving in the plane $(x^1, x^2, x^3 = 0) \in \mathbb{R}^3$. Dynamics is defined in terms of the single particle Schrödinger operator

$$H = \frac{1}{2}(p - A)^2 + V. \tag{1}$$

Where $B = \operatorname{rot} A$ is the external magnetic field, p denotes the momentum operator and V a one particle potential. H acts on functions in

$$\mathcal{H} = L^2\left(\mathbb{R}^2, dx^1\, dx^2 = \frac{1}{2i}\, dz\, d\bar{z}\right), \quad z = x^1 + ix^2.$$

A is a continuously differentiable vector potential and V a scalar potential in the Kato class.

Later we shall consider the situation where V is a random potential V_ω, $\omega \in \Omega$.

Physical states will always be zero temperature states of a non interacting, many body fermion system. In the single particle language this is a projector onto

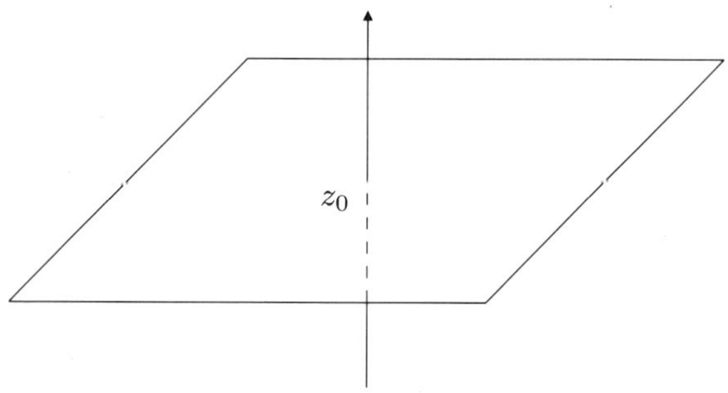

FIGURE 3. A magnetic flux piercing the plane at z_0.

all eigenstates below the Fermi energy of the Schrödinger operator in question. Hence a physical many body state corresponds to a spectral projector of the one particle Schrödinger operator H. The number of particles in the many body state is formaly given by Trace P which is of course infinit in most cases.

Now we come to the definition of charge deficiency. Let me motivate this concept first in physical terms. Consider some fermions moving in the plane in a state P. "Add addiabatically a magnetic flux unit through the point $z_0 \in \mathbb{R}^2$" (How this is done will be explained below). This produces a new fermionic state Q. It turns out to be unitarily conjugated to P.

In general Q will have a different charge since the adiabatic change of flux has pushed some particles off to infinity or sucked in some from infinity. This charge difference is superficially expressed by $\dim Q - \dim P$ which is in general badly defined. Fortunately there is a natural generalization of this expression in terms of the index of a pair of projections:

DEFINITION 1. Let P and Q be orthogonal projectors such that their difference is compact. Then

$$\operatorname{Ind}(P, Q) := \dim \operatorname{Ker}(P - Q - 1) - \dim \operatorname{Ker}(P - Q + 1). \tag{2}$$

That this is a natural generalization of the difference of dimensions is the content of

THEOREM 1. *Let P, Q be a pair of orthogonal projectors on a Hilbert space and suppose that $P - Q$ belongs to a trace ideal S_{n_0} for some $n_0 \in \mathbb{N}$. Then* $\operatorname{Ind}(P, Q) = \operatorname{Trace}(P - Q)^n$ *for all odd numbers $n \geq n_0$.*

The proof of this theorem is a simple consequence of the pretty algebraic structure of the algebra generated by the two projectors. This will be the subject of Section 3.

Now I want to explain the meaning of "adding adiabtically a flux unit" and try to give you an intuition why in this process some charge is possibly pushed to infinity. Consider the time dependent Schrödinger operator

$$H(t) := \frac{1}{2}\big(p - \Lambda(t)\operatorname{grad}\big(\arg(z - z_0)\big) - A\big)^2 + V \tag{3}$$

where p, A and V are as before and Λ is a switchfunction which interpolates between zero and one.

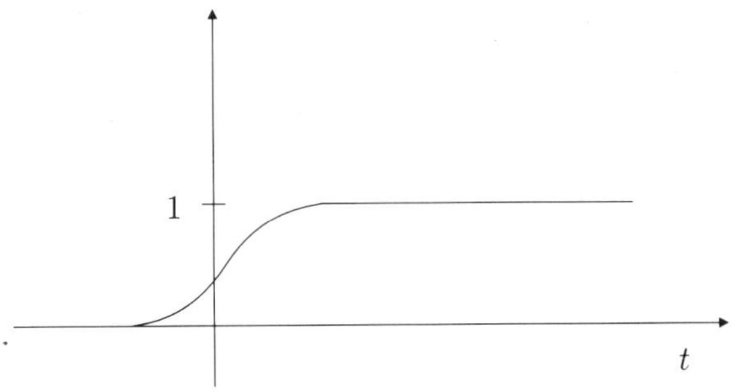

FIGURE 4. Typical switch function.

DEFINITION 2. The realvalued function Λ of $t \in \mathbb{R}$ is called a switch if it is continuously differentiable, monotonically nondecreasing and

$$\Lambda(\infty) = \Lambda(-\infty) + 1.$$

By construction $H(t)$ interpolates between $H(1)$ and UHU^{-1} where U is the multiplication operator by the function $u(z) = z/|z|$.

The vector potential $a_{z_0}(z) = \operatorname{grad}\bigl(\arg(z - z_0)\bigr)$ which is defined for all $z \neq z_0$ generates a unit singular magnetic flux tube piercing the plane $x_3 = 0$ at z_0:

$$\int_{|z-z_0|=r>0} dz\, a_{z_0}(z) = i \int_{|z-z_0|=2\pi} d\bigl(\arg(z - z_0)\bigr) = 2\pi i. \tag{4}$$

Imagine that the state of the system at $t = -\infty$ is P, where P denotes the spectral projection of H

$$P = E_\Delta(H), \quad \Delta = (-\infty, \varepsilon_F).$$

ε_F is the Fermi Energy. If we now evolve P according to the dynamics defined by $H(t)$ and if we do that on a very large time scale (in the adiabatic limit), the final state of the system is $Q = UPU^{-1}$, provided ε_F lies in a spectral gap of H [**9, 13, 5**]. This last condition is crucial. Hence $Q = UPU^{-1}$ is the state into which the system develops if you add very slowly to the state P one unit of magnetic flux. Hence adding one flux unit adiabatically defines a loop in the space of projectors modulo unitary conjugation.

Now I shall explain the physical picture which lies behind the words "charges are pushed to infinity". The time dependent vector potential in $H(t)$ generates an electric field $E(t) = -\dot{\Lambda}(t)a_{z_0}(z)$. It is azymutal and produces together with the magnetic field perpendicular to the 2-plane a Lorentz force acting on charged particles. The resulting drift velocity is radial (Figure 5).

Now let's return to the definition of charge deficiency.

DEFINITION 3. If P denotes a many body state of noninteracting fermions and U the unitary multiplication operator introduced previously, the charge deficiency is defined as the index of P and UPU^{-1}:

$$Q^D = \operatorname{Ind}(P, UPU^{-1})$$

The following theorem tells us that charge deficiency is well defined under natural conditions:

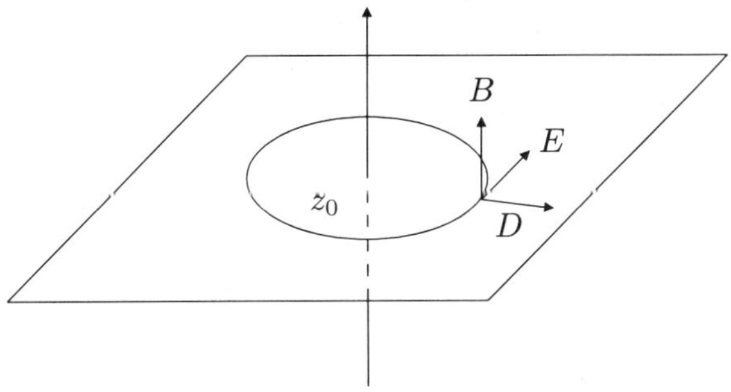

FIGURE 5. Adiabatic switch of magnetic flux produces electric field and drift.

THEOREM 2. *Let P be a projector which can be written in terms of an integral kernel $p(z,z')$ satisfying the following two conditions*:

(i) *$p(z,z')$ is continuous.*

(ii) *p decays off the diagonal like*

$$|p(z,z')| \leq \frac{\text{const}}{\left(1+|z-z'|\right)^{2+\varepsilon}}$$

for some $\varepsilon > 0$ and const $\in \mathbb{R}$.

Then $P - UPU^{-1}$ belongs to the trace ideal $\mathcal{S}_3$ and

$$Q^D = \int_{\mathbb{R}^6} p(z_1,z_2)p(z_2,z_3)p(z_3,z_1)\left(1-\frac{u(z_1)}{u(z_2)}\right)\left(1-\frac{u(z_2)}{u(z_3)}\right)\left(1-\frac{u(z_3)}{u(z_1)}\right).$$

The integral converges in the absolute sense.

The proof of this theorem requires some analysis [**4**]. For smooth p's with compact support it is in [**10, 11**].

3. The v. Neumann Algebra of Two Projectors

The v. Neumann algebra of two projectors has been studied some time ago by Araki and by Kato [**1, 14**]. It appeard again naturally in the context of the quantum Hall effect [**3**].

To investigate the v. Neumann algebra $R = \{P,Q\}''$ it's usefull to note that the pairs of orthogonal projectors can be parametrized by two selfadjoint operators A and B satisfying the two relations

$$A^2 + B^2 = 1 \tag{5}$$

$$\{A,B\} = 0 \tag{6}$$

A and B are expressed in terms of P and Q by the formulas $A = P - Q$, $B = 1 - P - Q$.

Let me summarize the general structural results about R following mainly reference [**8**]. The spectral projectors of A and B for the eigenvalues $+1$ and -1 are minimal central projectors. Their sum

$$E := E_A(1) + E_A(-1) + E_B(1) + E_B(-1) = E_A(0) + E_B(0)$$

projects onto $\ker([P,Q])$. From that one concludes further that R decomposes into an abelian part RE and a I_2 v. Neumann algebra $RE^{\perp}$,

$$R = RE + R(1-E). \tag{7}$$

The non commutative part $RE^{\perp}$ is generated by 3 matrix units

$$\tau_1 = \operatorname{sign} A, \quad \tau_2 = \operatorname{sign} B, \quad \tau_3 = i\tau_1\tau_2, \tag{8}$$

where A and B are now restricted to $E^{\perp}$. They satisfy the commutation relations of the Pauli matrices on range $E^{\perp}$

$$\tau_r\tau_s = i\varepsilon_{rst}\tau_t + 1\delta_{rs}. \tag{9}$$

Every element of $RE^{\perp}$ has a unique representation

$$a = \sum_{l=1}^{3} a_l(|A|)\tau_l \tag{10}$$

where the a_l's are functions of the central generator $|A|$. Hence $RE^{\perp}$ carries a quaternionic structure.

The index of (P,Q) is by definition a property of the abelian part RE only. In order to understand Theorem 1 we have to realize why the contribution to the trace of $(P-Q)^n$ which originates from $RE^{\perp}$ vanishes. This will be explained now.

By definition of the index one gets the formula

$$\operatorname{Ind}(P,Q) = \dim(\operatorname{Range} E_A(1)) - \dim(\operatorname{Range} E_A(-1)). \tag{11}$$

If n is odd and larger or equal than n_0 (see Theorem 1) we get for the trace of A^n the formula

$$\operatorname{Trace} A^n = \operatorname{Trace} A^n E + \operatorname{Trace} A^n E^{\perp} \tag{12}$$

$$= \operatorname{Ind}(P,Q) + \operatorname{Trace} A^n E^{\perp}. \tag{13}$$

Notice that we used the equality

$$E_A(0) = E_B(1) + E_B(-1) \tag{14}$$

which follows from (2). In order to prove Theorem 1 it remains to prove that the second term in (12) vanishes. This follows from the fact that τ_1 and τ_2 anticommute (9); hence

$$\tau_2 A\tau_2 = \tau_2|A|\tau_1\tau_2 \tag{15}$$

$$= -A \tag{16}$$

on range $E^{\perp}$. This proves Theorem 1.

On R there is a natural circle action generating a modular group of automorphism [**8**].

4. Adiabatic Charge Transport

In this section I introduce the adiabatic charge transport for non interacting fermions living in the noncompact configuration space $\mathbb{R}^2$. The key words will be Kubo's Formula, adiabatic connection and adiabatic curvature. The setup is modelled along the lines used previously in the analysis of the quantum Hall effect for interacting particle systems in multiple connected compact configuration space [**2**]. The physical mechanism which produces the charge transport in this case is qualitatively different from the one before. There we "added a flux unit", here we impose a voltage drop across the sample in the one direction which adds up in time to a flux quantum. Then the total charge is measured which moves across the 1-axis.

To put this into analytic terms let us consider the 2-parametric family of operators

$$\begin{aligned} H(\phi) &:= e^{i(\phi,\Lambda)} H e^{-i(\phi,\Lambda)}, \quad (\phi \in \mathbb{R}^2) \\ &= \frac{1}{2}(p - \phi_1 \operatorname{grad}\Lambda_1 - \phi_2 \operatorname{grad}\Lambda_2 - A)^2 + V \end{aligned} \tag{17}$$

where (ϕ, Λ) is the multiplication operator by $\phi_1\Lambda_1(x_1) + \phi_2\Lambda_2(x_2)$. Λ_1 and Λ_2 are switch-functions in the terminology introduced previously.

If we change ϕ_1 adiabatically in time from zero to one we get a family of time dependent Schrödinger operators

$$H_{\phi_2}(t) := \frac{1}{2}\big(p - \phi_1(t) \operatorname{grad}\Lambda_1 - \phi_2 \operatorname{grad}\Lambda_2 - A\big)^2 + V \tag{18}$$

parametrized by ϕ_2. The time dependant piece creates an electric field

$$E(t,x) = -\dot{\phi}_1(t) \operatorname{grad}\Lambda_1(x_1) \tag{19}$$

of the type mentioned before. The current operator is defined by the formula

$$J(t) = \frac{\partial H_{\phi_2}(t)}{\partial t}. \tag{20}$$

Charge transport can be computed in the adiabatic limit using the adiabatic theorem of quantum mechanics [**5**]. I will not explain this point since it is very much the same as in the case of particles in a configuration space of finite size. Let me just mention the main input which goes in this analysis: The Fermi energy ε_F has to be in a gap of the spectrum,

$$P(\phi) = E_{(-\infty,\varepsilon_F]}\big(H(\phi)\big) \tag{21}$$

where E denotes the spectral projector of the Hamiltonian (17) in question.

The adiabatic charge transport is

$$Q^H := (2\pi i) \lim_{L\to\infty} \operatorname{Trace} \chi_L \omega(\phi) \chi_L \tag{22}$$

where χ_L is the multiplication operator by the characteristic function of the square $[-L/2, L/2] \times [-L/2, L/2] \subset \mathbb{R}^2$. ω is given by the following expression

$$\omega(\phi) = P(\phi)\big[\partial_1 P(\phi), \partial_2 P(\phi)\big] P(\phi). \tag{23}$$

Notice that Trace $\chi_L \omega(\phi) \chi_L$ is infact ϕ-independent.

Trace ω has two nice interpretations. Firstly it is Kubo's formula as the following formal computation proves: consider the 2-Form on $\mathbb{R}^2 \widetilde{\omega} := \operatorname{Trace} PdPdP$ and compute it in terms of the resolvant of H:

$$\text{(24)} \quad \operatorname{Trace} PdPdP = \left(\frac{1}{2\pi i}\right)^2 \operatorname{Trace} P \int (H-z)^{-1} dH (H-z)^{-1} P^{\perp} \\ \times (H-\omega)^{-1} dH (H-\omega)^{-1} P$$

Using standard formal manipulations and Cauchy's formula one finds

$$\text{(25)} \quad \operatorname{Trace} PdPdP = \sum_{\varepsilon_\alpha < \varepsilon_F < \varepsilon_\beta} \frac{\langle \alpha | dH | \beta \rangle \langle \beta | dH | \alpha \rangle}{(\varepsilon_\alpha - \varepsilon_\beta)^2}..$$

Secondly Trace ω is the curvature of the adiabatic or canonical connection $\nabla = Pd$ on the vector bundle $E = \{ \langle \phi, \Psi \rangle \mid \phi \in \mathbb{R}^2, \Psi \in \operatorname{Range} P(\phi) \}$.

5. Translational Covariance

In this section it is shown that the two concepts of charge deficience and of charge transport coincide if the system posseses the property of translational covariance. The corner stone is a formula or rather two formulas for the area of a triangle.

The most natural setup for translational covariance is the case of ergodic random Schrödinger operators of the type

$$\text{(26)} \quad H_\omega = \frac{1}{2}(p-A)^2 + V_\omega .$$

DEFINITION 4. H_ω is called covariant if

$$\text{(27)} \quad V_\omega(x) = V_{a\omega}(x-a), \quad (x, a \in \mathbb{R}^2, \omega \in \Omega).$$

This implies for the integral kernel of the spectral projectors the following relation

$$\text{(28)} \quad p_\omega(x,y) = \mathcal{M}_a(x) p_{a\omega}(x-a, y-a) \mathcal{M}_a^{-1}(y),$$

where $\mathcal{M}_a(x)$ is a unitary multiplication operator by a smooth function.

This motivates the crucial assumption in

THEOREM 3. *Let P_ω be a spectral projector of the ergodic family of random Schrödinger operators H_ω as defined by eq.* (26). *Suppose the assumptions of Theorem 2 hold together with translational covariance* (28). *Then*

(i) $Q^D(\omega := \operatorname{Ind}(P_\omega, U P_\omega U^{-1})$ *is an integer and constant for a.e.* $\omega \in \Omega$;

(ii) $\langle Q^H \rangle := \int_\Omega d\mu(\omega) Q_H^\omega = Q^D(\omega)$ *a.e.* $\omega \in \Omega$.

I shall explain the proof of the theorem in the special case of a single deterministic Schrödinger operator. For this simplified case the assertion of the Theorem reduces to $Q^D = Q^H$. The argument contains however already the essential ingrediants.

The proof is based on two formulas for the area of a triangle: The first one is

$$\text{(29)} \quad \det(x,y) = \frac{1}{2\pi i} \int_{\mathbb{R}^2} dz \left(1 - \frac{u(z)}{u(z-x)}\right)\left(1 - \frac{u(z-x)}{u(z-y)}\right)\left(1 - \frac{u(z-y)}{u(z)}\right)$$

where u denotes the function $u(z) = z/|z|$ introduced previously. The second equation for the area of a triangle is

$$\det(x,y) = \int_{\mathbb{R}} dz\, \big(\Lambda_1(z_1+x_1) - \Lambda_1(z_1)\big)\big(\Lambda_2(z_2+y_2) - \Lambda_2(z_2+x_2)\big) \\ \{1 \leftrightarrow 2\} \tag{30}$$

which holds for any pair of switch functions Λ_1, Λ_2 in the terminology of Section 2 (Definition 1).

Now we compute the first of the two charges Q^D using the formula in Theorem 2 and translational covariance

$$p(x,y) = M_a(x)p(x-a, y-a)M_a^{-1}(x) \tag{31}$$

$$Q^D = \int dx\, dy\, dz\, p(x,y)p(y,z)p(z,x)\left(1 - \frac{u(x)}{u(y)}\right)\left(1 - \frac{u(y)}{u(z)}\right) \\ \times \left(1 - \frac{u(z)}{u(x)}\right). \tag{32}$$

After a change of variables suggested by the translational covariance, we can integrate over a two plane. The integral to do is just the one in (29); so we get the formula

$$Q^D = 2\pi i \int dx\, dy\, p(0,x)p(x,y)p(y,0)\det(x,y). \tag{33}$$

A similar computation can be carried through for

$$Q^H := -2\pi i \lim_{L\to\infty} \operatorname{Trace} \chi_L \omega \chi_L. \tag{34}$$

Consider first the adiabatic curvature ω. A direct computation shows that

$$\omega = \big([P,\Lambda_1]P^{\perp}[P,\Lambda_2] - (1 \leftrightarrow 2)\big) \tag{35}$$

which leads to the following integral for the diagonal of the integral kernel

$$\omega(x,x) = \int dy\, dz\, p(x,y)p^{\perp}(y,z)p(z,x) \\ \times \Big\{\big(\Lambda_1(y_1) - \Lambda_1(x_1)\big)\big(\Lambda_2(z_2) - \Lambda_2(x_2)\big) - (1 \leftrightarrow 2)\Big\}. \tag{36}$$

By covariance the integrand can be replaced by

$$p(0,y)p^{\perp}(y,z)p(z,0) \\ \times \Big\{\big(\Lambda_1(y_1+x_1) - \Lambda_1(x_1)\big)\big(\Lambda_2(z_2+x_2) - \Lambda_2(y_2+x_2)\big) - (1 \leftrightarrow 2)\Big\} \tag{37}$$

Hence integration over $x \in \mathbb{R}^2$ leads now to the second area formula (30). In the infinit volume limit this gives formula

$$Q^H = 2\pi i \int dx\, dy\, p(0,x)p(x,y)p(y,0)\det(x,y) \tag{38}$$

Hence the two charges are equal.

6. Concluding Remarks

We have made assumptions about continuity and decay of integral kernels of spectral projectors and it is natural to ask wether such properties can be expected to hold in typical cases. Fortunately this is so [**4**]. I will however not discuss this technical question in this lecture. The crucial property for decay is the spectral gap condition, which holds for instance for the case of constant magnetic field and sufficiently small potential V_ω.

In both approaches mentioned during this talk the adiabatic Theorem of quantum mechanics plays a predominent role since it turns the computation of charge deficiency Q^D and charge transport Q^H into a quantity of particular interest. In the first case Q^D is an index of Fredholm type in the second case Q^H is a Chern number. The appearence of a Chern number can be visualized as follows: Consider a one particle state. It is a solution of the time dependent Schrödinger equation

$$id_t\Psi\big(\phi_1(t),\phi_2\big) = H\big(\phi_1(t),\phi_2\big)\Psi\big(\phi_1(t),\phi_2\big) \tag{39}$$

where $H(\phi)$ is a Schrödinger operator of type (17). The initial conditions are such that $\Psi\big(\phi_1(-\infty),\phi_2\big)$ is in the range of $P(\phi_1 = 0,\phi_2)$ where P denotes the spectral projector needed for the definition of a multiparticle fermion state. $\phi_1(t)$ interpolates between zero and 2π.

The solution of (39) is a section of the trivial vector bundle $T \times \mathcal{H}$ where T denotes Flux space with coordinates $\phi = (\phi_1,\phi_2)$ and $\mathcal{H}$ is the Hilbert space $L^2(dz\,d\bar{z}/2i)$. In the adiabatic limit time evolution takes place in the subbundle $E = \big\{(\phi,\Psi) \mid \Psi \in \operatorname{Range} P(\phi)\big\}$. This vector subbundle can be nontrivial with nontrivial first Chern number (degree of the vectorbundle). It is remarkable that this Chern number is precisely the charge transport in the adiabatic limit [**5**]. That Chern numbers appear naturally in the context of particles in magnetic fields had been noticed before [**17**].

References

1. H. Araki, *On quasi free states of CAR and Bogoliubov automorphisms*, Publ. Res. Inst. Math. Sci. **6** (1970), 385–442.
2. J. E. Avron and R. Seiler *Quantization of the hall conductance for general multiparticle Schrödinger Hamiltonians*, Phys. Rev. Lett. **54** (1985), 259–262.
3. J. E. Avron, R. Seiler, and B. Simon, *The index of a pair of projections*, SFB 288 Preprint, no. 56, (to appear in J. Funct. Anal.).
4. ——, *Charge deficiency, charge transport and compairison of dimensions*, SFB 288 Preprint, no. 45, 1993, (to appear in Comm. Math. Phys.).
5. J. E. Avron, R. Seiler, and L. G. Yaffe, *Adiabatic theorems and applications to the quantum Hall effect*, Comm. Math. Phys. **110** (1987) 33–49.
6. J. Bellissard, *Ordinary quantum Hall effect and non-commutative cohomology*, Localization in Disordered Systems, (W. Weller and P. Zieche, eds.), Teubner, Leipzig, 1988.
7. M. Sh. Birman, *A proof of the Fredholm trace formula as an application of a simple embedding for kernels of integral operators of trace class in $L^2(\mathbb{R}^m)$*, Preprint, Department of Mathematics, Linkping University S-581 83 Linkping, Sweden
8. S. Borac, *On the algebra generated by two projectors*, SFB 288 Preprint, no. 103, (1994).
9. M. Born and V. Fock, *Beweis des Adiabatensatzes*, Z. Phys. **51** (1928), 165–169.
10. A. Connes, *Noncommutative differential geometry*, Inst. Hautes Études Sci. Publ. Math. **62** (1986), 257–360.
11. —— *Géometrie non commutative*, InterEdition, Paris, 1990.
12. J. Fröhlich and K. Gawedzki, *Conformal field theory and geometry of strings*, Mathematical Quantum Theory I: Field Theory and Many-Body Theory, (J. Feldman, R. Froese, L. Rosen,

eds.), CRM Proc. & Lecture Notes, vol. 7, Amer. Math. Soc., Providence, RI, 1994, (to appear).

13. T. Kato, *On the adiabatic theorem of quantum mechanics*, J. Phys. Soc. Japan **5** (1950), 435–439.
14. —— *Perturbation theory of linear operators*, Springer, Berlin, Heidelberg, New York, 1966.
15. M. Klein and R. Seiler, *Power law corrections to the Kubo formula vanish in quantum Hall systems*, Comm. Phys. Rev. Lett. **128** (1990), 141.
16. Q. Niu and D. J. Thouless, *Quantum Hall effect with realistic boundary conditions*, Phys. Rev. B **35** (1986), 2188.
17. S. P. Novikov, *Magnetic Bloch functions and vector bundles* Sov. Math. Dokl. **23** (1981), 298–303.
18. R. E. Prange and S. M. Girvin, *The quantum Hall effect.* Springer, 1987.
19. D. J. Thouless, M. Kohmoto, P. Nightingale, and M. den Nijs, *Quantum Hall conductance in a two dimensional periodic potential* Phys. Rev. Lett. **49** (1982), 40.

TECHNISCHE UNIVERSITÄT BERLIN, STRASSE DES 17.JUNI 136, 10623 BERLIN, GERMANY
E-mail address: seiler@math.tu-berlin.de

Centre de Recherches Mathématiques
CRM Proceedings and Lecture Notes
Volume 8, 1995

Number Theory, Classical Mechanics and the Theory of Large Atoms

L. A. Seco

A non-relativistic atom of nuclear charge Z fixed at the origin, and N quantized electrons at positions $x_i \in \mathbb{R}^3$ is described by the Hamiltonian

$$H_{Z,N} = \sum_{i=1}^{N} \left(-\Delta_{x_i} - \frac{Z}{|x_i|} \right) + \frac{1}{2} \sum_{i \neq j} \frac{1}{|x_i - x_j|}$$

which acts on the Hilbert space $\mathcal{H}$ of antisymmetric functions in $L^2(\mathbb{R}^{3N})$. The ground state energy of such a system is given by

$$E(Z) = \inf_{N \geq 0} E(Z,N), \quad E(Z,N) = \inf_{\substack{\phi \in \mathcal{H} \\ \|\phi\|=1}} \langle H_{Z,N}\phi, \phi \rangle.$$

As Z goes to infinity, the energy $E(Z)$ admits an asymptotic expansion of the form

$$E(Z) = -c_{\mathrm{TF}} Z^{7/3} + \frac{1}{8} Z^2 - c_s Z^{5/3} + O(Z^{5/3-a}), \quad a > 0.$$

The first term above was introduced by Thomas and Fermi in [**Th**], [**Fe**], and proved rigorously in [**LS**] (See also [**Li**] for a review of Thomas-Fermi theory). The Z^2 term was discovered by Scott in [**Sc**] and proved to be true in a series of papers by Hughes-Siedentop-Weikard, in [**Hu**], [**SW1**], [**SW2**] and [**SW3**]. Its generalization to molecules was obtained by Ivrii-Sigal [**IS**]. The $Z^{5/3}$ term was obtained by Schwinger in [**Sch**], and proved to be correct in [**FS1**], [**FS2**], [**FS3**], [**FS4**], [**FS5**], [**FS6**], [**FS7**] and [**FS8**].

In view of (11–18), it is naturally conjectured (see [**Feff**]) that the next term in the energy asymptotics for $E(Z)$ above is given by the following sum

$$\Psi_Q(Z) = \sum_{l=1}^{l_{\mathrm{TF}}} \frac{2l+1}{\left(\int \left(V_{\mathrm{TF}}^Z(r) - l(l+1)/r^2\right)_+^{-1/2} dr\right)/\pi} \times \mu\left(\frac{1}{\pi} \int \left(V_{\mathrm{TF}}^Z(r) - \frac{l(l+1)}{r^2} \right)_+^{1/2} dr \right)$$

1991 *Mathematics Subject Classification.* Primary: 35J10, 81V70; Secondary: 35Q40, 81V55, 81Q15.

This reasearch was supported by a N.A.T.O research grant no. CRG921184, by NSERC grants no. OGP0121848 and EQPEQ336, by a CICYT grant and by a Connaught Fellowship.

This is the final form of the paper.

where $\mu(x)(x,\mathbb{Z})^2 - 1/12$, V^Z_{TF} is the Thomas-Fermi potential for an atom with charge Z (see [**Li**]), which satisfies the perfect scaling condition

$$V^Z_{\mathrm{TF}}(r) = Z^{4/3} V^1_{\mathrm{TF}}(Z^{1/3} \cdot r)$$

where we have

$$V^1_{\mathrm{TF}}(r) = \frac{y(a \cdot r)}{r}, \quad a = \left(\frac{3\pi}{2}\right)^{2/3}$$

and y is the Thomas-Fermi function, solution of the Thomas-Fermi equation

$$\left.\begin{aligned} y''(r) &= \frac{y^{3/2}(r)}{r^{1/2}} \\ y(0) &= 1 \\ \lim_{r \to \infty} y(r) &= 0 \end{aligned}\right\}.$$

Finally, l_{TF} is the greatest integer such that $V^Z_{\mathrm{TF}}(r) - l(l+1)/r^2$ is positive somewhere.

The book of Englert ([**En**]; see also references thereof) contains a discussion of oscillatory terms in the asymptotics of $E(Z)$.

It was proved in [**CFS1**] and [**CFS2**] that this sum Ψ_Q corresponds to a sum of classical data of a certain classical hamiltonian, which would then suggest that the expansion for $E(Z)$ is a trace formula which one would expect from a path integral picture.

The purpose of this paper is to describe the analysis involved in understanding the sum $\Psi_Q(Z)$ as a function of Z, which turns out to be an adaptation of a well-known method in analytic number theory developed mostly by Van der Corput to understand the number of lattice points in a large circle. We begin with a few remarks about analytic number theory.

1. Number Theory

Consider sums of the form

$$S(\lambda) = \sum_{l=1}^{[\lambda]} f\left(\frac{l}{\lambda}\right) \cdot \left(\lambda \cdot \phi\left(\frac{l}{\lambda}\right)\right)$$

where λ is a large number, μ is a periodic function with average 0, f is an amplitude function which can be viewed as constant and ϕ is a smooth function which satisfies the crucial non-degeneracy condition

$$|\phi''(x)| \geq c_0 > 0.$$

Particular cases of sums of this kind give rise to two well-known problems in analytic number theory, namely

1. $f \equiv 1$, $\mu(x) = e^{2\pi i x}$, $\phi(x) = x^2$. In this case, $S(\lambda)$, for λ integer, corresponds to the Gauss sums. The value of S is then known explicitly, and satisfies the estimate

$$S(\lambda) = O(\lambda^{1/2}).$$

2. $f \equiv 1$, $x - [x] - 1/2$, $\phi(x) = \sqrt{1-x^2}$. In this case, S is related to the error $E(\lambda)$ in the lattice point problem for the circle in $\mathbb{R}^2$, which can is defined as follows: take a large circle on $\mathbb{R}^2$ of radius λ, and denote by $N(\lambda)$ the number of lattice points in $\mathbb{Z}^2$ which fall inside this circle. Then

$$E(\lambda) = N(\lambda) - \pi\lambda^2$$

and it is an old problem in number theory to prove that

$$E(\lambda) = O(\lambda^\alpha)$$

for the best posssible value of α. It was observed very early, by Gauss and Dirichlet, that one can take $\alpha = 1$ which is an obvious geometric fact, and is also obviously satisfied by $S(\lambda)$. Different probabilistic approaches (as the one by Cramer, for instance) indicate that α above will not be smaller than $1/2$. What follows is a *brief* historic overview of the estimates for α (see [**GK**] for details).

$$\alpha = \begin{cases} 1, & \text{Gauss-Dirichlet, 1849.} \\ 2/3 = 0.666\ldots, & \text{Voronoi 1904, Hardy, 1917.} \\ 66/100 = 0.66, & \text{Van der Corput 1922.} \\ 163/247 = 0.659919\ldots, & \text{Walfisz 1927.} \\ 27/41 = 0.6585\ldots, & \text{Nieland-Van der Corput 1928.} \\ 15/23 = 0.6521\ldots, & \text{Tichmarsh 1935.} \\ 13/20 = 0.65, & \text{Loo Keng Hua 1942.} \\ 24/37 = 0.6486\ldots, & \text{Kolesnik-Yin Wen Lin 1962.} \\ 35/54 = 0.6481\ldots, & \text{Kolesnik 1971.} \\ 278/429 = 0.648018\ldots, & \text{Kolesnik 1985.} \\ 7/11 = 0.636636\ldots, & \text{Iwaniec-Mozzochi 1988, Huxley 1992.} \end{cases}$$

Note now that the perfect scaling condition of the Thomas-Fermi potential shows that our sum Ψ_Q is (almost exactly) of the form $S(\lambda)$ as defined above, where $\mu(x)(x,\mathbb{Z})^2 - 1/12$, $\lambda = Z^{1/3}$, and

$$\phi(x) = \int \left(\frac{y(r)}{r} - \frac{x^2}{r^2} \right)_+^{1/2} dr.$$

The proof of the non-degeneracy condition for ϕ was done in [**FS8**], and it has the peculiarity that it is a computer assisted proof.

A natural question then arises: what is the level of difficulty in analysing the size of Ψ_Q? Is is as simple as the analysis of the gauss sums above? Or so hard as the analysis of the lattice point problem?

A method devised by Van der Corput (or at least, a variant of it), in his attempts to understand the lattice point problem provides the answer: we compute our sum using Poisson summation, and then we expand each Fourier integral using stationary phase. In doing this, we end up with a sum in which μ is replaced by its Fourier coefficients $\hat{\mu}(n)$. If they decrease fast enough (like $|n|^{-3/2}$, it so happens), the sum is bounded by $\lambda^{1/2}$. In our case, $\mu(n) \sim |n|^{-2}$, therefore, after realizing that the size of our amplitude function is $Z^{4/3}$, we can conclude that $\Psi \sim Z^{4/3+1/6}$.

2. Classical Mechanics

The Van der Corput method does more than merely tell us the size of Ψ. A mechanical analysis of the result given by the stationary phase expansion shows that the function Ψ is (to leading order) a sum of classical data associated to a certain classical hamiltonian, which displays the relationship between classical and quantum mechanics reminiscent of the Feynmann path integral representation of Schrödinger operators.

In order to explain our main result, consider a classical particle with mass 1/2, moving in the negative radial potential given by $-V_{\rm TF}^Z$. Such motion is planar, and we consider the closed orbits at energy 0 which arise for angular momentum M; all such orbits for a fixed M can be obtained by rotations of each other. On each of those orbits, the particle "oscillates" in the sense that the distance to the origin $r(t)$ varies between $r_{\rm min}(M)$ and $r_{\rm max}(M)$ and closes after passing $n(M)$ times through, say, $r_{\rm max}(M)$. We also consider the integer $l(M)$, the winding number of the orbit around 0, the action $S(M)$, the period $T(M)$, and the following locally defined quantity: for a closed trajectory with angular momentum M, and n oscillations, and given ε small, consider a trajectory (not necessarily closed) with angular momentum $M+\varepsilon$ which, after $n(M)$ oscillations between succesive $r_{\rm max}(M+\varepsilon)$, misses to close by an angle $2\pi\alpha_M(\varepsilon)$, where we take α between 0 and 1. Then, we define

$$D(M) = \lim_{\varepsilon\to 0} \varepsilon^{-1}\alpha_M(\varepsilon)$$

In this classical formalism, the non-degeneracy condition $\phi''(x) \neq 0$ stated amounts to the fact that $D(M) \neq 0$ for all closed trajectories. This means that closed trajectories are isolated once we factor out the trivial symmetry given by the rotation group. Its role in the proof is similar to the non-vanishing curvature of the sphere in the circle problem.

With this notation, our main result is as follows:

THEOREM 1.

$$\Psi_Q(Z) = \Psi_0(Z) + o(Z^{3/2})$$

where

$$\Psi_0(Z) = 2\pi\cdot Z^{3/2}\cdot \sum_{\substack{\text{closed trajectories}\\ \text{at energy } 0}} \delta\cdot\frac{\hat\mu(n)}{n}\cdot\frac{M}{T}\cdot\left|D(M)\right|^{-1/2}\cdot e^{i\left(Z^{1/3}S-\pi\cdot(l+(\operatorname{sign} n)/4)\right)}$$

where $\delta = 1$ except in the case of a perfect circular trajectory, when we have $\delta = 1/2$. Furthermore, the sum is absolutely convergent.

We now state precisely the result mentioned in the previous section for the size of $\Psi_Q(Z)$

THEOREM 2. *There are universal constants K (large) and κ_0 (small but strictly positive), such that*

$$\left|\psi_Q(Z)\right| \le K\cdot Z^{3/2}, \qquad \int_{Z_0}^{Z_0+Z} \left|\psi_Q(z)\right|^2\frac{dz}{z^3} \ge \kappa_0\cdot Z$$

whenever $Z \ge KZ_0^{2/3}$. Furthermore,

$$\liminf_{\substack{Z\to\infty\\ Z=1,2,3,\ldots}} \left|Z^{-3/2}\psi_Q(Z)\right| \neq 0$$

The lower bound requires some extra work, and hinges on the fact that a certain number is not zero: the relevance of the non-vanishing of this number is analog to the non vanishing of the $L(\chi, 1)$ in Dirichlet's theorem on the number of primes in arithmetic progressions.

It is interesting to note that the size of the error term above $o(Z^{3/2})$ depends on whether a certain number is rational or not. More precisely, let $r_{\max}$ be the radius of the circular closed trajectory corresponding to the maximal angular momentum $M_{\max}$ at energy 0. Then, if

$$\sqrt{1 - \frac{1}{2} r_{\max} \cdot M_{\max}}$$

is rational, then the error term can be seen to be $O(Z^{3/2-\varepsilon})$. Otherwise, specially if this number is very well approximated by rationals (say, a Louiville number), then in general one cannot improve the o-result. However, if in the sum over closed trajectories above one includes also the trajectories with complex period, then the error term is always of size $O(Z^{3/2-\varepsilon})$.

The proofs of Theorems 1 and 2 can be found in [**CFS2**].

The book of Gutzwiller [**Gu**] contains a discussion of the interplay between classical and quantum mechanics in relation with trace formulas.

References

[CFS1] Córdoba, A., Fefferman, C., Seco, L., *A trigonometric sum relevant to the non-relativistic theory of atoms*, P.N.A.S., to appear.

[CFS2] ——, *Weyl sums and atomic energy oscillations*, Rev. Mat. Iberoamericana, to appear.

[En] Englert, B. G., *Semiclassical Theory of the Atom*, Lecture Notes in Phys., vol 301, Springer Verlag.

[Feff] Fefferman, C., *Atoms and analytic number theory*, Proc. Amer. Math. Soc. (1991).

[FS1] Fefferman, C. and Seco, L., *The Ground-state energy of a large atom*, Bull. Amer. Math. Soc. **23** (1990), no. 2, 525–530.

[FS2] ——, *Eigenvalues and eigenfunctions of ordinary differential operators*, Adv. Math., to appear.

[FS3] ——, *The eigenvalue sum for a one-dimensional potential*, Adv. Math., to appear.

[FS4] ——, *The density in a one-dimensional potential*, Adv. Math., to appear.

[FS5] ——, *The eigenvalue sum for a three-dimensional radial potential*, Adv. Math., to appear.

[FS6] ——, *The density in a three-dimensional radial potential*, Adv. Math., to appear.

[FS7] ——, *On the Dirac and Schwinger corrections to the ground-state energy of an atom*, Adv. Math., to appear.

[FS8] ——, *A periodicity of the Hamiltonian flow in the Thomas-Fermi potential*, Rev. Mat. Iberoamericana, to appear.

[Fe] Fermi, E., *Un metodo statistico per la determinazione di alcune priorieta dell'atome* Rend. Accad. Naz. Lincei **6** (1927), 602–607.

[GK] Graham, S. W. and Kolesnik, G., *Van der Corput's Method of Exponential Sums*, London Math. Soc. Lecture Note Ser., vol. 126, Cambridge Univ. Press.

[Gu] Gutzwiller, M., *Chaos in Classical and Quantum Mechanics*, Springer Verlag, 1990.

[Hu] Hughes, W., *An atomic energy lower bound that agrees with Scott's correction*, Adv. in Math. **79** (1990), 213–270.

[IS] Ivrii, V. and Sigal, I. M., *Asymptotics of the ground state energies of large Coulomb systems*, Ann. Math. **138** (1993), no. 2, 243–335.

[Li] Lieb, E., *Thomas-Fermi and related theories of atoms and molecules*, Rev. of Modern Phys. bf 53 (1981) no. 4, 603–641.

[LS] Lieb, E. and Simon, B., *Thomas-Fermi theory of atoms, molecules and solids*, Adv. Math. **23** (1977), 22–116.

[Sch] Schwinger, J., *Thomas-Fermi model: the second correction* Phys. Rev. A, **24** (1981), no. 5, 2353–2361.

[Sc] Scott, J. M. C., *The binding energy of the Thomas-Fermi atom*, Philos. Mag. **43** (1952), 859–867,

[SW1] Siedentop, H., Weikard, R., *On the Leading Energy Correction for the statistical model of the atom: interacting case* Comm. Math. Phys. **112** (1987), 471–490,

[SW2] ——, *On the Leading Correction of the Thomas-Fermi model: lower bound* and an appendix by A. M. K. Müller. Invent. Math. **97** (1989), 159–193.

[SW3] ——, (1990) *A new phase space localization technique with applications to the sum of negative eigenvalues of Schrödinger operators*, Ann. Sci. École Norm. Sup. **24** (1991), 215–225.

[Th] Thomas, L. H. *The calculation of atomic fields* Proc. Cambridge Philos. Soc. **23** (1927), 542–548.

DEPT. OF MATHEMTAICS, UNIVERSITY OF TORONTO, 100 ST. GEORGE ST., TORONTO, ONTARIO, M5S 1A1, CANADA

E-mail address: seco@math.toronto.edu

Centre de Recherches Mathématiques
CRM Proceedings and Lecture Notes
Volume 8, 1995

Bound for the Atomic Ground State Density at the Nucleus

Heinz Siedentop

ABSTRACT. The atomic ground state density at the nucleus is bounded from above by a certain explicitly given positive constant times the third power of the nuclear charge.

1. Introduction

The Hamiltonian of an atom of N electrons with q spin states each and a fixed nucleus of charge Z located at the origin is given by

$$H_{N,Z} = \sum_{\nu=1}^{N} \left(-\Delta_\nu - \frac{Z}{|\mathfrak{r}_\nu|} \right) + \sum_{\substack{\mu,\nu=1 \\ \mu<\nu}}^{N} \frac{1}{|\mathfrak{r}_\mu - \mathfrak{r}_\nu|} \tag{1}$$

self-adjointly realized in $\bigwedge_{\nu=1}^{N} (L^2(\mathbb{R}^3) \otimes \mathbb{C}^q)$. Furthermore we write ψ for an eigenfunction that belongs to the bottom of the spectrum of $H_{N,Z}$, i.e., a ground state eigenfunction and

$$\begin{aligned} \rho_\psi(\mathfrak{r}) = & \sum_{\sigma_1,\ldots,\sigma_N=1}^{q} \sum_{\nu=1}^{N} \\ & \int_{\mathbb{R}^{3(N-1)}} \left|\psi(\mathfrak{r}_1,\sigma_1;\ldots;\mathfrak{r}_{\nu-1},\sigma_{\nu-1};\mathfrak{r},\sigma_\nu;\mathfrak{r}_{\nu+1},\sigma_{\nu+1};\ldots;\mathfrak{r}_N,\sigma_N)\right|^2 \\ & \qquad d\mathfrak{r}_1 \ldots d\mathfrak{r}_{\nu-1} d\mathfrak{r}_{\nu+1} \ldots d\mathfrak{r}_N \end{aligned} \tag{2}$$

for the corresponding density.

A quantity of particular interest is the ground state density $\rho_\psi(0)$ at the nucleus. Recently Narnhofer [**4**] argued that one might expect for an atom $\rho_\psi(0) = O(Z^3)$. In this talk we will outline a proof of this conjecture. In fact we can prove

THEOREM 1. *Let ρ_ψ be a ground state density of $H_{N,Z}$ and $N = Z$, then*

$$\rho_\psi(0) \le \frac{\pi}{48} q Z^3 + \text{const}\, Z^{161/54}.$$

This value is not in disagreement with Lieb's Strong Scott Conjecture [**3**] which is actually older and stronger than Narnhofer's: according to Lieb the scaled atomic density $\rho_\psi(\mathfrak{r}/Z)/Z^3$ should converge to the corresponding quantity of the bare Schrödinger operator $H_{N,Z}^0$ which equals $H_{N,Z}$ except for the omission of the second sum, the electron-electron interaction. If one assumes this convergence to be pointwise this would predict $\rho_\psi(0) = \zeta(3)/(8\pi) q Z^3 + o(Z^3)$ (Lieb [**3**, (7.35)]). We

---

1991 *Mathematics Subject Classification.* Primary: 35J10; Secondary: 81Q10.
This final version of this paper will be published elsewhere.

are now explicitly able to see that our theorem does support this conjecture, since $\pi/48$ which is about 0.065 is bigger than $\zeta(3)/8\pi$ which is about 0.048. On the other hand it shows that our estimate is rather good—we loose only using (3)—but presumably not yet sharp.

2. Outline of the Proof

The starting point of our proof is the inequality

$$\rho_\psi(0) \leq \frac{Z}{4\pi}\int_{\mathbb{R}^3} \frac{\rho_\psi(\mathfrak{r})}{|\mathfrak{r}|^2}\, d\mathfrak{r} \tag{3}$$

of Hoffmann-Ostenhof et al. [**1**]. The evaluation of the right hand side will be done by controlling the ground state energy up to second order in Z which in turn requires control of the ground states on distances Z^{-1} and $Z^{-1/3}$ from the origin. To do so we pursue a strategy similar to [**6**].

For the sake of notational simplicity we formulate our theorem and this outline of the proof for the most canonical case only, namely neutral atoms, i.e., $N = Z$.

First we define for $\epsilon \in [0, 1/4)$

$$H^\epsilon_{N,Z} = H_{N,Z} - \sum_{\nu=1}^{N} \frac{\epsilon}{|\mathfrak{r}_\nu|^2}.$$

With this notation we may write for positive ϵ

$$\begin{aligned} \int_{\mathbb{R}^3} \frac{\rho_\psi(\mathfrak{r})}{|\mathfrak{r}|^2}\, d\mathfrak{r} &= \frac{1}{\varepsilon}\left[(\psi, H_{N,Z}\psi) - (\psi, H^\epsilon_{N,Z}\psi)\right] \\ &\leq \frac{1}{\varepsilon}\left[E_{TF}(N,Z) + \frac{q}{8}Z^2 - \inf\sigma(H^\epsilon_{N,Z}) + \text{const}\, Z^{47/24}\right] \end{aligned} \tag{4}$$

where we use [**5**, Theorem 1]. (We denote by $E_{TF}(N,Z)$ the infimum of the Thomas-Fermi functional

$$\mathcal{E}_{TF}(\rho) = \int_{\mathbb{R}^3} \frac{3}{5}\gamma\rho(\mathfrak{r})^{5/3} - \frac{Z}{|\mathfrak{r}|}\rho(\mathfrak{r})\, d\mathfrak{r} + D(\rho,\rho)$$

on $\mathcal{I} = \left\{\rho \in L^1(\mathbb{R}^3) \cap L^{5/3}(\mathbb{R}^3) \mid \rho \geq 0, \int \rho \leq N\right\}$ where $\gamma = (6\pi^2/q)^{2/3}$ and

$$D(\sigma,\rho) = \frac{1}{2}\int_{\mathbb{R}^3}\int_{\mathbb{R}^3} \frac{\overline{\sigma(\mathfrak{r})}\rho(\mathfrak{s})}{|\mathfrak{r}-\mathfrak{s}|}\, d\mathfrak{r}\, d\mathfrak{s}$$

is the usual Coulomb scalar product.)

To proceed we need to estimate the infimum $\inf\sigma(H^\epsilon_{N,Z})$ of the spectrum of $H^\epsilon_{N,Z}$ from below. To this end we begin—following Lieb [**3**]—by proving

LEMMA 1. *The kinetic energy of any (approximate) ground state of $H^\epsilon_{N,Z}$ fulfills the following estimate for $N = Z \to \infty$:*

$$\int_{\mathbb{R}^{3N}} |\nabla\psi_\epsilon|^2 = O(Z^{7/3})$$

uniformly in $\varepsilon \in [0, 1/4)$.

Moreover using Lieb's inequality on the correlation (Lieb [**2**]) we have

$$(5)\quad \left(\psi_\epsilon, \sum_{\substack{\mu,\nu=1\\ \mu<\nu}}^{N} \frac{1}{|\mathfrak{r}_\mu - \mathfrak{r}_\nu|}\psi_\varepsilon\right) \geq \int_{\mathbb{R}^3}\int_{\mathbb{R}^3} \frac{\rho_{TF}(\mathfrak{r})\rho_{\psi_\epsilon}(\mathfrak{r}')}{|\mathfrak{r}-\mathfrak{r}'|}\,d\mathfrak{r}\,d\mathfrak{r}' - \text{const}\int_{\mathbb{R}^3} \rho_{\psi_\epsilon}(\mathfrak{r})^{4/3}\,d\mathfrak{r} - D(\rho_{TF},\rho_{TF}).$$

This implies for any (approximate) ground state ψ_ϵ of $H^\epsilon_{N,Z}$ using the Schwarz inequality, Lemma 1, and the Lieb-Thirring inequality

$$(6)\qquad \int \rho_{\psi_\epsilon}^{4/3} \leq \sqrt{\int \rho_{\psi_\epsilon}^{5/3}}\sqrt{\int \rho_{\psi_\epsilon}} \leq \text{const}\, Z^{5/3}.$$

Thus

$$(7)\quad \inf\sigma(H^\epsilon_{N,Z}) \geq \inf\sigma\left(\sum_{\nu=1}^{N} \underbrace{1\otimes\cdots\otimes 1}_{\nu-1 \text{ factors}} \otimes h^{\epsilon,TF}_{N,Z} \otimes \underbrace{1\otimes\cdots\otimes 1}_{N-\nu \text{ factors}}\right) - D(\rho_{TF},\rho_{TF}) - \text{const}\, Z^{5/3}$$

where

$$(8)\qquad h^{\epsilon,TF}_{N,Z} = -\Delta - \frac{\varepsilon}{|\cdot|^2} - \frac{Z}{|\cdot|} + \rho_{TF} \star \frac{1}{|\cdot|}.$$

The first term on the right hand side can be estimated by the sum of all negative eigenvalues of (8) which are in turn given by

$$\begin{aligned} h^\epsilon_l \varphi_{n,l}(r) &:= \left(-\frac{d^2}{dr^2} + \frac{l(l+1)-\varepsilon}{r^2} - \frac{Z}{r} + \int_0^\infty \frac{\sigma(r')}{\max\{r,r'\}}\,dr'\right)\varphi_{n,l}(r) \\ &= \epsilon^Z_{n,l}\varphi_{n,l}(r) \end{aligned}$$

($l = 0,1,2,\dots$) using separation of variables. Here $\sigma(r) = 4\pi r^2 \rho_{TF}(\mathfrak{r})$ for $r = |\mathfrak{r}|$. (Note that ρ_{TF} is spherically symmetric.)

We treat small l, i.e., $l < L = [Z^{4/27}]$, and large l, i.e., $l \geq L$, differently:

SMALL l. We estimate h^ϵ_l from below by

$$-\frac{d^2}{dr^2} + \frac{l(l+1)-\varepsilon}{r^2} - \frac{Z}{r}.$$

The eigenvalues of this operator are

$$-\frac{Z^2}{4(\lambda_{l,\epsilon}+n)^2}$$

where $n = 1, 2, \ldots$ and $\lambda_{l,\epsilon}$ is the positive root of $\lambda(\lambda+1) = l(l+1) - \epsilon$; more explicitly $\lambda_{l,\epsilon} = \sqrt{(l+1/2)^2 - \epsilon} - 1/2$. Thus

$$\begin{aligned}
&-\sum_{l=0}^{L-1} q(2l+1)\,\mathrm{tr}(h_l^\varepsilon)_- \\
&\geq -\sum_{l=0}^{L-1} q(2l+1) \sum_{n=1}^{\infty} \frac{Z^2}{4(\lambda_{l,\epsilon}+n)^2} \\
&= -\frac{Z^2}{4} \sum_{l=0}^{L-1} q(2l+1) \sum_{n=1}^{\infty} \left(\frac{1}{(l+n)^2} + \frac{\varepsilon}{(\sqrt{(l+1/2)^2} - 1/2 + n)^3 \sqrt{(l+1/2)^2}} \right. \\
&\qquad \left. + \frac{\epsilon^2}{4} \frac{3((l+1/2)^2 - \xi)^{1/2} + \sqrt{(l+1/2)^2 - \xi} - 1/2 + n}{(\sqrt{(l+1/2)^2 - \xi} - 1/2 + n)^4 ((l+1/2)^2 - \xi)^{3/2}} \right)
\end{aligned}$$

for some ξ with $0 < \xi < \epsilon$. We obtain a lower bound by replacing ξ by any number bigger or equal to ϵ and smaller than $1/4$.

Note that the sum over n is convergent for each term separately. Moreover, for the last two terms the summation over l when extended to infinity is convergent, too. The term of interest is the one that is homogeneous of degree one in ϵ:

$$\begin{aligned}
(9) \qquad -q\frac{Z^2}{2} \sum_{l=0}^{L-1} \sum_{n=1}^{\infty} \frac{\varepsilon}{(l+n)^3} &\geq -q\frac{Z^2}{2} \sum_{l=0}^{\infty} \sum_{n=1}^{\infty} \frac{\varepsilon}{(l+n)^3} \\
&= -q\frac{Z^2}{2} \sum_{\mu=1}^{\infty} \frac{\epsilon}{\mu^2} = -\frac{\pi^2}{12} q Z^2 \epsilon.
\end{aligned}$$

LARGE l. We follow the treatment in [**6**] which yields

$$(10) \qquad \sum_{l=L}^{\infty} q(2l+1)\,\mathrm{tr}(h_l^\varepsilon)_- = \sum_{l=L}^{\infty} q(2l+1)\,\mathrm{tr}(h_l^0)_- + O(Z^{53/27})$$

uniformly in ϵ.

Summation over l analogously to [**6**] yields

$$(11) \quad \inf \sigma(H_{N,Z}^\epsilon) \geq E_{TF}(N,Z) + \left(\frac{1}{8} - \frac{\pi^2}{12}\right) q Z^2 \epsilon - \mathrm{const}(Z^2\epsilon^2 + Z^{53/27}).$$

Picking now $\varepsilon = Z^{-1/54}$ and putting together (3), (4), and (11) yields the desired result.

Acknowledgement. It is a pleasure to thank Thomas Hoffmann-Ostenhof for the invitation to the Schrödinger Institute in Vienna where the present work was done and for telling me about his lower and upper bound in the bosonic case which actually generated my interest in the subject.

References

1. Maria Hoffmann-Ostenhof, Thomas Hoffmann-Ostenhof, and Walter Thirring, *Simple bounds to the atomic one-electron density at the nucleus and to expectation values of one-electron operators*, J. Phys. B **11** (1978), no. 19, L571–L575.
2. Elliott H. Lieb, *A lower bound for Coulomb energies*, Phys. Lett. A **70** (1979), 444–446.
3. ———, *Thomas-Fermi and related theories of atoms and molecules*, Rev. Modern Phys. **53** (1981), 603–641.

4. Heide Narnhofer, *Towards control over electron density in heavy atoms*, Preprint, UWThPh-1993-15, 1993.
5. Heinz Siedentop and Rudi Weikard, *On the leading energy correction for the statistical model of the atom: Interacting case*, Comm. Math. Phys. **112** (1987), 471–490.
6. ——, *A new phase space localization technique with application to the sum of negative eigenvalues of Schrödinger operators*, Ann. Sci. École Norm. Sup. **24** (1991), no. 2, 215–225.

MATEMATIKK, NTH, N-7034 TRONDHEIM, NORWAY

Current address: Matematisk institutt, Universitetet i Oslo, p.b. 1053, N-0316 Oslo, Norway

E-mail address: heinz@math.uio.no

Centre de Recherches Mathématiques
CRM Proceedings and Lecture Notes
Volume 8, 1995

Spectral Analysis of N-Body Stark Hamiltonians

Ira Herbst and Erik Skibsted

1. Results

In this note we present recent results on absence of pure point and singular continuous spectra for a large class of N-body Stark Hamiltonians [**HMS**]. The model is the following. We consider a system of N ν-dimensional particles labelled $1, \dots, N$ with masses, charges, positions and momenta denoted by m_i, q_i, x_i, and p_i, respectively. The interaction consists of two parts. One part (the external part) is due to the presence of an electric field $\mathcal{E}$, while the other part (the internal interaction) is given as a sum of pair potentials each one, $v_{ij} \in C^1(\mathbb{R}^v)$, assumed to obey the decay condition

$$|v_{ij}(y)| + |\nabla v_{ij}(y)| = o(1) \text{ for } y \to \infty. \tag{1.1}$$

Then the total Hamiltonian reads

$$\widetilde{H} = \sum_{i=1}^{N} \left(\frac{p_i^2}{2m_i} - q_i \mathcal{E} \cdot x_i \right) + V;$$
$$V(x) = \sum_{1 \le i < j \le N} v_{ij}(x_i - x_j), \quad x = (x_1 \dots, x_N),$$

on $L^2(\mathbb{R}^{\nu N})$. To separate out the center of mass motion we introduce the metric

$$x \cdot y = \sum_{i=1}^{N} 2m_i x_i \cdot y_i; \quad x = (x_1, \dots, x_N), y = (y_1, \dots, y_N).$$

Corresponding to the decomposition $\mathbb{R}^{\nu N} = X^{\perp} \oplus X$ where

$$X = \left\{ x = (x_1, \dots, x_N) \;\middle|\; x_i \in \mathbb{R}^v, \sum_{i=1}^{N} m_i x_i = 0 \right\}$$

the total Hamiltonian takes the form

$$\widetilde{H} = H_0 \otimes I + I \otimes H.$$

1991 *Mathematics Subject Classification.* Primary: 35J10; Secondary: 81Q10.

The final version of this paper will be submitted for publication elsewhere.

The operator H (defined on $L^2(X)$) is the Hamiltonian we shall be concerned with. It takes the form

$$H = -\Delta - E \cdot x + V;$$
$$E = \left(\left(\frac{q_1}{2m_1} - \frac{Q}{2M}\right)\mathcal{E}, \dots, \left(\frac{q_N}{2m_N} - \frac{Q}{2M}\right)\mathcal{E}\right),$$
$$Q = \sum_{i=1}^{N} q_i, M = \sum_{i=1}^{N} m_i.$$

We shall here present three results. In addition to the condition (1.1) we need the assumption that $q_i/m_i \neq q_j/m_j$, for some pair i, $j \leq N$ (i.e. that $E \neq 0$). The first is

THEOREM 1.1. *Suppose* (1.1) *and* $E \neq 0$. *Then the pure point spectrum,* $\sigma_{pp}(H)$, *is empty.*

For the second one we need the assumption

$$\partial^\beta v_{ij} \in L^\infty(\mathbb{R}^\nu); \quad |\beta| = 2, \tag{1.2}$$

on the second order distributional derivatives. The statement is

THEOREM 1.2. *Suppose* (1.1), (1.2) *and* $E \neq 0$. *Then the singular continuous spectrum,* $\sigma_{sc}(H)$, *is empty.*

For the third statement we need the notations $\langle p \rangle$ and $\langle x \rangle$ for the selfadjoint operators on $L^2(X)$ given by $(1-\Delta)^{1/2}$ and multiplication by $(1+x\cdot x)^{1/2}$, respectively. It reads

THEOREM 1.3. *Suppose* (1.1), (1.2) *and* $E \neq 0$. *Then*

(1) $\forall \delta > 1/2$:

$$\sup_{\operatorname{Im} z \neq 0} \|\langle p \rangle^{-\delta}(H-z)^{-1}\langle p \rangle^{-\delta}\| < \infty,$$

(2) $\forall \delta' > 1/4$:

$$\sup_{\operatorname{Im} z\neq 0, \operatorname{Re} z \leq 0} (1+|\operatorname{Re} z|)^\nu \|\langle x \rangle^{-\delta'}(H-z)^{-1}\langle x \rangle^{-\delta'}\| < \infty; \quad \nu < \frac{4\delta'-1}{4\delta'+1},$$

while

$$\sup_{\operatorname{Im} z\neq 0, \operatorname{Re} z \geq 0} (1+|\operatorname{Re} z|)^\nu \|\langle x \rangle^{-\delta'}(H-z)^{-1}\langle x \rangle^{-\delta'}\| < \infty; \quad \nu < -\frac{1}{2}.$$

Moreover in both cases the (two) boundary values

$$\lim_{\operatorname{Im} z \downarrow(\uparrow) 0} \langle p \rangle^{-\delta}(H-z)^{-1}\langle p \rangle^{-\delta}, \qquad \lim_{\operatorname{Im} z \downarrow(\uparrow) 0} \langle x \rangle^{-\delta'}(H-z)^{-1}\langle x \rangle^{-\delta'}$$

exist in $\mathcal{B}(L^2(X))$ *and are attained locally uniformly.*

The results above generalize some recently obtained results [**Sk**], [**T**]. The first of these papers needed an additional geometric condition that was verified for $N = 3$ and for "atoms" for $N > 3$. The second needed a stronger condition than (1.1) namely the short range condition given by replacing the right hand side by $O(|y|^{-1/2-\varepsilon})$. Previous to these works is [**Si**] that contains Theorem 1.1 for Born Oppenheimer molecules. Here the condition of repulsion (between electrons) was important in the proof.

As was the case for the above mentioned works we prove our results by commutator methods. As for Theorem 1.1 we apply a variant of the method of Froese and Herbst [**FH**] and for Theorem 1.3 we apply the Mourre method [**M**]. We shall in the next section outline some ingredients of the proof of Theorem 1.1. An immediate consequence of the proof is the so called Mourre estimate that allows us to apply the abstract theory of [**M**]. In this sense the proof of Theorem 1.3 is the easy part. It will not be discussed. As for Theorem 1.2 it follows immediately from Theorem 1.3. We remark that our methods can be extended to include local singularities of at most Coulomb type, see [**HMS**].

All results hold in the framework of generalized Schrödinger operators. As a matter of fact the proofs are most conveniently given in this framework. In order to fix notations we give in the following a detailed account of it. We consider a finite dimensional real vector space X with an inner product $x \cdot y$, and a finite family $\{X_a \mid a \in \mathcal{A}\}$ of subspaces of X. It is assumed that this family is closed with respect to intersection and there exist $a_{\min}, a_{\max} \in \mathcal{A}$ such that $X_{a_{\min}} = X$ and $X_{a_{\max}} = 0$. The orthogonal complement of X_a in X is denoted by X^a. We order $\mathcal{A}$ by writing $a_2 \subset a_1$ if $X^{a_2} \subset X^{a_1}$. For given a we define $\#a$ as the largest $n \in \mathbb{N}$ for which there exists an increasing sequence $a = a_n \subsetneqq a_{n-1} \subsetneqq \cdots \subsetneqq a_1 = a_{\max}$. The largest such n is $N := \#a_{\min}$. The orthogonal projections onto X_a and X^a are denoted by Π_a and Π^a, respectively. The corresponding components for $x \in X$ are denoted x_a and x^a. We use similar notations for the components of the momentum operator $p = -i\nabla = (p_1, \ldots, p_{\dim X})$. We put $\langle x \rangle = (1 + |x|^2)^{1/2}$; $|x|^2 = x \cdot x$. Let $E \in X$. Then the Hamiltonian

$$H = -\Delta - E \cdot x + V; \quad V = \sum_{a \in \mathcal{A}} V_a,$$

is essential selfadjoint on $C_0^\infty(X) \subset L^2(X)$ under the following assumption on the potentials (cf. (1.1)). For all $a \in \mathcal{A}$ the real potential $v_a(x^a) \in C^1(X^a)$ obeys

$$\partial^\beta V_a(x^a) = o(1) \text{ for } x^a \to \infty; \quad |\beta| \leq 1. \tag{1.3}$$

By convention $\partial^\beta V_{a_{\min}} = 0$.

For $a \neq a_{\min}$ we introduce the sub-Hamiltonian

$$H^a = (p^a)^2 - E^a \cdot x^a + V^a; \quad V^a = \sum_{b \subset a} V_b,$$

and

$$H_a = p^2 - E \cdot x + V^a = (p_a)^2 - E_a \cdot x_a + H^a.$$

For $R, \varepsilon > 0$ the notations $\chi(\cdot < R)$ and η_ε stand for the characteristic functions of the intervals $(-\infty, R)$ and $(-\varepsilon, \varepsilon)$, respectively.

2. Proofs

In this section we outline the main ingredients of the proof of Theorem 1.1 (for generalized Schrödinger operators). This is done by stating a series of lemmas and often omitting proofs. The most important are Lemmas 2.2 and 2.3. The proof of

Theorem 1.1 is by induction with respect to $\#a$ (starting from above). Throughout the section we assume the statement of the theorem for sub-Hamiltonians, i.e.

$$a \neq a_{\max}, \quad E^a \neq 0 \Rightarrow \sigma_{pp}(H^a) = \emptyset. \tag{2.1}$$

The first result concerns basic domain questions. Its proof is elementary.

LEMMA 2.1. (1) *The domain $D(H) \subseteq H^2_{\text{loc}}$, the local Sobolev space of order 2. Moreover*

$$p\langle x\rangle^{-1/2}(H-i)^{-1} \text{ and } p^2\langle x\rangle^{-1}(H-i)^{-1}$$

are bounded.

(2) *Suppose $0 \leq h \in C^\infty(X)$ and $\psi \in \mathcal{D}(H)$ are given such that $hH\psi$, $h\psi$, $E \cdot xh\psi$, $(\partial_j h)\psi \in L^2(X)$ for any component $\partial_j h$ of ∇h. Then*

$$hp^2\psi, p_j h\psi, hp_j\psi \in L^2(X).$$

The proof of the next result is more complicated. It is based on (2.1) and a result of one of the authors [**H**, Proposition 2.1].

LEMMA 2.2. *Suppose $R > 0$, $a_{\min} \neq b \subset a \neq a_{\max}$ and $E^b \neq 0$. Then*

$$\|\eta_\varepsilon(H^a - \lambda)\chi(|x^b|_. < R)\| \to 0 \text{ for } \varepsilon \to 0,$$

uniformly in $\lambda \in \mathbb{R}$.

To state the next result we introduce smooth functions $f\colon X \to \mathbb{R}_+$ and $\theta\colon \mathbb{R}_+ \to \mathbb{R}$ (depending on parameters) as follows. Let

$$f(x) = f_\kappa(x) = \kappa\langle x\rangle + E\cdot x; \quad \kappa \gg 1,$$

and

$$\theta(t) = \theta_\kappa(t) = \mu \log\left(1 + \frac{\gamma t}{\kappa\mu}\right) + \frac{\alpha}{\kappa}t; \quad \mu > 1, \gamma > 0, \alpha \geq 0.$$

LEMMA 2.3. *Let $\lambda \in \mathbb{R}$ and $\alpha_0 > 0$. Then for a (small) $\delta > 0$ and all sufficiently large κ there exist a constant $C > 0$ and a compact operator K such that*

$$\forall \mu > 1, \quad \gamma + |\alpha - \alpha_0| \leq \delta,$$
$$\phi \in C_0^\infty(X)\colon C\|e^{\theta(f)}(H-\lambda)\phi\| \geq \|e^{\theta(f)}\phi\| - \|Ke^{\theta(f)}\phi\|.$$

The previous estimate leads to

LEMMA 2.4. *Suppose $(H-\lambda)\psi = 0$; $\lambda \in \mathbb{R}, \psi \in \mathcal{D}(H)$, and that*

$$\alpha_0 := \sup\{\alpha \geq 0 | e^{\alpha\langle x\rangle}\psi(x) \in L^2(X)\} > 0.$$

Then

$$\alpha_0 = \infty.$$

PROOF. Suppose the statement is false. Then we come to a contradiction by the following use of Lemma 2.3. Let $\delta > 0$ be given in accordance with the lemma. Then we choose any $\alpha < \alpha_0$ such that $\alpha_0 < \alpha + \delta/2$ and put $\gamma = \delta/2$. It remains to specify κ. We choose it so large that in addition to the estimate of the lemma the conditions

$$\alpha\left(1 + \frac{|E|}{\kappa}\right) < \alpha_0 \text{ and } \alpha_0 < \left(\alpha + \frac{\delta}{2}\right)\left(1 - \frac{|E|}{\kappa}\right) \tag{2.2}$$

hold. Now we can use Lemma 2.1 (2) and the first of the above conditions to show by an approximation argument that ϕ in the estimate can be replaced by ψ; i.e. that

$$\|Ke^{\theta(f)}\psi\| \geq \|e^{\theta(f)}\psi\|. \tag{2.3}$$

Since the right hand side goes to infinity as $\mu \to \infty$ (by the second condition in (2.2)) we come to a contradiction by letting $\mu \to \infty$ in (2.3) recalling that a compact operator takes a weakly convergent sequence to a strongly convergent one. □

"PROOF OF THEOREM 1.1". We can now show that under the assumption of Lemma 2.4 $\psi = 0$. This proves Theorem 1.1 except for the fact that α_0 could be zero. This case will not be discussed, but we remark that it can be treated (i.e. excluded) by similar means.

By using Lemmas 2.1(2) and 2.4 the above statement follows readily from the following formula valid for any $\phi \in C_0^\infty(X)$:

$$\begin{aligned}\|e^{\alpha E\cdot x}\{(H-\lambda) - V\}\phi\|^2 &= \|\{p^2 - E\cdot x - (\lambda + \alpha^2|E|^2)\}e^{\alpha E\cdot x}\phi\|^2 \\ &\quad + 4\alpha^2\|E\cdot pe^{\alpha E\cdot x}\phi\|^2 + 2\alpha|E|^2\|e^{\alpha E\cdot x}\phi\|^2; \quad \lambda \in \mathbb{R}, \alpha \geq 0.\end{aligned}$$

Because of Lemma 2.4 we can apply it to $\phi = \psi$. This gives the condition

$$\|e^{\alpha E\cdot x}V\psi\|^2 \geq 2\alpha|E|^2\|e^{\alpha E\cdot x}\psi\|^2,$$

which clearly holds for all $\alpha \geq 0$ only if $\psi = 0$. □

"PROOF OF LEMMA 2.3". We give the main ingredients but stop short of a complete proof. With f and θ as in the lemma we abbreviate

$$A_f = \frac{(\nabla f\cdot p + p\cdot\nabla f)}{2}, \quad \theta = \theta(f), \quad \theta' = \left(\frac{d}{dt}\theta\right)(f), \quad \theta'' = \left(\frac{d^2}{dt^2}\theta\right)(f),$$

and compute the commutator

$$\begin{aligned}i\big[H, e^\theta A_f e^\theta\big] &= e^\theta\{T_1 + \cdots + T_4\}e^\theta, \\ T_1 = 4A_f\theta' A_f, \quad T_2 &= 2pf^{(2)}p, \quad T_3 = (E - \nabla V)\cdot\nabla f, \\ T_4 &= -\frac{1}{2}\Delta^2 f + \nabla f\cdot\nabla\big((\theta'^2 - \theta'')|\nabla f|^2\big).\end{aligned}$$

By further computations

$$T_1, T_2 \geq 0, \quad T_4 = O\big(\langle x\rangle^{-1}\big) \text{ for } x \to \infty.$$

The latter statement is uniform with respect to the parameters μ, γ and α (cf. (2.8) below).

To deal with the term T_3 we notice that

$$\nabla V\cdot\nabla\langle x\rangle = o(1) \text{ for } x \to \infty.$$

Hence

$$T_3 = E \cdot \kappa \frac{x}{\langle x \rangle} + \varepsilon_0 - \sum_{E^a \neq 0} \nabla V_a(x^a) \cdot E + o(1); \quad \varepsilon_0 = |E|^2.$$

For the first term we have

$$E \cdot \kappa \frac{x}{\langle x \rangle} = \mathrm{Re}\left(\frac{\kappa}{\langle x \rangle}\{p^2 + V - H\}\right) \geq -\mathrm{Re}\left(\frac{\kappa}{\langle x \rangle} H\right) + o(1).$$

Putting the above computations together we finally obtain

$$(2.4) \qquad \begin{gathered} i[H, e^\theta A f e^\theta] = e^\theta\{\dots\}e^\theta; \\ \{\dots\} \geq T_1 + \varepsilon_0 - \sum_{b:\ E^b \neq 0} h_b - \mathrm{Re}\left(\frac{\kappa}{\langle x \rangle} H\right) + h; \\ h_b(x) = \nabla V_b(x^b) \cdot E, \quad h(x) = o(1). \end{gathered}$$

We emphasize that $h = o(1)$ holds uniformly with respect to the parameters μ, γ and α. Under this condition the last two terms on the right hand side of the inequality are easy to treat in the procedure described below. We shall here focus on the third term.

For $\phi \in C_0^\infty(X)$ and any given linear operator T we abbreviate the expectation $\langle e^\theta \phi, T e^\theta \phi \rangle = \langle T \rangle$. We proceed by taking the expectation of both sides of the inequality (2.4) in states $\phi \in C_0^\infty(X)$. As for the left hand side we use

$$\langle \phi, i[H, e^\theta A_f e^\theta]\phi \rangle \leq 3^{-1}\langle T_1 \rangle + \|(\theta')^{-1/2} e^\theta (H - \lambda)\phi\|^2.$$

When subtracting the term $3^{-1}\langle T_1 \rangle$ we obtain the positive term $2/3\langle T_1 \rangle$ on the right hand side. This term (and $\langle \varepsilon_0 I \rangle$) will be used to estimate the third term on the left hand side of (2.4) now to be discussed.

By choosing R large enough we have

$$(2.5) \qquad -\sum_{b:\ E^b \neq 0} h_b(x) \geq -\sum_{b:\ E^b \neq 0} \chi\big(|x^b| < R\big) h_b(x) - \frac{\varepsilon_0}{6}.$$

In order to deal with the first term on the right hand side of (2.5) we put $\bar{\lambda} = \lambda + (\alpha_0)^2$, and let $\{j_a\}$, $a \neq a_{\max}$, be a partition of unity (i.e. $\sum_a j_a(x) = 1$) of bounded measurable functions on X such that for any $R > 0$ and $b \not\subset a$

$$\chi\big(|x^b| < R\big) j_a = 0 \text{ for } |x| \text{ large}.$$

We decompose

$$(2.6) \qquad \begin{gathered} \chi\big(|x^b| < R\big) = \eta_\varepsilon(H - \bar{\lambda})\chi\big(|x^b| < R\big) + \mathcal{E}_b; \\ \mathcal{E}_b = (H - \bar{\lambda})\zeta_\varepsilon(H - \bar{\lambda})\chi\big(|x^b| < R\big), \quad \zeta_\varepsilon(t) = t^{-1}\big(1 - \eta_\varepsilon(t)\big). \end{gathered}$$

The first term can be written in the form

$$\eta_\varepsilon(H - \bar{\lambda})\chi\big(|x^b| < R\big) = \sum_{b \subset a} \eta_\varepsilon(H_a - \bar{\lambda})\chi\big(|x^b| < R\big) j_a + K; \quad K \text{ compact}.$$

But by Lemma 2.2

$$\big\|\eta_\varepsilon(H_a - \bar{\lambda})\chi\big(|x^b| < R\big)\big\| \to 0 \text{ for } \varepsilon \to 0.$$

Hence for ε small enough

$$(2.7)\qquad -\operatorname{Re}\sum_{b:\,E^b\neq 0}\eta_\varepsilon(H-\bar\lambda)\chi\big(|x^b|<R\big)h_b\geq -\frac{\varepsilon_0}{6}I-K;\quad K \text{ compact},$$

and thus it suffices to look at the contribution from the second term on the right hand side of (2.6) (for fixed $\varepsilon>0$).

We estimate

$$\begin{aligned}|\langle\mathcal{E}_b h_b\rangle| &\leq \|(H-\bar\lambda)e^\theta\phi\|C\|e^\theta\phi\|\\ &\leq C'\|(H-\bar\lambda)e^\theta\phi\|^2+\varepsilon'\|e^\theta\phi\|^2\\ &\leq 2C'\big\|(H-\lambda-(\theta'^2-\theta'')|\nabla f|^2+2i\theta' A_f)e^\theta\phi\big\|^2\\ &\qquad +2C'\Big\|\big(\{(\alpha_0)^2-(\theta'^2-\theta'')|\nabla f|^2\}+2i\theta' A_f\big)e^\theta\phi\Big\|^2+\varepsilon'\|e^\theta\phi\|^2;\\ &\qquad\qquad \varepsilon'>0, C'=C'_{\varepsilon,\varepsilon'}.\end{aligned}$$

As for the first term we use the fact that

$$\big(H-\lambda-(\theta'^2-\theta'')|\nabla f|^2+2i\theta' A_f\big)e^\theta=e^\theta(H-\lambda).$$

To estimate the second term we use

$$(2.8)\qquad \Big|\theta'(t)-\frac{\alpha}{\kappa}\Big|\leq\frac{\gamma}{\kappa},\quad \big|\theta''(t)\big|\leq\frac{\gamma}{\kappa t},\quad |\theta'''(t)|\leq\frac{2\lambda}{\kappa t^2}$$

to obtain the bounds

$$\big|(\alpha_0)^2-(\theta'^2-\theta'')|\nabla f|^2\big|\leq C\big(\langle x\rangle^{-1}+\kappa^{-1}+\delta\big)$$

and

$$\theta'\leq C\kappa^{-1}.$$

In both cases C is independent all parameters (constrained to $\gamma+|\alpha-\alpha_0|\leq\delta$). We apply the above estimates for κ large and δ small. (Notice the bound $\|2i\theta' A_f e^\theta\phi\|^2\leq\sup\theta'\langle T_1\rangle$.) As a result

$$\begin{aligned}(2.9)\qquad -\Big\langle\operatorname{Re}\sum_{b:\,E^b\neq 0};\quad \mathcal{E}_b h_b\Big\rangle\\ \geq -\tilde\varepsilon\|e^\theta\phi\|^2-C\|e^\theta(H-\lambda)\phi\|^2-\Big\langle\frac{T_1}{3}\Big\rangle+\big\langle O(\langle x\rangle^{-2})\big\rangle;\quad \tilde\varepsilon=\frac{\varepsilon_0}{6}.\end{aligned}$$

By combining (2.5), (2.6), (2.7) and (2.9) we finally obtain that for a constant $C>0$ and compact K (both independent of μ, γ and α)

$$\begin{aligned}(2.10)\qquad \Big\langle-\sum_{b:\,E^b\neq 0}h_b(x)\Big\rangle+C\|e^\theta(H-\lambda)\phi\|^2\\ \geq -\frac{\varepsilon_0}{2}\|e^\theta\phi\|^2-\Big\langle\frac{T_1}{3}\Big\rangle-\|Ke^\theta\phi\|^2+\big\langle\tilde h\big\rangle;\quad \tilde h=o(1).\end{aligned}$$

By (2.8) we can choose $C' \geq C + \sup(\theta'^{-1})$ independent of μ, λ and α (constrained to $\gamma + |\alpha - \alpha_0| \leq \delta$). Hence by (2.4), (2.10) and a previous remark we obtain the (uniform) bound

$$C'\|e^\theta(H-\lambda)\phi\|^2 \geq \frac{\varepsilon_0}{2}\|e^\theta\phi\|^2 + \left\langle\frac{T_1}{3}\right\rangle - \|Ke^\theta\phi\|^2 + \left\langle\tilde{h} + h - \mathrm{Re}\left(\frac{\kappa}{\langle x\rangle}H\right)\right\rangle,$$

which is a main intermediate result. □

References

[FH] R. Froese, I. Herbst, *Exponential bounds and absence of positive eigenvalues for N-body Schrödinger operators*, Comm. Math. Phys. **87** (1982), 429–447.

[H] I. Herbst, *Temporal exponential decay for the Stark effect in atoms*, J. Funct. Anal. **48** (1982), 224–251.

[HMS] I. Herbst, J. S. Möller, and E. Skibsted, *Spectral analysis of N-body Stark Hamiltonians*, Aarhus Univ., 1994, (preprint).

[M] E. Mourre, *Absence of singular continuous spectrum for certain selfadjoint operators*, Comm. Math. Phys. **91** (1981), 391–408.

[Si] I. M. Sigal, *Stark effect in multielectron systems: non-existence of bound states*, Comm. Math. Phys. **122** (1989), 122.

[Sk] E. Skibsted, *Absolute spectral continuity for N-body Stark Hamiltonians*, Ann. Inst. H. Poincaré Phys. Théor. **61** (1994), no. 3, ??–??.

[T] H. Tamura, *Spectral analysis for N-particle systems with Stark effect: Non-existence of bound states and principle of limiting absorption*, Ibaraki Univ, 1993, (preprint).

Department of Mathematics, University of Virginia, Charlottesville, VA 22903, USA

E-mail address: iwh@virginia.edu

Matematisk Institut Aarhus University, Ny Munkegade 8000 Aarhus C, Denmark

E-mail address: skibsted@mi.aau.dk

Centre de Recherches Mathématiques
CRM Proceedings and Lecture Notes
Volume **8**, 1995

Ground State Resonances for Hydrogen in an Intense Magnetic Field

Richard Froese and Roger Waxler

Today we would like to speak about a continuing study of the quantum mechanics describing a hydrogen atom in a constant, and very large, magnetic field. Explicitly we have been studying the system whose Hamiltonian is given formally by the expression

$$H = \frac{1}{2m_e}\left(-i\hbar\nabla + \frac{eB}{c}\hat{e}_3 \times x\right)^2 - \frac{e^2}{|x|} \tag{1}$$

where m_e is the mass of the electron, e is negative of the charge of an electron, $\hbar$ is Planck's constant divided by 2π, c is the speed of light, $\hat{e}_3 = (0,0,1)$ and $x = (x_1, x_2, x_3)$. H is the Hamiltonian which describes a very simple physical system, the simplest atom, hydrogen, in a constant magnetic field in the approximation in which the atom is taken to consist of an electron without spin interacting with an infinitely massive and structureless proton. We expect H to be physically meaningful for fields which are not so large that relativistic effects cannot be ignored. One expects relativistic effects to be small for fields B for which the length scale induced by the magnetic field, often called the Larmour radius,

$$\rho_{\text{mag}} = \left(\frac{\hbar c}{eB}\right)^{1/2},$$

is much larger than the Compton wavelength of an electron [**9**]. This implies that

$$\left(\frac{\hbar c}{eB}\right)^{1/2} \gg \frac{\hbar}{m_e c}$$

so that we should restrict our attention to fields for which

$$B \ll \frac{m_e^2 c^3}{\hbar e} \approx 4 \cdot 10^{13}\,\text{Gauss}.$$

On the other hand, by large field we mean the limit in which the Larmour radius is much smaller than the length scale of the hydrogen atom, the Bohr radius,

$$\rho_B = \frac{\hbar^2}{m_e e^2}.$$

1991 *Mathematics Subject Classification.* Primary: 35P15; Secondary: 81V45.
An expanded version of the manuscript will be published in Reviews in Mathematical Physics.

Explicitly, we can handle fields for which

$$\left(\frac{\hbar c}{eB}\right)^{1/2} \ll \frac{\hbar^2}{m_e e^2}$$

which implies that

$$B \gg \frac{m_e^2 e^3 c}{\hbar^3} \approx 3 \cdot 10^9 \text{ Gauss.}$$

This last condition, combined with the condition that the problem remain non-relativistic, implies that we must restrict ourselves to fields satisfying $3 \cdot 10^9$ Gauss $\ll$ $B \ll 4 \cdot 10^{13}$ Gauss. Thus, there is a large range of magnetic fields for which the Hamiltonian (1) provides a reasonable model for the physical situation and for which our asymptotics are valid. We point out however that field strengths of 10^9 Gauss are, quite literally, astronomical. The largest stationary fields yet produced in a laboratory are of the order of 10^6 Gauss [**7**]. The effects produced by fields larger than 10^9 Gauss have been studied by astrophysicists interested in the behavior of atoms near the surface of neutron stars where one expects to find large magnetic fields of the order of 10^{12} Gauss or even larger (for more information see the reviews by Ruderman [**9**]).

Introducing the dimensionless parameter

$$\lambda = \sqrt{2}\, \frac{m_e e^2}{\hbar^2} \left(\frac{eB}{\hbar c}\right)^{-1/2} = \left(\frac{2 m_e^2 e^3 c}{\hbar^3 B}\right)^{1/2},$$

changing to cylindrical coordinates and scaling appropriately we find that

$$H = \frac{m_e e^4}{\hbar^2} \bigoplus_{m=-\infty}^{\infty} H^{(m)},$$

where

$$H^{(m)} = \lambda^{-2}\left(-\frac{\partial^2}{\partial r^2} - \frac{1}{r}\frac{\partial}{\partial r} + \frac{m^2}{r^2} + r^2 + 2m\right) - \frac{\partial^2}{\partial z^2} - \frac{1}{\sqrt{\lambda^2 r^2 + z^2}} \tag{2}$$

acts in $\mathrm{L}^2([0,\infty), r\,dr) \otimes \mathrm{L}^2(R, dz)$. Note that large B means small λ. The spectrum of $H^{(m)}$ in the small λ limit has been rigorously studied by Avron, Herbst, and Simon [**1**] and by Froese and Waxler [**4**]. Avron, Herbst, and Simon showed that for each $n \in \{0, 1, 2, \dots\}$ there is a threshold at $\lambda^{-2}(m + |m| + 2n + 1)$ at which the multiplicity of the essential spectrum increases. Further, they studied the ground state and showed that the ground state eigenvalue, ε_0, of $H^{(m)}$ has the asymptotic form

$$\varepsilon_0 = -\left(\ln\left(\frac{1}{\lambda}\right) - \ln\left(\ln\left(\frac{1}{\lambda}\right)\right) + \mathcal{O}(1)\right)^2 + \frac{1}{\lambda^2}(m + |m| + 1). \tag{3}$$

Froese and Waxler then showed that the rest of the eigenvalues of $H^{(m)}$ have the asymptotic form

$$\varepsilon_k^{(\pm)} = -\frac{1}{4k^2} + \frac{\alpha_k^{(\pm)}}{k^3} + \frac{1}{\lambda^2}(m + |m| + 1) \tag{4}$$

where the coefficients $\alpha_k^{(\pm)}$ have been estimated and satisfy $0 < \alpha_k^{(-)} < \alpha_k^{(+)} < k^3\big(-1/(4(k+1)^2) + 1/(4k^2)\big)$. Further, these eigenvalues are simple and the eigenfunction corresponding to $\varepsilon_k^{(-)}$, $\phi_k^{(-)}(r,z)$, is odd with respect to $z \to -z$ while the eigenfunction corresponding to $\varepsilon_k^{(+)}$, $\phi_k^{(+)}(r,z)$, is even with respect to $z \to -z$. Note that there is an accumulation point at $1/(\lambda^2)\big(m + |m| + 1\big)$.

Our intention here is to discuss resonances, in the sense of dilation analyticity **[2]** **[3]** **[6]** **[10]**, of $H^{(m)}$ **[5]**. In the theory of dilation analyticity resonances arise as complex-valued eigenvalues of a non-selfadjoint operator related to the Hamiltonian by a formal complex scaling. In the present context we consider the operator

$$H_\theta^{(m)} = \lambda^{-2}\left(-\frac{\partial^2}{\partial r^2} - \frac{1}{r}\frac{\partial}{\partial r} + \frac{m^2}{r^2} + r^2 + 2m\right) - e^{-2i\theta}\frac{\partial^2}{\partial z^2} - \frac{1}{\sqrt{\lambda^2 r^2 + e^{2i\theta} z^2}},$$

acting in $\mathrm{L}^2\big([0,\infty), r dr\big) \otimes \mathrm{L}^2(R, dz)$, obtained by letting $z \to e^{i\theta} z$ in (2). Eigenvalues of $H_\theta^{(m)}$ which have non-zero imaginary parts are called resonances.

By a relative compactness argument [**1**], [**4**] the essential spectrum of $H_\theta^{(m)}$ is the same as that of

$$\lambda^{-2}\left(-\frac{\partial^2}{\partial r^2} - \frac{1}{r}\frac{\partial}{\partial r} + \frac{m^2}{r^2} + r^2 + 2m\right) - e^{-2i\theta}\frac{\partial^2}{\partial z^2}.$$

Thus the continuous spectrum of $H_\theta^{(m)}$ is

$$\sigma_c(H_\theta^{(m)}) = \bigcup_{J=0}^{\infty}\left(\frac{1}{\lambda^2}\big(m + |m| + 2J + 1\big) + e^{-2i\theta}[0,\infty)\right).$$

We have shown that, given J, if λ is small enough then there is a resonance, $\varepsilon_0^{(J)}$, of $H_\theta^{(m)}$ which satisfies

$$\text{(5)}\quad \operatorname{Re}\varepsilon_0^{(J)} = -\left(\ln\left(\frac{1}{\lambda}\right) - \ln\left(\ln\left(\frac{1}{\lambda}\right)\right) + \mathcal{O}(1)\right)^2 + \frac{1}{\lambda^2}\big(m + |m| + 2J + 1\big)$$

and

$$\text{(6)}\qquad \operatorname{Im}\varepsilon_0^{(J)} = -c_J\lambda\left(\ln\left(\frac{1}{\lambda}\right) + \mathcal{O}\left(\frac{\ln(\ln 1/\lambda))}{\ln(1/\lambda)}\right)\right)$$

where the numbers c_J are positive and have been determined. We conjecture that for each k there are resonances whose imaginary parts go to zero as $\lambda \downarrow 0$ and whose real parts satisfy

$$\operatorname{Re}\varepsilon_k^{(J\pm)} = -\frac{1}{4k^2} + \frac{\alpha_k^{(J\pm)}}{k^3} + \frac{1}{\lambda^2}\big(m + |m| + 2J + 1\big)$$

where $0 < \alpha_k^{(J-)} < \alpha_k^{(J+)} < k^3\big(-1/(4(k+1)^2) + 1/(4k^2)\big)$. We have, to date, established the truth of this conjecture only for bounded k. Note that the position of these resonances is given, to leading order, by their real parts, and that, other than the shift by $(2J)/(\lambda^2)$, these positions are approximately given by (3) and (4). Explicitly

$$\lim_{\lambda\downarrow 0}\left|\varepsilon_0^{(J)} - \frac{2J}{\lambda^2} - \varepsilon_0\right| \to 0$$

and we expect that

$$\lim_{\lambda \downarrow 0} \left| \varepsilon_k^{(J\pm)} - \frac{2J}{\lambda^2} - \varepsilon_k^{(\pm)} \right| \to 0.$$

Thus, we call the $\varepsilon_0^{(J)}$ ground state resonances and the $\varepsilon_k^{(J\pm)}$ excited state resonances.

In the rest of this talk we would like to discuss our calculation of the ground state resonances. In the limit we are considering, the small λ limit, the Larmour radius is much smaller than the Bohr radius. One expects that in the plane perpendicular to the magnetic field the effect of the magnetic field on the electron dominates that of the proton. The atomic physics should take place mostly in the direction parallel to the field **[8]** **[9]** **[1]**. Explicitly, if $g_n^{(m)}(r)$ are the eigenfunctions of the pure magnetic field part of $H^{(m)}$ (or, for that matter, $H_\theta^{(m)}$),

$$\left(-\frac{\partial^2}{\partial r^2} - \frac{1}{r}\frac{\partial}{\partial r} + \frac{m^2}{r^2} + r^2 + 2m \right) g_n^{(m)}(r) = \big(m + |m| + 2n + 1\big) g_n^{(m)}(r),$$

then the eigen-functions of $H^{(m)}$ **[4]** and the resonant eigenfunctions that we have found, the eigen-functions of $H_\theta^{(m)}$ with $J > 0$ **[5]**, are approximately of the form by $g_J^{(m)}(r) f(z)$ for some $f \in \mathrm{L}(R, dz)$. To extract this one dimensional behavior we have used the following trick. We use the functions $g_n^{(m)}(r)$ as a basis for $\mathrm{L}^2\big([0,\infty), rdr\big)$ and write $H_\theta^{(m)}$ as a matrix with respect to this basis. If we let $v_{nn'}(\lambda, z) = \langle g_n^{(m)} (\lambda^2 r^2 + z^2)^{-1/2}, g_{n'}^{(m)} \rangle$,

$$\begin{aligned} a_\theta^{(J)} &= H_{\theta\ JJ}^{(m)} \\ &= -e^{-2i\theta} \frac{d^2}{dz^2} - v_{JJ}(\lambda, e^{2i\theta} z^2) + \frac{1}{\lambda^2}\big(m + |m| + 2J + 1\big), \\ b_{\theta n}^{(J)} &= -v_{Jn}(\lambda, e^{2i\theta} z^2) \end{aligned}$$

and $M_\theta^{(J)}$ be the matrix for $H_\theta^{(m)}$ with the Jth row and column deleted then $H_\theta^{(m)}$ has a resonance, ε, iff there is a $\psi \in \mathrm{L}^2\big([0,\infty), rdr\big) \otimes \mathrm{L}^2(R, dz)$ with $\langle \psi, g_n^{(m)} \rangle = f_n(z)$ such that

$$\begin{pmatrix} a_\theta^{(J)} - \varepsilon & b_\theta^{(J)t} \\ b_\theta^{(J)} & M_\theta^{(J)} - \varepsilon \end{pmatrix} \begin{pmatrix} f_J(z) \\ f_\perp(z) \end{pmatrix} = \begin{pmatrix} (a_\theta^{(J)} - \varepsilon) f_J(z) + b_\theta^{(J)t} f_\perp(z) \\ b_\theta^{(J)} f_J(z) + (M_\theta^{(J)} - \varepsilon) f_\perp(z) \end{pmatrix} = \begin{pmatrix} 0 \\ 0 \end{pmatrix}$$

where

$$f_\perp(z) = \begin{pmatrix} f_0(z) \\ f_1(z) \\ \vdots \\ f_{J-1} \\ f_{J+1} \\ \vdots \end{pmatrix}.$$

If ε is chosen so that it is not too close to any of the rays in $\sigma_c(H_\theta^{(m)})$ other than $(m + |m| + 2J + 1)\lambda^2 + e^{-2i\theta}[0, \infty)$ then we have shown that $M_\theta^{(J)} - \varepsilon$ is invertible.

Explicitly, if

$$\left|\operatorname{Im} e^{2i\theta}\left(\frac{1}{\lambda^2}(m+|m|+2J+1)-\varepsilon\right)\right| < \frac{3\sin(2\theta)}{2\lambda^2}$$

then the resolvent expansion for $(M_\theta^{(J)}-\varepsilon)^{-1}$ about

$$\left(-e^{-2i\theta}\frac{d^2}{dz^2}-\varepsilon\right)^{-1}\delta_{nn'},$$

with $n, n' \neq J$, converges. In particular $\left\|(M_\theta^{(J)}-\varepsilon)^{-1}\right\| \leq \text{const } \lambda^2$. It follows that

$$f_\perp = -(M_\theta^{(J)}-\varepsilon)^{-1} b_\theta^{(J)} f_J$$

so that $H_\theta^{(m)}$ has a resonance, ε, iff there is an $f_J \in \mathrm{L}^2(R, dz)$ with

$$0 = \left(a_\theta^{(J)} - \varepsilon - b_\theta^{(J)t}(M-\varepsilon)^{-1} b_\theta^{(J)}\right) f_J.$$

Letting $\delta = \varepsilon - \big(m+|m|+2J+1\big)/\lambda^2$ and

$$h_\theta(\delta) = -e^{-2i\theta}\frac{d^2}{dz^2} - v_{JJ}(\lambda, e^{2i\theta}z^2) - b_\theta^{(J)t}\big(M_\theta^{(J)}-\varepsilon\big)^{-1} b_\theta^{(J)}$$

we have that $H_\theta^{(m)}$ has a resonance, $\varepsilon = \big(m+|m|+2J+1\big)/\lambda^2 + \delta$, iff $h_\theta(\delta) - \delta$ has non-trivial kernel.

$h_\theta(\delta)$ is a integro-differential operator in one dimension, the dimension parallel to the magnetic field. The effects from the plane perpendicular to the field are all encoded in $b_\theta^{(J)t}\big(M_\theta^{(J)}-\varepsilon\big)^{-1} b_\theta^{(J)}$ and don't contribute to leading order in λ. However, it turns out that the imaginary parts of the resonances come from the contribution of $b_\theta^{(J)t}\big(M_\theta^{(J)}-\varepsilon\big)^{-1} b_\theta^{(J)}$. Thus we can't be satisfied with leading order estimates of the eigenvalues of $H_\theta^{(m)}$, as we were in the case of the spectrum, we must here estimate higher order terms.

Now we restrict the range of ε further by insisting that $\operatorname{Re}\delta < -1$, putting ε into the region in which we expect to find the Jth ground state resonance, and proceed, essentially as in §4 of [**4**], to find an appropriate eigenvalue condition for $H_\theta^{(m)}$. Since we are investigating the small λ limit it is tempting to try to expand about the $\lambda = 0$ limit of $h_\theta(\delta)$. Unfortunately, the $\lambda \downarrow 0$ limit is singular. Considering the form of $H_\theta^{(m)}$ it is not difficult to convince oneself that, formally, one should have

$$\begin{aligned} h_\theta^{lim} &= \lim_{\lambda\downarrow 0} h_\theta(\delta) \\ &= -e^{-2i\theta}\frac{d^2}{dz^2} - \frac{e^{-i\theta}}{|z|}. \end{aligned}$$

More explicitly, one can show that

$$\left|v_{nn'}(\lambda, e^{2i\theta}z^2) - \frac{e^{-i\theta}}{|z|}\delta_{nn'}\right| \leq \text{const } \lambda^2 \frac{1}{|z|^3}$$

and that $\left|v_{nn'}(\lambda, e^{2i\theta}z^2)\right| \leq \text{const } 1/\lambda$. The domain of h_θ^{lim} is not a core for $h_\theta(\delta)$. In more colloquial terms, because of the singularity that develops at $z = 0$ the domain of $h_\theta(\delta)$ changes drastically at $\lambda = 0$. To get around this problem we note that $h_\theta(\delta)$, acting in $\mathrm{L}^2(R, dz)$, is unitarily equivalent to $h_{\theta N}(\delta) \oplus h_{\theta D}(\delta)$ acting in

$\mathrm{L}^2\big([0,\infty),dz\big)$, where N and D refer to Neumann and Dirichlet boundary conditions at $z=0$. One finds that the $\lambda=0$ limit of $h_{\theta N}(\delta)$ is singular while the $\lambda=0$ limit of $h_{\theta D}(\delta)$ is not. If we let

$$h_{\theta 0}=\left(-e^{-2i\theta}\frac{d^2}{dz^2}-\frac{e^{-i\theta}}{z}\right)_D$$

in $\mathrm{L}^2\big([0,\infty),dz\big)$ then we can control $h_{\theta D}(\delta)$ by using $h_{\theta 0}$ as a first approximation. We find, as expected (the ground state should be in the Neumann part of the spectrum), that $h_{\theta D}(\delta)-\delta$ has no kernel in the range to which δ has been restricted.

To control the Neumann operator we study $\big(h_{\theta N}(\delta)+1\big)^{-1}$ rather than $h_{\theta N}(\delta)$ directly. We then use the following fact about second order differential operators on the half line. If

$$h_{\theta D}^{(1)}=-\left(e^{-2i\theta}\frac{d^2}{dz^2}+v_{JJ}(\lambda,e^{2i\theta}z^2)\right)_D$$

and

$$h_{\theta N}^{(1)}=-\left(e^{-2i\theta}\frac{d^2}{dz^2}+v_{JJ}(\lambda,e^{2i\theta}z^2)\right)_N$$

then both $\big(h_{\theta D}^{(1)}+1\big)^{-1}$ and $\big(h_{\theta N}^{(1)}+1\big)^{-1}$ exist and

$$(h_{\theta N}^{(1)}+1)^{-1}-(h_{\theta D}^{(1)}+1)^{-1}=R_\theta$$

where R_θ is a rank one operator with integral kernel

$$R_\theta(z_1,z_2)=-\frac{e^{2i\theta}}{\psi^{(+)\prime}(0)\psi^{(+)}(0)}\psi^{(+)}(e^{i\theta}z_1)\psi^{(+)}(e^{i\theta}z_2).$$

The function $\psi^{(+)}$ is the unique element of $\mathrm{L}^2\big([0,\infty),dz\big)$ which solves the equation

$$\left(-\frac{d^2}{dz^2}-v_{JJ}(\lambda,z^2)+1\right)\psi^{(+)}(z)=0.$$

Letting

$$F_\theta(\delta)=b_\theta^{(J)t}(M_{\theta N}^{(J)}-\varepsilon)^{-1}b_\theta^{(J)}\big(h_{\theta N}^{(1)}+1\big)^{-1}$$

we find that $\big\|F_\theta(\delta)\big\|\le \text{const }\lambda/\ln(1/\lambda)$ and that

$$\big(h_{\theta N}(\delta)+1\big)^{-1}=((h_{\theta N}^{(1)}+1)^{-1}+R_\theta)(1-F_\theta)^{-1}.$$

It follows that $h_{\theta N}(\delta)-\delta$ has non-trivial kernel iff

$$(h_{\theta D}^{(1)}+1)^{-1}+R_\theta-\frac{1}{\delta+1}(1-F_\theta)$$

does. Now we use the specific form of R_θ. In general, if A is an invertible operator on some vector space and κ is an element of that vector space then $A+\kappa\otimes\kappa$ has non-trivial kernel iff there is some vector $v\notin\ker(\kappa\otimes\kappa)$ with $(A+\kappa\otimes\kappa)v=0$. Subtracting $(\kappa\otimes\kappa)v=\kappa\langle v,\bar\kappa\rangle$ from both sides, multiplying by A^{-1}, taking an inner product with $\bar\kappa$ and dividing by $\langle v,\bar\kappa\rangle$ we find that $A+\kappa\otimes\kappa$ has non-trivial kernel iff $1=-\langle A^{-1}\kappa,\bar\kappa\rangle$. Thus we obtain the resonance condition

$$1=\frac{1}{\psi^{(+)\prime}(0)\psi^{(+)}(0)}\left\langle\left((h_{\theta D}^{(1)}+1)^{-1}-\frac{1}{\delta+1}(1-F_\theta)\right)^{-1}\psi_\theta^{(+)},\psi_{-\theta}^{(+)}\right\rangle. \tag{7}$$

The remainder of the work goes into solving the resonance condition (7). First we show that in the range to which δ has been restricted there is a unique solution, δ_1, of the resonance condition with F_θ set equal to zero,

$$\psi^{(+)\prime}(0)\psi^{(+)}(0) = \left\langle \left((h_{\theta D}^{(1)} + 1)^{-1} - \frac{1}{\delta_1 + 1} \right)^{-1} \psi_\theta^{(+)}, \psi_{-\theta}^{(+)} \right\rangle .$$

Further, $\delta_1 \in R$ and

$$\delta_1 = -\left(\ln\left(\frac{1}{\lambda}\right) - \ln\left(\ln\left(\frac{1}{\lambda}\right)\right) + \mathcal{O}(1)\right)^2 . \tag{8}$$

We then solve (7) by expanding about $\delta = \delta_1$. We find that there is a solution, δ, to (7) which satisfies $|\delta - \delta_1| \le \text{const}\ \lambda\big(\ln 1/\lambda\big)^2$. This establishes (5). Further, we show that

$$\operatorname{Im}\delta = -c_J\lambda \left(\sqrt{-\delta_1} + \mathcal{O}\left(\frac{1}{\sqrt{-\delta_1}}\right)\right) \tag{9}$$

where

$$c_J = 2^{1/2} \sum_{n=0}^{J-1} \frac{1}{\sqrt{J-n}} \left(\int_0^\infty v_{Jn}(1, z^2) \cos\big(\sqrt{2(J-n)}\ z\big)\, dz \right)^2 .$$

Note that c_J is positive and depends only on J (the explicit λ in $v_{Jn}(\lambda, z^2)$ has been scaled out of the integral and is set to 1 in this last expression). (9), together with (8), establishes (6).

References

1. J. E. Avron, I. W. Herbst, and B. Simon, *Schrödinger operators with magnetic fields* III. *atoms in homogeneous magnetic field*, Comm. Math. Phys. **79** (1981), 529–572; *Strongly bound states of hydrogen in intense magnetic field*, Phys. Rev. A **20** (1979), 2287–2296.
2. E. Balslev and J. M. Combes, *Spectral Properties of Schrödinger operators with dilation analytic potentials*, Comm. Math. Phys. **22** (1971), 280–294.
3. H. L. Cycon, R. G. Froese, W. Kirsch, B. Simon, *Schrödinger Operators*, Springer-Verlag, 1987.
4. R. Froese and R. Waxler, *The Spectrum of a hydrogen atom in an intense magnetic field*, Rev. Math. Phys. (accepted).
5. ——, *Ground state resonances of a hydrogen atom in an intense magnetic field*, Rev. Math. Phys. (accepted).
6. W. Hunziker, *Resonances, metastable states and exponential decay laws in perturbation theory*, Comm. Math. Phys. **132** (1990), 177–188.
7. H. Jones, *Beyond 45 tesla*, Europhys. News **22** (1991), 154–155.
8. L. D. Landau and E. M. Lifshitz, *Quantum Mechanics*, 3rd edition, Pergamon, 1977.
9. M. Ruderman, *Matter and magnetospheres in superstrong magnetic fields*, Role of Magnetic Fields in Physics and Astrophysics, (V. Canuto, ed.), Ann. New York Acad. Sci., vol. 257, 1975; *Matter in Superstrong Magnetic Fields*, Physics of Dense Matter, (C. J. Hanson, ed.), D. Reidel, 1974, 117–131.
10. B. Simon, *Resonances and complex scaling: a rigorous overview*, Internat. J. Quant. Chem. **14** (1978), 529–542.

DEPT. OF MATHEMATICS, UNIVERSITY OF BRITISH COLUMBIA, VANCOUVER, B.C. V6T 1Y4, CANADA

E-mail address: rfroese@wespe.math.ubc.ca

DEPARTMENT OF MATHEMATICS, S.U.N.Y. AT BUFFALO, BUFFALO, N.Y. 14214-3093, USA

E-mail address: rwax@newton.math.buffalo.edu

Centre de Recherches Mathématiques
CRM Proceedings and Lecture Notes
Volume 8, 1995

On the Scattering Matrix for N-Particle Hamiltonians

D. Yafaev

Our goal here is to derive stationary representations for the scattering matrix of the N-particle Schrödinger operator. We consider arbitrary incoming or outgoing channels of scattering. In particular, it is shown that the scattering matrix is a strongly continuous function of the spectral parameter (energy).

1. Similarly to the two-particle case, the scattering theory for $N, N \geq 3$, interacting quantum particles with a Hamiltonian H can be formulated as the assertion about existence and completeness of some wave operators $W^{\pm}(H, \widehat{H}; J)$. All possible channels of scattering (determined by decompositions of the N-particle system into non-interacting clusters of particles) are naturally taken into account in the framework of an auxiliary "large" Hilbert space $\widehat{\mathcal{H}}$ in which some "effective free" Hamiltonian $\widehat{H}$ acts. The operator $\widehat{H}$ is constructed in terms of the kinetic energy operators of relative motion of clusters and energies of bound states of these clusters. The "identification" J is constructed in terms of the corresponding eigenfunctions. Thus scattering theory in the N-particle case is naturally formulated in a couple of (different) spaces.

However, results of the abstract scattering theory are definitely not applicable to the triple $\{\widehat{H}, H; J\}$ which impedes to obtain stationary representations for the scattering matrix $S = S(J)$. Actually, one of the ingredients of the problem is a correct choice of an auxiliary identification $\tilde{J}$ such that $\{\widehat{H}, H; \tilde{J}\}$ satisfies already standard assumptions of the smooth scattering theory. In particular, this gives a representation for $\widetilde{S} = S(\tilde{J})$ in terms of boundary values of the resolvent $R(z)$ of H and the spectral family of $\widehat{H}$.

Let us explain a construction of $\tilde{J}$ on the example of the channel of scattering, where all particles are asymptotically free both before and and after interaction. In this channel we choose $\tilde{J}$ as a first order differential operator $M^{(0)}$ with coefficients which are homogeneous functions of degree zero. It is required that these coefficients vanish in neighbourhoods of "singular" directions, where pair potentials do not tend to zero.

Thus we first construct the scattering matrix $\widetilde{S}$ corresponding to the auxiliary identification $\tilde{J}$ and then find its relation to the physical scattering matrix S. Our restriction on coefficients of $M^{(0)}$ implies that we have a reasonably good representation for the physical scattering amplitude of the free channel only away from singular directions.

Our representations for the scattering matrix require a detailed study of boundary values of the resolvent $R(z)$ as $z = \lambda \pm i\varepsilon$, $\varepsilon \to 0$. In complement to the limiting

1991 *Mathematics Subject Classification.* Primary: 81U20; Secondary: 47F05.
A detailed paper is being published in "Integral Equations and Operator Theory".

absorption principle we prove uniform boundedness of operators $G_a R(z) G_b^*$, where G_a, G_b (indices a, b label decompositions of N particles into clusters) are suitable first order differential operators with coefficients vanishing as $|x|^{-1/2}$ at infinity.

Note that a time-dependent definition determines the scattering matrix for almost all energies only. This is not quite satisfactory from the physics point of view. Using stationary representations we show that the scattering matrix is actually well-defined for all energies and is a continuous (in the strong operator sense) function of the energy.

We follow here a non-perturbative approach of [**6**], [**3**] and [**7**] to the N-particle scattering theory. Actually, our constructions rely heavily on the paper [**7**] where it is shown that methods of the smooth scattering theory can be applied to the N-particle problem. In the framework of the Faddeev's approach [**2**] to the three-particle scattering theory stationary representations for the scattering matrix were obtained in [**9**]. Note that the representations of [**9**] and of this paper do not coincide.

2. Let us briefly recall some basic definitions of the scattering theory. Let H_j, $j = 1, 2$, be a couple of self-adjoint operators in Hilbert spaces $\mathcal{H}_j$, $j = 1, 2$, respectively, and let $J\colon \mathcal{H}_1 \to \mathcal{H}_2$ be an H_1-bounded operator. Denote by $E_j(\Lambda) = E(\Lambda; H_j)$ the spectral family of H_j. Suppose that an interval Λ is such that the operator H_1 is absolutely continuous on the subspace $\mathcal{H}_1(\Lambda) = E_1(\Lambda)\mathcal{H}_1$. Suppose also that $JE_1(\Lambda) \in \mathcal{B}$ ($\mathcal{B}$ is the class of bounded operators). The wave operator for the pair H_1, H_2, "identification" J and interval Λ is defined by the relation

$$W^{\pm} = W^{\pm}(H_2, H_1; J, \Lambda) = \operatorname*{s-lim}_{t \to \pm J\infty} \exp(iH_2 t) J \exp(-iH_1 t) E_1(\Lambda) \tag{1}$$

under the assumption that this strong limit exists.

Suppose now that both wave operators (1) exist. Then the scattering operator for the triple H_1, H_2, J and the interval Λ is defined as

$$\mathbf{S} = \mathbf{S}(H_2, H_1; J, \Lambda) = W^+(H_2, H_1; J, \Lambda)^* W^-(H_2, H_1; J, \Lambda). \tag{2}$$

To introduce the scattering matrix we consider a diagonalization of the operator H_1 on Λ. In view of applications we suppose that the operator H_1 has the absolutely continuous spectrum of constant multiplicity on Λ. Thus we consider a representation of the space $\mathcal{H}_1(\Lambda)$ as the L_2-space of vector-functions with values in some auxiliary Hilbert space $\mathbf{N}$ ($\dim \mathbf{N}$ equals the multiplicity of the spectrum):

$$\mathcal{H}_1(\Lambda) \leftrightarrow L_2(\Lambda; \mathbf{N}). \tag{3}$$

Let F_1 be the operator which maps unitarily the left-hand side onto its right-hand side. We require that $(F_1 H_1 f)(\lambda) = \lambda (F_1 f)(\lambda)$ for $f \in \mathcal{H}_1(\Lambda)$ and, consequently,

$$\left\| (F_1 f)(\lambda) \right\|_{\mathbf{N}}^2 = \frac{d\big(E_1(\lambda) f, f\big)}{d\lambda}, \qquad \text{a.e. } \lambda \in \Lambda.$$

Since the operator (2) commutes with H_1, it acts in (3) as multiplication by an operator-function $S(\lambda) = S(\lambda; H_2, H_1; J)\colon \mathbf{N} \to \mathbf{N}$, called the scattering matrix. In other words,

$$(\mathbf{S} f, g) = \int_\Lambda \big(S(\lambda)(F_1 f)(\lambda), (F_1 g)(\lambda) \big)_{\mathbf{N}} \, d\lambda$$

for any $f, g \in \mathcal{H}_1(\Lambda)$. The scattering matrix is defined for a.e. $\lambda \in \Lambda$ and is bounded by $\|J\|^2$.

To describe a stationary formula for the scattering matrix it is necessary to introduce auxiliary wave operators

$$W^{+}\big(H_1, H_1; E_1(\Lambda)J^*JE_1(\Lambda)\big). \tag{4}$$

They commute with H_1 and hence act as multiplication by operator-functions $w^{\pm}(\lambda)\colon \mathbf{N} \to \mathbf{N}$ in the representation (3). We put

$$\mathbf{B}_{+}(z) = J^*\mathbf{V} - \mathbf{V}^*R_2(z)\mathbf{V}, \quad \mathbf{B}_{-}(z) = \mathbf{V}^*J - \mathbf{V}^*R_2(z)\mathbf{V},$$

where

$$\mathbf{V} = H_2J - JH_1. \tag{5}$$

Let a formal operator $Z_1(\lambda)$ be defined by $Z_1(\lambda)f = (F_1 f)(\lambda)$. The scattering matrix satisfies two equalities (for each of the signs "+" and "−")

$$S(\lambda) = w^{\pm}(\lambda) - 2\pi i Z_1(\lambda)\mathbf{B}_{\pm}(\lambda + i0)Z_1^*(\lambda), \tag{6}$$

which we call its stationary representations. If $J = I$, then $w^{+} = w^{-}$, $\mathbf{B}_{+} = \mathbf{B}_{-}$ and, consequently, two representations (6) coincide. In the general case they are different and we give both since neither of them is better than another.

Of course, representations (6) are only formal. Below we give assumptions on $\mathbf{V}$ which guarantee that the operators $Z_1(\lambda)\mathbf{B}_{\pm}(\lambda+i0)Z_1^*(\lambda)$ are correctly defined, bounded and depend continuously (in the weak sense) on $\lambda \in \Lambda$.

It turns out that though the notation $Z_1(\lambda)$ is only formal the product

$$Z_1(\lambda)K^* =: Z_1(\lambda; K), \tag{7}$$

where $K\colon \mathcal{H}_1 \to \mathcal{K}$ ($\mathcal{K}$ is an auxiliary Hilbert space), can be correctly defined. This requires, of course, some conditions on the operator K which can be verified in applications. Let us accept the following

DEFINITION 1. Let K be an H_1-bounded operator. Suppose that there exists a bounded weakly continuous in $\lambda \in \Lambda$ operator $Z_1(\lambda; K)\colon \mathcal{K} \to \mathbf{N}$ such that for every $f \in \mathcal{K}$ and a.e. $\lambda \in \Lambda$ the equality

$$(F_1K^*f)(\lambda) = Z_1(\lambda; K)f$$

holds. In this case we say that the operator (7) is correctly defined. We set $U_1(\lambda; K) = Z_1(\lambda; K)^*$.

Now we are able to rewrite representations (6) in a correct form. The proof of the following statement (under fairly more general assumptions) can be found in [**1**] or [**8**], Theorem 5.5.3.

THEOREM 2. *Let the perturbation* (5) *satisfy* $\mathbf{V} = K_2^*\mathcal{V}K_1$, *where* $K_j\colon \mathcal{H}_j \to \mathcal{K}$, K_1 *is* H_1*-bounded,* K_2 *is* $|H_2|^{1/2}$*-bounded and* $\mathcal{V} \in \mathcal{B}$. *Assume that the operators* $Z_1(\lambda; K_1)$, $Z_1(\lambda; K_2J)$ *are correctly defined and their adjoints* $U_1(\lambda; K_1)$, $U_1(\lambda; K_2J)$ *are strongly continuous in* $\lambda \in \Lambda$. *Suppose also that the operator-function*

$$\mathcal{R}_2(z; K_2) := K_2R_2(z)K_2^*$$

is weakly continuous in z, $\Re z \in \Lambda$, $\pm\Im z \geq 0$. *Put*

$$\mathcal{A}^+(\lambda) = Z_1(\lambda; K_2 J)\mathcal{V}U_1(\lambda; K_1), \quad \mathcal{A}^-(\lambda) = \mathcal{A}^+(\lambda)^*, \tag{8}$$

$$\mathbf{A}(\lambda) = Z_1(\lambda; K_1)\mathcal{V}^*\mathcal{R}_2(\lambda + i0; K_2)\mathcal{V}U_1(\lambda; K_1) \tag{9}$$

and

$$A^\pm(\lambda) = \mathcal{A}^\pm(\lambda) - \mathbf{A}(\lambda). \tag{10}$$

Then the wave operators (1) *and* (4) *exist and the scattering matrix* $S(\lambda)$ *satisfies for* $\lambda \in \Lambda$ *the relations*

$$S(\lambda) = w^\pm(\lambda) - 2\pi i A^\pm(\lambda).$$

In particular, the operator-functions $S(\lambda) - w^+(\lambda)$ *and* $S(\lambda) - w^-(\lambda)$ *are weakly continuous in* $\lambda \in \Lambda$.

Remark. Theorem 2 can be naturally extended to the case where $\mathbf{V}$ is a finite sum $\mathbf{V} = \sum_n K_{2,n}^* \mathcal{V}_n K_{1,n}$. This case can be reduced to the former one if one introduces "vector" operators $K_j = \bigoplus_n K_{j,n}$ and a diagonal matrix operator $\mathcal{V} = \operatorname{diag}\{\mathcal{V}_n\}$.

3. The N-particle Schrödinger operator is the self-adjoint operator $H = T + V$ in the Hilbert space $\mathcal{H} = L_2(\mathbb{R}^d)$, where $T = -\Delta$ and V is multiplication by a real function $V(x)$. Suppose that some finite number α_0 of subspaces X^α of $X := \mathbb{R}^d$ is given and let x^α, x_α be the orthogonal projections of $x \in X$ on X^α and its orthogonal complement $X_\alpha = X \ominus X^\alpha$, respectively. We assume that

$$V(x) = \sum_{\alpha=1}^{\alpha_0} V^\alpha(x^\alpha),$$

where V^α are decaying real functions of variables x^α. Clearly, $V^\alpha(x^\alpha)$ tends to zero as $|x| \to \infty$ outside of any conical neighbourhood of X_α. However $V^\alpha(x^\alpha)$ is constant on planes parallel to X_α. Due to this property the structure of the spectrum of H is much more complicated than in the two-particle case.

Let us introduce the set of all linear sums

$$X^a = X^{\alpha_1} + X^{\alpha_2} + \cdots + X^{\alpha_k}$$

of subspaces X^{α_j}. The zero subspace $X^0 = \{0\}$ is included in this set and X itself is excluded. We consider also the set of orthogonal complements $X_a := X \ominus X^a$, which consists of all intersections of subspaces X_α. Let x^a and x_a be the orthogonal projections of $x \in X$ on the subspaces X^a and X_a, respectively. In the multiparticle terminology, index a parametrizes decompositions of an N-particle system into noninteracting clusters; x^a is the set of "internal" coordinates of clusters, x_a describes their relative motion.

The scattering matrix is, of course, well-defined only for short-range potentials. However, our estimates on the resolvent hold as well true in the case, where potentials contain long-range parts. Therefore we distinguish two types of assumptions on functions V^α. The first of them is related to the short-range case and the second—to the general one. Let us define operators $T^\alpha = -\Delta^\alpha$ in the space $\mathcal{H}^\alpha = L_2(X^\alpha)$. Derivatives of V^α are understood below in the sense of distributions. We usually use the same notation for a function and the operator of multiplication by this function.

ASSUMPTION 3. Operators $V^\alpha(T^\alpha+I)^{-1/2}$ are compact in the space $\mathcal{H}^\alpha$ and operators

$$\big(|x^\alpha|+1\big)^\rho V^\alpha(T^\alpha+I)^{-1/2}$$

are bounded in $\mathcal{H}^\alpha$ for some $\rho>1$.

ASSUMPTION 4. Functions V^α admit representations $V^\alpha=V_s^\alpha+V_l^\alpha$, where the short-range V_s^α and long-range V_l^α parts satisfy the following conditions. Operators $V_s^\alpha(T^\alpha+I)^{-1}$, $V_l^\alpha(T^\alpha+I)^{-1}$, $|\nabla V_l^\alpha|(T^\alpha+I)^{-1}$ are compact in the space $\mathcal{H}^\alpha$ and operators $\big(|x^\alpha|+1\big)^\rho V_s^\alpha(T^\alpha+I)^{-1}$, $\big(|x^\alpha|+1\big)^\rho|\nabla V_l^\alpha|(T^\alpha+I)^{-1}$ are bounded in $\mathcal{H}^\alpha$ for some $\rho>1$.

Let us define for each a a potential

$$V^a=\sum_{X^\alpha\subset X^a}V^\alpha,$$

which does not depend on x_a, and the remaining part $V_a=V-V^a$, which tends to zero outside of any conical neighbourhood of X_a. Let the operator H^a be defined in the space $\mathcal{H}^a=L_2(X^a)$ by the equality $H^a=T^a+V^a$, $T^a=-\Delta^a$. The union over all a of their point spectra

$$\Upsilon_0=\bigcup\sigma^{(p)}(H^a)$$

is called the set of thresholds for the operator H. *Below Λ is always an arbitrary bounded interval such that $\bar\Lambda\cap\Upsilon=\emptyset$*, where $\bar\Lambda$ is the closure of Λ and $\Upsilon=\Upsilon_0\cup\sigma^{(p)}(H)$. Let Q be multiplication by $(x^2+1)^{1/2}$. The following basic result (see [**4, 5**]) of spectral theory of multiparticle Hamiltonians is called the limiting absorption principle.

PROPOSITION 5. *Let Assumption 4 hold. Then for any $r>1/2$ the operator $Q^{-r}(T+I)^{1/2}R(z)(T+I)^{1/2}Q^{-r}$ is continuous in topology of the norm with respect to z for $\Re z\in\Lambda$, $\pm\Im z\geq 0$.*

Let us give the precise formulation of the scattering problem for N-particle Hamiltonians. It can be conveniently performed in terms of the scattering theory in a couple of spaces. Our considerations are localized on an interval Λ. Let us introduce eigenvalues $\lambda^{a,n}$ of the operator H^a lying below Λ:

$$\lambda^{a,n}<\inf\Lambda. \tag{11}$$

Denote by $\psi^{a,n}$ a normalized eigenvector of the operator H^a corresponding to an eigenvalue $\lambda^{a,n}$. If some eigenvalues coincide, the corresponding eigenvectors are supposed to be orthogonal. We usually write $\mathbf{a}$ instead of a couple $\{a,n\}$.

Let us introduce a space

$$\widehat{\mathcal{H}}=\bigoplus_{\mathbf{a}}\mathcal{H}_{\mathbf{a}},\quad \mathcal{H}_{\mathbf{a}}=\mathcal{H}_a=L_2(X_a), \tag{12}$$

of all scattering channels and a "free" Hamiltonian

$$\widehat{H}=\bigoplus_{\mathbf{a}}T_{\mathbf{a}},\quad T_{\mathbf{a}}=T_a+\lambda^{\mathbf{a}},\quad T_a=-\Delta_a, \tag{13}$$

in this space. Here and below the sums are taken over all $\mathbf{a}$ satisfying (11). We define an identification $J\colon \widehat{\mathcal{H}} \to \mathcal{H}$ by the relations

$$J = \sum_{\mathbf{a}} J^{\mathbf{a}}, \quad J^{\mathbf{a}} f_{\mathbf{a}} = f_{\mathbf{a}} \otimes \psi^{\mathbf{a}}, \tag{14}$$

where the tensor product is determined by the decomposition $\mathcal{H} = \mathcal{H}_a \otimes \mathcal{H}^a$. In particular, $T_0 = T$, $\lambda^0 = 0$ and $J^0 = I$. Note that $H_a = T_a \otimes I + I \otimes H^a$ and hence

$$H_a J^{\mathbf{a}} = J^{\mathbf{a}} T_{\mathbf{a}}. \tag{15}$$

The basic result of the scattering theory for N-particle Schrödinger operators can be formulated as follows.

THEOREM 6. *Let Assumption 3 hold. Then the wave operators $W^{\pm}(H, \widehat{H}; \hat{J}, \Lambda)$ exist, are isometric and complete.*

This theorem implies, of course, that the scattering operator $\mathbf{S}$ defined by (2) is unitary in the space $\widehat{\mathcal{H}}(\Lambda) = E(\Lambda; \widehat{H})\widehat{\mathcal{H}}$ and commutes with the operator $\widehat{H}$. Clearly, $\mathbf{S}$ is the matrix operator with components

$$\mathbf{S}_{\mathbf{ab}} = (W_{\mathbf{a}}^{+})^* W_{\mathbf{b}}^{-} : \mathcal{H}_b \to \mathcal{H}_a, \quad W_{\mathbf{a}}^{\pm} = W^{\pm}(H, T_{\mathbf{a}}; J^{\mathbf{a}}, \Lambda).$$

To define the corresponding scattering matrix we consider a diagonal for $\widehat{H}$ representation of the space $\widehat{\mathcal{H}}(\Lambda)$ in a direct integral:

$$\hat{\mathcal{H}}(\Lambda) \leftrightarrow L_2(\Lambda; \mathbf{N}), \tag{16}$$

where

$$\mathbf{N} = \bigoplus_{\mathbf{a}} \mathbf{N}_{\mathbf{a}}, \quad \mathbf{N}_{\mathbf{a}} = \mathbf{N}_a = L_2(\mathbb{S}^{d_a - 1}).$$

Let us introduce a unitary mapping $\widehat{F} = \bigoplus_{\mathbf{a}} F_{\mathbf{a}}$ of the left-hand side of (16) onto its right-hand side. Explicitly, $F_{\mathbf{a}}$ is given by the standard formula

$$(F_{\mathbf{a}} f)(\lambda; \omega_a) = (\kappa^{\mathbf{a}})^{-1+d_a/2} 2^{-1/2} (\Phi_a f)(\kappa^{\mathbf{a}} \omega_a), \quad \kappa^{\mathbf{a}} = (\lambda - \lambda^{\mathbf{a}})^{1/2},$$

where

$$(\Phi_a f)(\xi_a) = (2\pi)^{-d_a/2} \int \exp(-i\langle \xi_a, x_a \rangle) f(x_a)\, dx_a$$

is the Fourier transform in the space $L_2(\mathbb{R}^{d_a})$ (the symbol $\langle \cdot, \cdot \rangle$ denotes the scalar product in different Euclidean spaces). Then $(\widehat{F}\widehat{H}f)(\lambda) = \lambda(\widehat{F}f)(\lambda)$ for $f \in \hat{\mathcal{H}}(\Lambda)$.

Once the spectral representation of $\widehat{H}$ is fixed, the scattering matrix $S(\lambda)\colon \mathbf{N} \to \mathbf{N}$ can be introduced by the general time-dependent definition of p. 297. Now $S(\lambda)$ is a "matrix" operator with components $S_{\mathbf{a},\mathbf{b}}(\lambda)$ acting from the space $L_2(\mathbb{S}^{d_b-1})$ into the space $L_2(\mathbb{S}^{d_a-1})$. We emphasize that $S(\lambda)$ is defined and unitary for a.e. $\lambda \in \Lambda$ only. Our goal below is to obtain stationary representations for it. When considering $S_{\mathbf{a},\mathbf{b}}(\lambda)$ we can put $J = J_{\mathbf{a}} + J_{\mathbf{b}}$ if $\mathbf{a} \neq \mathbf{b}$ and $J = J_{\mathbf{a}}$ if $\mathbf{a} = \mathbf{b}$. Thus, without loss of generality, we may suppose that the sums (12)–(14) contain a finite number of terms only.

4. To justify the representations for the scattering matrix we need to reinforce the estimates of [**7**] called the radiation conditions-estimates there. In contrast to [**7**], where estimates were formulated in terms of the unitary group $\exp(-iHt)$, we study the resolvent of H here.

Let ∇_a be the gradient in the variable x_a (i.e. $\nabla_a u$ is the orthogonal projection of ∇u on the subspace X_a) and let $\nabla_a^{(s)}$,

$$(\nabla_a^{(s)} u)(x) = (\nabla_a u)(x) - |x_a|^{-2}\langle(\nabla_a u)(x), x_a\rangle x_a,$$

be its orthogonal projection in X_a on the plane orthogonal to the vector x_a. Let

$$\mathbf{X}_a(\varepsilon) = \{|x_a| > (1-\varepsilon)|x|\}, \quad \varepsilon \in (0,1),$$

be a conical neighbourhood of $X_a \setminus \{0\}$. We put

$$\mathbf{\Gamma}_a(\varepsilon) = X \Big\backslash \bigcup_{X_a \not\subset X_b} \mathbf{X}_b(\varepsilon).$$

Clearly, $\mathbf{\Gamma}_a(\varepsilon)$ gets larger as ε decreases and every closed cone $\mathbf{\Gamma}_a$ such that $\mathbf{\Gamma}_a \cap X_b = \{0\}$ if $X_a \not\subset X_b$ belongs to a cone $\mathbf{\Gamma}_a(\varepsilon)$ for sufficiently small ε. In particular, $\mathbf{\Gamma}_0$ is the whole space with some neighbourhoods of all subspaces X_α removed from it. Denote by $\chi(\mathcal{Y})$ the characteristic function of a set $\mathcal{Y} \subset \mathbb{R}^d$. Let us introduce the operator

$$G_a(\varepsilon) = \chi(\mathbf{\Gamma}_a(\varepsilon))Q^{-1/2}\nabla_a^{(s)} \tag{17}$$

which acts from the space $\mathcal{H}$ into $\mathcal{H} \otimes \mathbb{C}^{d_a}$. Our resolvent estimates are formulated in the following

THEOREM 7. *Let Assumption 4 hold. Choose any $\varepsilon > 0$ and set $G_a = G_a(\varepsilon)$. Then for arbitrary a and b*

$$\|G_a R(z) G_b^*\| \le C, \quad \Re z \in \Lambda.$$

COROLLARY 8. *Let $\Re z \in \Lambda$, $\pm\Im z \ge 0$ and $r > 1/2$. Then the operator-function $Q^{-r}(T+I)^{1/2}R(z)G_b^*$ is strongly continuous and operator-functions $G_a R(z)(T+I)^{1/2}Q^{-r}$, $G_a R(z) G_b^*$ are weakly continuous in z.*

We need also some results on strong continuity of some operator-functions related to the spectral family of the "free" operator $\widehat{H}$. In some sense these results can be considered as generalizations of Sobolev theorems about traces of functions on a sphere or, more precisely, of the corresponding dual assertions.

Let us introduce operators

$$\big(U_{\mathbf{a}}(\lambda)u\big)(x_a) = \kappa^{(d_a-2)/2} 2^{-1/2}(2\pi)^{-d_a/2} \int \exp\big(i\kappa\langle\omega_a, x_a\rangle\big) u(\omega_a)\, d\omega_a,$$
$$\omega_a \in \mathbb{S}^{d_a-1}, \quad \kappa = (\lambda - \lambda^{\mathbf{a}})^{1/2} > 0,$$

and

$$G_b(\varepsilon) J^{\mathbf{a}} U_{\mathbf{a}}(\lambda) \colon \mathbf{N}_a \to \mathcal{H}. \tag{18}$$

THEOREM 9. *Let Assumption 4 hold. Then for arbitrary a, b and $\varepsilon > 0$ the operator* (18) *is bounded and strongly continuous in $\lambda > \lambda^{\mathbf{a}}$.*

5. In this section we construct an auxiliary identification $\tilde{J}$ such that the triple $\{\widehat{H}, H, \tilde{J}\}$ satisfies the assumptions of Theorem 2. This gives immediately stationary representations for the scattering matrix $\widetilde{S}(\lambda)$ corresponding to $\tilde{J}$. A verification of the assumptions of Theorem 2 is based on the estimates of p. 4. We return back to the "physical" identification J in the next part.

Let us introduce a differential operator

$$M^{(a)} = 2^{-1} \sum_{j=1}^{d} (m_j^{(a)} D_j + D_j m_j^{(a)}), \quad m_j^{(a)} = \partial m^{(a)} / \partial x_j, \tag{19}$$

with a "generating" function $m^{(a)}$ obeying the following conditions:

1° $m^{(a)}(x)$ is a real C^∞-function, which is homogeneous of degree 1 for $|x| \geq 1$ and $m^{(a)}(x) = 0$ for $|x| \leq 1/2$.

2° Let b be arbitrary. If $x \in \mathbf{X}_b(\varepsilon)$ and $|x| \geq 1$, then $m^{(a)}(x) = m^{(a)}(x_b)$, i.e. $m^{(a)}(x)$ does not depend on x^b.

3° Let $X_a \not\subset X_b$. If $x \in \mathbf{X}_b(\varepsilon)$, then $m^{(a)}(x) = 0$.

Functions $m^{(a)}$ with such such properties were constructed in [**7**]. We emphasize that ε is a fixed positive number which can be chosen arbitrary small.

Let us consider the "effective perturbation" (cf. (5)) for the pair H_a, H relative to the identification $M^{(a)}$:

$$\mathbf{V}^{(a)} := HM^{(a)} - M^{(a)}H_a = [T, M^{(a)}] + [V^a, M^{(a)}] + V_a M^{(a)}, \tag{20}$$

Properties of different terms in the right-hand side were considered in [**7**]. They are formulated in the following three assertions.

PROPOSITION 10. *Let Assumption 3 hold and let $m^{(a)}$ satisfy the conditions 1°, 2°. Then*

$$[V^a, M^{(a)}] = (T+I)^{1/2} Q^{-r} B_a Q^{-r} (T+I)^{1/2},$$

where $2r = \rho > 1$ and $B_a \in \mathcal{B}$.

PROPOSITION 11. *Let Assumption 3 hold and let $m^{(a)}$ satisfy the conditions 1°, 3°. Then*

$$V_a M^{(a)} = (T+I)^{1/2} Q^{-r} \widetilde{B}_a Q^{-r} (T+I)^{1/2},$$

where $2r = \rho > 1$ and $\tilde{B}_a \in \mathcal{B}$.

PROPOSITION 12. *Let $m^{(a)}$ satisfy the conditions 1°, 2°. Then*

$$[T, M^{(a)}] = \sum_b G_b^* B_{a,b} G_b + Q^{-3/2} B_a^{(0)} Q^{-3/2}, \quad B_{a,b} \in \mathcal{B}, \quad B_a^{(0)} \in \mathcal{B},$$

where the operators G_b are defined by (17).

Recall that the operator $\widehat{H}$ and its spectral representation were constructed in p. 296. Let us introduce formal operators $Z_{\mathbf{a}}(\lambda)$, $Z_{\mathbf{a}}(\lambda) f = (F_{\mathbf{a}} f)(\lambda)$. Then (also formally) $U_{\mathbf{a}}(\lambda) = Z_{\mathbf{a}}(\lambda)^*$. We set

$$\mathbf{Z}_a(\lambda) = Z_{\mathbf{a}}(\lambda)(J^{\mathbf{a}})^*, \quad \mathbf{U}_a(\lambda) = \mathbf{Z}_a(\lambda)^* = J^{\mathbf{a}} U_{\mathbf{a}}(\lambda).$$

Let us define the identification $\tilde{J}$ by the relation (cf. (14))

$$\tilde{J} = \sum_{\mathbf{a}} \tilde{J}^{\mathbf{a}}, \quad \tilde{J}^{\mathbf{a}} = M^{(a)} J^{\mathbf{a}}. \tag{21}$$

We check that the triple $\{\widehat{H}, H, \tilde{J}\}$ satisfies on Λ the assumptions of Theorem 2.

Let us consider $H\tilde{J} - \tilde{J}\widehat{H}$. In virtue of (15)

$$\widetilde{\mathbf{V}}^{(\mathbf{a})} := H\tilde{J}^{\mathbf{a}} - \tilde{J}^{\mathbf{a}}T_{\mathbf{a}} = (HM^{(a)} - M^{(a)}H_a)J^{\mathbf{a}} = \mathbf{V}^{(a)}J^{\mathbf{a}}.$$

According to the equality (20) and Propositions 10–12 this operator admits the representation

$$\widetilde{\mathbf{V}}^{(\mathbf{a})} = (T+I)^{1/2}Q^{-r}B_aQ^{-r}(T+I)^{1/2}J^{\mathbf{a}} + \sum_b G_b^*B_{a,b}G_bJ^{\mathbf{a}}, \tag{22}$$

where $r > 1/2$, B_a, $B_{a,b}$ are bounded operators and the sum is taken over all b. The operator (22) can be factorized as $\widetilde{\mathbf{V}}^{(\mathbf{a})} = K^*\mathcal{V}K_{\mathbf{a}}^{(0)}$, where $\mathcal{V} \in \mathcal{B}$, K is a "vector" operator with components $Q^{-r}(T+I)^{1/2}$ and G_b (all b), $K_{\mathbf{a}}^{(0)}$ is a "vector" operator with components $Q^{-r}(T+I)^{1/2}J^{\mathbf{a}}$ and $G_bJ^{\mathbf{a}}$.

The weak continuity of the operator $KR(z)K^*$ in z, $\Re z \in \Lambda$, $\pm\Im z \geq 0$, is a consequence of Proposition 5 and Corollary 8. The assumptions of Theorem 2 with respect to the "free" operator $T_{\mathbf{a}}$ require boundedness and strong continuity of the operators $K_{\mathbf{a}}^{(0)}U_{\mathbf{a}}(\lambda)$ and $K_{\mathbf{a}}\tilde{J}^{\mathbf{a}}U_{\mathbf{a}}(\lambda)$. This is equivalent to the same statements about the operators

$$Q^{-r}(T+I)^{1/2}J^{\mathbf{a}}U_{\mathbf{a}}(\lambda), \quad G_bJ^{\mathbf{a}}U_{\mathbf{a}}(\lambda),$$
$$Q^{-r}(T+I)^{1/2}M^{(a)}J^{\mathbf{a}}U_{\mathbf{a}}(\lambda), \quad G_bM^{(a)}J^{\mathbf{a}}U_{\mathbf{a}}(\lambda),$$

acting from the space $\mathbf{N_a}$ into $\mathcal{H}$. Boundedness and strong continuity of these operators are consequences of the standard Sobolev (dual) trace theorem and of Theorem 9.

Thus we have obtained the following:

THEOREM 13. *Let Assumption* 3 *hold. Suppose that $\tilde{J}$ is defined by* (19), (21), *where $m^{(a)}$ obeys the conditions* 1°–3°. *Then the triple $\{\widehat{H}, H, \tilde{J}\}$ satisfies on Λ the assumptions of Theorem* 2.

Theorem 2 gives immediately stationary representations for the scattering matrix $\widetilde{S}(\lambda) = S(\lambda; H, \widehat{H}, \tilde{J})$ or for its components $\widetilde{S}_{\mathbf{a,b}}(\lambda)$. Actually, following (8)–(10), we set

$$\mathcal{A}_{\mathbf{a,b}}^{+}(\lambda) = \mathbf{Z_a}(\lambda)M^{(a)}\mathbf{V}^{(b)}\mathbf{U_b}(\lambda), \quad \mathcal{A}_{\mathbf{a,b}}^{-}(\lambda) = \mathbf{Z_a}(\lambda)(\mathbf{V}^{(a)})^*M^{(b)}\mathbf{U_b}(\lambda), \tag{23}$$

$$\mathbf{A_{a,b}}(\lambda) = \mathbf{Z_a}(\lambda)(\mathbf{V}^{(a)})^*R(\lambda+i0)\mathbf{V}^{(b)}\mathbf{U_b}(\lambda) \tag{24}$$

and

$$A_{\mathbf{a,b}}^{\pm}(\lambda) = \mathcal{A}_{\mathbf{a,b}}^{\pm}(\lambda) - \mathbf{A_{a,b}}(\lambda). \tag{25}$$

We note that the representation (20) together with analytical results of Propositions 10–12 allow us to rewrite the right-hand sides of (23), (24) as combinations of bounded operators. This gives a precise sense to these definitions. We emphasize that operators (23) are strongly and the operator (24) is weakly continuous in λ.

The auxiliary wave operators

$$\mathbf{w}_{\mathbf{a,b}}^{\pm} = W^{\pm}(T_{\mathbf{a}}, T_{\mathbf{b}}; E_{\mathbf{a}}(\Lambda)(J^{\mathbf{a}})^*M^{(a)}M^{(b)}J^{\mathbf{b}}E_{\mathbf{b}}(\Lambda)) \tag{26}$$

act as multiplications by operator-functions $w^{\pm}_{\mathbf{a},\mathbf{b}}(\lambda)\colon \mathbf{N_b} \to \mathbf{N_a}$. They can be calculated explicitly (see Proposition 17 below). Combining Theorems 2 and 13 we arrive at the following

THEOREM 14. *Under the assumptions of Theorem 13 the scattering matrix $\widetilde{S}_{\mathbf{a},\mathbf{b}}(\lambda)$ satisfies the following two equalities*

$$\widetilde{S}_{\mathbf{a},\mathbf{b}}(\lambda) = w^{\pm}_{\mathbf{a},\mathbf{b}}(\lambda) - 2\pi i A^{\pm}_{\mathbf{a},\mathbf{b}}(\lambda). \tag{27}$$

6. The operator $S_{\mathbf{a},\mathbf{b}}(\lambda)$ can (at least formally) be represented by its kernel parametrized by angular variables $\omega_a \in \mathbb{S}^{d_a-1}$ and $\omega_b \in \mathbb{S}^{d_b-1}$. As it is physically natural to expect, these kernels acquire additional singularities as ω_a approaches some plane $X_{a'} \subset X_a$, $X_{a'} \neq X_a$ or ω_b approaches some plane $X_{b'} \subset X_b$, $X_{b'} \neq X_b$. We concentrate here on the simplest case when ω_a and ω_b are separated from all such "singular" planes $X_{a'}$ and $X_{b'}$, respectively.

To derive representations for components $S_{\mathbf{a},\mathbf{b}}(\lambda)$ of the scattering matrix $S(\lambda) = S(\lambda; H, \widehat{H}, J)$ with the physical identification J we find a relation connecting $\widetilde{S}(\lambda)$ and $S(\lambda)$. First, we get rid of operators $M^{(a)}$ in the definition of wave operators

$$W^{\pm}(H, H_a; M^{(a)}, \Lambda) = W^{\pm}(H, H_a)W^{\pm}(H_a, H_a; M^{(a)}, \Lambda) \tag{28}$$

(this equality is called the multiplication theorem for wave operators). It turns out that the operators $W^{\pm}(H_a, H_a; M^{(a)}, \Lambda)J^{\mathbf{a}}$ can be calculated explicitly. Let us define on X_a the function

$$\underline{m}^{(a)}(x_a) = |x_a| m^{(a)}\big(x_a |x_a|^{-1}\big).$$

This function coincides with the trace of $m^{(a)}(x)$ on the subspace X_a if $|x_a| \geq 1$ and it is extended by homogeneity (of degree 1) to all $x_a \neq 0$. Denote by χ_Λ the characteristic function of Λ.

We start with the case $a = 0$ and set

$$\mathbf{\Omega}_0^{\pm} = \mathbf{\Omega}_0^{\pm}(m^{(0)}) = W^{\pm}(H_0, H_0; M^{(0)}, \Lambda).$$

These operators commute with H_0. Recall that

$$\big(\exp(-iH_0 t) f\big)(x) \sim (2it)^{-d/2} \exp(ix^2/4t) \hat{f}(x/2t), \quad \hat{f} = \Phi_0 f,$$

as $t \to \pm\infty$ (the sign " $\sim$ " means that the difference between left and right sides tends to zero in $L_2(\mathbb{R}^d)$).

LEMMA 15. *Let $m^{(0)}$ be an arbitrary smooth homogeneous function of degree 1. Then the operator $\Phi_0 \mathbf{\Omega}_0^{\pm} \Phi_0^*$ is multiplication by the function $\pm \underline{m}^{(0)}(\pm\xi)\chi_\Lambda(\xi^2)$.*

In the general case we have:

PROPOSITION 16. *Let $m^{(a)}$ be an arbitrary smooth homogeneous function of degree 1 and $m^{(a)}(x) = m^{(a)}(x_a)$ in some conical neighbourhood of X_a for $|x| \geq 1$. Then*

$$W^{\pm}(H_a, H_a; M^{(a)}, \Lambda) J^{\mathbf{a}} = \mathbf{\Omega}_{\mathbf{a}}^{\pm}(m^{(a)}) \otimes J^{\mathbf{a}},$$

where the operator $\Phi_a \mathbf{\Omega}_{\mathbf{a}}^{\pm}(m^{(a)}) \Phi_a^*$ acts as multiplication by the function

$$\pm \underline{m}^{(a)}(\pm\xi_a)\chi_\Lambda(\lambda^{\mathbf{a}} + \xi_a^2).$$

In particular, this assertion allows us to calculate the operators (26).

PROPOSITION 17. *If* $\mathbf{a} \neq \mathbf{b}$, *then* $\mathbf{w}^{\pm}_{\mathbf{a},\mathbf{b}} = 0$. *If* $\mathbf{a} = \mathbf{b}$, *then* $\mathbf{w}^{\pm}_{\mathbf{a},\mathbf{a}} = \boldsymbol{\Omega}^{\pm}_{\mathbf{a}}(m^{(a)})^2$, *where* $\Phi_a \boldsymbol{\Omega}^{\pm}_{\mathbf{a}}(m^{(a)})\Phi^*_a$ *is the same as in Proposition* 16.

Furthermore, with a help of (28) we find a relation between the scattering operators

$$\boldsymbol{\Omega}^{+}_{\mathbf{a}}(m^{(a)})\mathbf{S}_{\mathbf{a},\mathbf{b}}\boldsymbol{\Omega}^{-}_{\mathbf{b}}(m^{(b)}) = \widetilde{\mathbf{S}}_{\mathbf{a},\mathbf{b}},$$

corresponding to the identifications J and $\tilde{J}$. In terms of the scattering matrices it means that for a.e. $\lambda \in \Lambda$

$$m^{+}_{\mathbf{a}}(\lambda)S_{\mathbf{a},\mathbf{b}}(\lambda)m^{-}_{\mathbf{b}}(\lambda) = \tilde{S}_{\mathbf{a},\mathbf{b}}(\lambda), \tag{29}$$

where $m^{\pm}_{\mathbf{a}}(\lambda)$ is multiplication by $\pm(\lambda - \lambda^{\mathbf{a}})^{1/2} m^{(a)}(\pm\omega_a)$.

We can rewrite the representations (27) for $\tilde{S}_{\mathbf{a},\mathbf{b}}(\lambda)$ taking into account Proposition 17. According to (29) this gives representations for the "physical" scattering matrix $S_{\mathbf{a},\mathbf{b}}(\lambda)$ cut-off by the operators $m^{\pm}_{\mathbf{a}}(\lambda)$.

THEOREM 18. *Under the assumptions of Theorem* 13

$$m^{+}_{\mathbf{a}}(\lambda)S_{\mathbf{a},\mathbf{b}}(\lambda)m^{-}_{\mathbf{b}}(\lambda) = m^{\pm}_{\mathbf{a}}(\lambda)^2\delta_{\mathbf{a},\mathbf{b}} - 2\pi i A^{\pm}_{\mathbf{a},\mathbf{b}}(\lambda), \quad a.e.\, \lambda \in \Lambda, \tag{30}$$

where $A^{\pm}_{\mathbf{a},\mathbf{b}}(\lambda)$ *is defined by equalities* (23)–(25), $\delta_{\mathbf{a},\mathbf{a}} = I$ *and* $\delta_{\mathbf{a},\mathbf{b}} = 0$ *if* $\mathbf{a} \neq \mathbf{b}$.

Because of the property (3) of functions $m^{(a)}$ and $m^{(b)}$, the equality (30) gives a representation for the sesquilinear form of $S_{\mathbf{a},\mathbf{b}}(\lambda)$ only on functions vanishing in neighbourhoods of singular directions (see the beginning of this part).

The continuity of $(S_{\mathbf{a},\mathbf{b}}(\lambda)f_{\mathbf{b}}, f_{\mathbf{a}})$ on dense sets of elements $f_{\mathbf{a}}, f_{\mathbf{b}}$ together with the bound $\|S_{\mathbf{a},\mathbf{b}}(\lambda)\| \leq 1$ ensure the weak continuity of $S_{\mathbf{a},\mathbf{b}}(\lambda)$. In view of the unitarity of the scattering matrix $S(\lambda)$ the weak continuity can be replaced here by the strong one. Let us formulate the result obtained.

THEOREM 19. *Let Assumption* 3 *hold. For any* $\mathbf{a}$, $\mathbf{b}$ *the operator* $S_{\mathbf{a},\mathbf{b}}(\lambda)$ *is strongly continuous in* $\lambda \in \Lambda$. *Thus the representation* (30) *holds for all* $\lambda \in \Lambda$.

Finally we note that in contrast to the two-particle problem in the general case one can not expect continuity of $S_{\mathbf{a},\mathbf{b}}(\lambda)$ in the topology of the norm.

References

1. M. Sh. Birman and D. R. Yafaev, *A general scheme in stationary scattering theory*, Problemy Mat. Fiz. **12** (1987), 89–117 (Russian); Amer. Math. Soc. Transl. (2) **157** (1993), 87–112.
2. L. D. Faddeev, *Mathematical Aspects of the Three Body Problem in Quantum Scattering Theory*, Trudy MIAN **69**, 1963. (Russian)
3. G. M. Graf, *Asymptotic completeness for N-body short-range quantum systems: A new proof*, Comm. Math. Phys. **132** (1990), 73–101.
4. E. Mourre, *Absence of singular spectrum for certain self-adjoint operators*, Comm. Math. Phys. **78** (1981), 391–400.
5. P. Perry, I. M. Sigal and B. Simon, *Spectral analysis of N-body Schrödinger operators*, Ann. of Math. **144** (1981), 519–567.
6. I. M. Sigal and A. Soffer, *The N-particle scattering problem: Asymptotic completeness for short-range systems*, Ann. of Math. **126** (1987), 35–108.
7. D. R. Yafaev, Radiation conditions and scattering theory for N-particle Hamiltonians, Comm. Math. Phys. **154** (1993), 523–554.
8. ——, *Mathematical Scattering Theory*, Amer. Math. Soc., 1992.
9. ——, *On the multichannel scattering theory in two spaces*, Theoret. and Math. Phys. **37** (1978), 867–874.

Département de Mathématiques, Université de Rennes 1, Campus de Beaulieu, 35042 Rennes Cedex, France

E-mail address: dimitri.yafaev@univ-rennes1.fr